# Of Sound Mind

## Life and Work of a Naval Scientist

*by*
*Alan Parsons*

© Copyright Alan Parsons, 2019

ISBN: 978-0-244-17984-7

Cover design taken from an original by Michel Boisrayon

*Big whirls have little whirls that feed on their velocity,*

*and little whirls have lesser whirls and so on to viscosity.*

L F Richardson (1922)

To Lucienne,

the love of my life.

# CONTENTS

PREFACE

BOOK ONE
Parsons Genealogy (~1700 – 1943)

BOOK TWO
Childhood (1943 – 1954)

BOOK THREE
Youth (1954 – 1964)

BOOK FOUR
Marriage (1964 – 1968)

BOOK FIVE
Work (1968 – 2018)

# Preface

I have written this volume to record my Parsons ancestry, some memories of my early life, and my work as a naval scientist.

With regard to my ancestry, the information I have discovered for the period before 1841 was obtained mainly from on-line transcriptions of Parish Registers for West Somerset covering baptisms, marriages and burials, and from various army records obtainable on-line through National Archives. The information for the period after 1841 was obtained mainly from on-line records of the censuses held every 10 years. So almost everything I have discovered was obtainable directly from the Internet in the comfort of an easy chair.

There are some exceptions: several birth, marriage and death certificates that I had to apply for and receive by post; documents and reference books providing background to the army career of my great-great-grandfather Thomas Parsons, supplied by The Keep Military Museum in Dorchester, a research centre for the Devon and Dorset regiments including the 39th Regiment of Foot in which my great-great-grandfather served; additional information provided by a distant cousin, Dorothy Parsons; and scraps of information left behind by deceased relatives. I have chosen not to provide references for this material. I wanted to create a narrative, not an academic document.

With regard to my own life, I have purposely included only outline memories of my daughters and their families. This is partly because I did not want to embarrass them, but mainly because the objective was to write about things concerning my early life and my work in underwater acoustics that perhaps they did not know about, and that possibly one day they and others might find interesting.

*Alan Parsons*
*Wantage*
*February 2019*

# BOOK ONE
# Parsons Genealogy (~1700 – 1943)

# Parsons Origins

The Parsons surname is fairly common in South West England, particularly in the counties of Dorset and Somerset. The records show that our family of Parsons has its origins in Somerset.

The Parsons surname was originally the occupational name for the servant of a parish priest or parson, or possibly the child of a parson, but not the parson himself. It is derived from the old French word *persone* meaning parish priest or devout man regarded as the representative *person* of the parish. The phonetic change from -*er*- to -*ar*- was a regular development in Medieval English.

In the 18th and 19th centuries there were two main concentrations of Parsons in Somerset, one centred around Bicknoller and Williton, both near Minehead, and the other around Kingston St Mary near Taunton. According to the records our family of Parsons originated from the latter area of Kingston St Mary, but an earlier connection with Minehead is possible.

Kingston St Mary, often referred to simply as Kingston, is a small village in the Quantock countryside just three miles north of Taunton. It boasts a village green, a post office, a handsome 15th century church of St Mary's complete with an adjoining vicarage, manor house and grange, a single pub called The Swan, and several thatched cottages in honey-coloured stone with traditional old-English country gardens. Its residents are now mainly commuters or wealthy retired couples.

Kingston St Mary is a model country village, but in the 1800s and earlier it was rather different and much busier than today. Kingston St Mary was one of many thriving agricultural communities populated largely by farm workers. In those days 1 in 3 of the population of England worked on the land. This was to reduce 1 in 10 by 1900.

The nearby town of Taunton became an important market centre that grew rapidly in the 19th century. The railway reached Taunton in 1842, and in 1843 the county court was moved there from Ilchester. Many local industries flourished in Taunton including silk making, brewing, iron founding, and notably the making of shirt collars.

# John Parsons (~1722 – 1776)
# Jacob Parsons (1760 – 1840)

It should be noted that the dates of birth / death obtained from early Parish Registers are actually dates of baptism / burial.

My g4 (great-great-great-great-) grandfather is John Parsons. He was born circa 1722 and probably worked throughout his life as an agricultural labourer.

John Parsons married Mary Tamlin in the parish church of Kingston St Mary on 1 February 1747/8[1]. It is likely that neither John nor Mary was born in Kingston St Mary, but they seem to have lived most of their lives in the parish.

The origins of John and Mary are speculative. In the available on-line transcriptions of baptismal registers for West Somerset there is only one John Parsons found before 1740. This John Parsons was baptised in the parish of Bicknoller, near Minehead, on 9 February 1722. His father was William Parsons. If this is the right John Parsons then his father William is my g5 grandfather.

The maiden name of John's wife Mary was Tamlin, not a common name. Possible relatives are Richard and Mary Tamlin (at different times spelt Tamplin or Tamling) who were married in the parish church of Kingston St Mary on 18 July 1714. Richard and Mary came from Cothelstone about three miles north-west of Kingston St Mary. The baptisms of three of their children are recorded: Richard on 28 December 1716, Thomas on 13 April 1722, and Elizabeth on 5 May 1728. It is possible that Richard Tamlin was related to our family because

---

[1] 1747 in the old calendar, 1748 in the new. This ambiguity exists for dates before the 1752 calendar change when Great Britain switched from the Julian to the continental Gregorian Calendar. Before 1752 the year began on 25 Mar. As a result of the change, the day following 31 Dec 1751 was 1 Jan 1752, not 1 Jan 1751 as under the old system. Thus 1751 was a short year running from 25 Mar to 31 Dec. Also, 11 days were omitted from the 1752 calendar. The day after 2 Sep 1752 was 14 Sep 1752. All this must have caused considerable confusion.

there are no other Tamlins recorded as having lived in the vicinity of Kingston St Mary at that time.

My established g4 grandparents John and Mary Parsons had eleven children born at two-year intervals between 1749 and 1770. This regular rate of child bearing was generally considered to be healthy for the mother. The names of the children starting with the eldest were Benjamin, John, Samuel, Abraham, Isaac, Betty, Jacob, Thomas, Mary, James, and William. Abraham and Isaac died in infancy. Their father John died on 23 January 1776.

John and Mary's seventh child, Jacob, is my g3 grandfather. He was born on 14 February 1760 and died on 31 October 1840. Jacob lived all his life in Kingston St Mary working as a carpenter. He married Jane Trott at the parish church on 8 June 1789. Jane was born in the village on 23 February 1766.

Jacob and Jane had five children all born between 1790 and 1799. Their names starting with the eldest were Mary, John, Jacob, Thomas and Sarah. The fourth child, Thomas, is my g2 grandfather. Thomas was baptised on 12 February 1797.

# Thomas Parsons (1797 – 1864)

Thomas is my g2 (great-great) grandfather. Quite a lot is known about his life, mainly from meticulous army records. According to Kingston St Mary's Parish Register he was baptised on 12 February 1797. Thomas Parsons spent his childhood in Kingston St Mary, his place of birth.

On 4 January 1816 Thomas enlisted in Taunton as a private in the 39th Dorsetshire Regiment of Foot, just six months after the battle of Waterloo. The army recorded that he joined up on his 17th birthday whereas, according to the record of his baptism, he was almost 19 years old. Perhaps he did not know his age. Because he was recorded as under-18 his first year in the army would not count as reckonable service for pension purposes.

The army was very different from today with no permanent regimental headquarters or barracks. Few soldiers were kept at home, the majority being stationed overseas. Unlike the Navy, officers were not recruited on merit but bought their commissions. Soldiers were enticed to join by the recruiting sergeants, there being no conscription until the First World War. Pay was low, conditions poor, and flogging was the main form of discipline. Nevertheless, conditions in the army were generally better at that time than those of agricultural labourers or the unemployed. One married soldier in every 12 was allowed to take his wife and family with him to help with the regiment's cooking and sewing.

Upon enlistment in 1816 Thomas was immediately sent to France to serve in the Army of Occupation. The task was to consolidate the recent victory over Napoleon and to keep the peace among the French populace. Thomas' regiment comprising some 800 men was stationed in NW France at Bethencourt (where Henry V crossed the Somme prior to the battle of Agincourt exactly 400 years earlier, and near the site of the Battle of the Somme exactly 100 years later). At the end of 1817 the regiment moved to Berles, then on to St Omer and Valenciennes.

At Wellington's instigation, the original five-year posting of the Army of Occupation in France was reduced to three. In 1818 the 39th Regiment received orders to proceed to Ireland. Most of the men had served continuously whilst in France, but the new posting precluded home leave for them to see their families. This was because the whole of Ireland at that time was part of the United Kingdom, and therefore counted as a "home posting".

The regiment left France on 30 October 1818, sailing from Calais to Dover. The inspecting officer, General Brisbane, reported that the men were in good health and had been well-behaved in France towards the local people. He stated that the regiment was "a very fine battalion in a high state of discipline and fit for any service".

From Dover the regiment marched the 140 miles to Portsmouth arriving on 11 November 1818. After five weeks in Portsmouth they sailed on 17 December for Cork in Ireland, arriving on Christmas Eve. From Cork they marched 180 miles to the regiment's final destination of Castlebar in County Mayo, arriving on 7 January 1819. It is difficult to comprehend the amount of marching these soldiers endured, always in full kit.

Thomas served with the regiment in Ireland for the next 6½ years with periods of service in several places including Castlebar, Dublin, Cork, Tralee, Limerick and Buttevant. The regiment was frequently split into small detachments of between 12 and 30 men. Their tasks were to suppress political insurrection and generally to act as a police force. This involved a wide range of duties such as dealing with cattle rustlers, illicit whiskey distilling, and troubles between Unionists and Catholics. In January 1822 some regimental detachments were employed in Cork to suppress a partial insurrection of the Whiteboys. The men were not trained for this type of police work and the local Catholic population resented their presence.

In early October 1824 the regiment marched to Buttevant in County Cork. Here the men were finally brought together as a single unit, having been working continuously in small disjointed detachments for the whole of their service in Ireland, with the exception of a few months while stationed in Dublin.

It was probably in Dublin that Thomas met his first wife Susannah McCauley (also known as Susan). They married on Boxing Day, 26 December 1820 at the Catholic Church of St Nicholas Within, Dublin. Nothing is known about Susannah's family. She was probably from Dublin and almost certainly a Catholic. There must have been a stigma attached to her marrying a Protestant soldier.

The birth of their first child, Ann, in 1824/5 probably took place in Ireland. According to Ann's age given in later census records she was born between June 1824 and May 1825

On 10 July 1825 the regiment received orders to proceed to Australia to oversee convicts and assist with the exploration and development of this new country. Australia, like Ireland, needed a strong military presence to maintain discipline. On 19 September 1825 the regiment embarked from Cork to Chatham to make preparations.

In the 17 months between 6 December 1825 and 5 May 1827 small parties of officers and men from the six service companies of the 39$^{th}$ Regiment of Foot departed at irregular intervals from Chatham in convict ships, escorting no less than 18 groups of convicts to Sydney, New South Wales. The journey typically took 4 to 6 months. The regimental headquarters under the command of Colonel Lindesay left in the last of these detachments. The four depôt companies remained in Canterbury for a further 9 months, eventually sailing in small detachments between 1 February 1828 and 30 August 1830. During this long window of transfer to Australia, the regiment was dispersed across the world with the chain of command greatly stretched.

It appears that Thomas sailed in a service company detachment leaving Chatham in December 1826, probably with his wife Susannah and daughter Ann. Dorothy Parsons[2]

---

[2] Dorothy Margaret Parsons is my second cousin once removed: that is, my great-great-grandfather Thomas Parsons is her great-grandfather. So Dorothy precedes me by one generation, hence the "once removed". She

of Cobham suggests the precise departure date was 29 December 1826 (original source unknown). This date is consistent with Thomas' service records, which state that he spent 5 years and 6 months in Australia prior to his service in India which began on 10 October 1832. So he must have arrived in Australia in April or May 1827. This suggests a departure date from Chatham in December 1826, in accordance with the information supplied by Dorothy Parsons.

From the records of the convict vessels leaving England and Ireland around the time of Thomas' departure, his most likely itinerary is that in late December 1826 he and his family first sailed from Chatham to Cork in Ireland where they collected a consignment of convicts, leaving Cork on the *Mariner* on 14 January 1827. This is the best fit with the available facts.

The *Mariner*'s Captain was Robert Nosworthy. The Surgeon Superintendent was Patrick McTernan, his first appointment on a convict ship. He kept a journal from 3 November 1826 to 5 June 1827 in which he recorded a daily weather report for the entire voyage and included a Hospital Diet Table. There was not a single case of scurvy. The *Mariner* sailed via the Cape of Good Hope where a dozen prisoners convicted of crimes at the Cape were embarked. The total number of convicts on board was then 161. The *Mariner* departed the Cape on 28 March 1827 arriving in Sydney on 23 May, a voyage of 129 days in all. The Guard from the 39th Foot was commanded by Captain Sturt. When the Guard were landed they marched to their quarters at noon through George Street, Sydney, with drums beating and fifes playing.

---

was born on 15 October 1924 in Surbiton and lived for many years in Cobham, Surrey. I discovered her by chance after responding to an advert she placed in the personal column of the magazine of the Somerset & Dorset Genealogy Society, asking for information about an ancestor. Up to this point her side of the family had been totally ignorant of the existence of my side, and *vice versa*. Dorothy provided me with much useful information.

First impressions on reaching Sydney are recorded in the reminiscences of Lieutenant William Coke of the 39th Foot who arrived a few months before Thomas. The following is an extract from his writings:

*On the fortunate termination of our perilous voyage in the Convict Ship we were all, as may naturally be supposed, anxious to get on shore at Sydney as soon as possible. Having landed, we marched up the street, and were met by the officers of the Buffs [His Majesty's 3rd Regiment of Foot], to which regiment I was to be attached till the headquarters came out. On our way we passed an almost naked native, lolling listlessly with a spear in his hand, and though he seemed to take no notice of anything, yet, nevertheless, nothing escaped his sleepy eye. Farther on we came to a guard house with a verandah round it, and a little beyond this on the same side, we passed the guard and entered the barrack yard. In front were the men's barracks, and the officers' quarters were chiefly placed at right angles to these – on the left sloping down towards the guard house and wall. They consisted of one floor, with a very broad portico or verandah, supported by strong pillars on dwarf walls, whereon, late at night, the officers used to sit and chat with one another. On these stones, during the hot winds, the thermometer often stood at 105° at midnight.*

Thomas served 5½ years in Sydney and Bathurst, New South Wales, Australia. His work would have included garrison duties, overseeing convicts during their working day, and exploration. The regiment also provided a mounted police force to keep law and order. In addition, detachments operated for short periods in Hobart, Van Diemen's Land (now Tasmania) where another penal colony had been established.

Thomas is listed on extant army payroll musters where he is identified as a member of No 4 Company (Plate 3). A Company comprised 100 men under the command of a Captain supported by a Lieutenant and an Ensign (later renamed sub-Lieutenant). Several Companies were grouped to form a Regiment. His rate of pay was 7d (seven pennies) per day, which included an additional 1d per day for completion of 7 years service. His pay would eventually rise to 8d per day on completion of 14 years service. Out of this money he had to pay for his food.

The regiment included some distinguished officers whom Thomas would have known. One of these was the renowned explorer Captain Charles Sturt, later to have a New South Wales University named after him. Captain Sturt's major achievement was his exploration of inland Australia. He was particularly associated with explorations in 1828-30 of the Macquarie, Lachlan, Murrumbidgee, Darling and Murray River system. These principal rivers now serve the communities and agricultural industries of the region where the Charles Sturt University has been established. As noted previously, Captain Sturt had sailed to Australia in *Mariner*, the same ship in which Thomas probably sailed. It is possible that Thomas made some small contribution to Captain Sturt's achievements.

Thomas and Susannah had a son John, born in Sydney on 4 November 1828 and baptised on 29 January 1829. John sadly died aged 4 months. He was buried on 12 March 1829. The baptismal and burial services both took place at St Philip's Church, the original Parish Church of the City of Sydney.

On 26 January 1831 their second daughter Elizabeth was born. She was baptised on 6 February in St Philip's Church, Sydney.

On 5 May 1832 the 39$^{th}$ Regiment of Foot received orders to leave for India, a standard destination for regiments ending their term in Australia. A soldier was usually given the option of staying on in Australia to settle. Many in the regiment took this option. In fact, a total of 59 soldiers from the 39$^{th}$ did so, and some prospered. But Thomas declined and settled for service in India.

The regiment sailed from Sydney, Australia, in two detachments. The first detachment comprised six companies and left on 21 July 1832 to arrive in Madras on 22 September.

The second detachment left on 3 December 1832 to arrive in Madras on 21 February 1833. Thomas left in the first detachment, probably with his wife and two daughters.

On arriving in India Thomas was again hospitalised for reasons unknown. On recovery he proceeded to Poonamallee and finally to Bangalore, the regional headquarters of the British administration. This was to be Thomas' base for the rest of his time in India.

Although Australia had been hot and unhealthy, India was even more inhospitable. The men's military clothing was not suited to the monsoon conditions. The regiment found India to be unsanitary. In March 1833 a serious outbreak of cholera resulted in the death of 1 officer, 4 sergeants, 42 rank and file, 2 women and 11 children.

The family entries in the 1841 and 1851 England censuses include a son William Parsons. According to his recorded age, William must have been born to Susannah in India in the window between July 1832 and March 1833, but no mention of this event has been found in the Indian records.

However, a son James was born at Bangalore on 30 March 1834 and baptised on 16 April 1834, but no further reference to James has been found. Possibly James died in infancy. Certainly, his mother Susannah died soon after the birth leaving Thomas a widower. Susannah was buried at Bangalore on 26 May 1834.

In 1834 the regiment was involved in active field operations against the Rajah of Coorg, a tyrant defying British military authority and giving refuge to rebels. The regiment was commended for obtaining the surrender of the Rajah at his capital Mercara with minimal casualties to themselves.

Thomas' service in India was curtailed as a result of an ailment leading to his honourable discharge. Such discharges due to illness or injury were not uncommon. The report issued at Bangalore on 10 November 1837 reads as below. It was signed by Thomas' commanding officer Lieutenant Colonel Thomas Poole and the Assistant Surgeon Robert Davis (both to die in India of cholera within the next 18 months). The

Regimental Board in Bangalore met one week later to ratify the discharge.

No. 116 Private Thomas Parsons H. M. 39th Regiment
Age by records 38 $^{10}/_{12}$ years
Certified that the above named Thomas Parsons has suffered from repeated attacks of fistula in ano *[anal fistula probably due to an abscess]*, the result of a breaking up of his constitution, and his *[he is]* unfit to discharge his military duties in consequence thereof.
Thomas Parsons bears the character of an excellent soldier in the Regiment, and his conduct whilst a patient in Hospital has always been exemplary, and he contributed every thing in his power to accelerate his cure.
This is a case of debility contracted in the Service, and not attributable to vice, design or intemperance.

Thomas' service in India officially ended on 30 November 1837. He had to sign an attestation that he had received all his dues up to that point. Being unable to write, Thomas made his mark with an 'X'.

It was another six months before Thomas finally arrived back in England, landing at Gravesend on 8 June 1838. He was dispatched to the General Hospital at Chatham for a further medical examination. The Chief Medical Officer confirmed his unfitness to continue in service as a result of him being "permanently incapacitated". His final discharge date was 25 July 1838. At this time he was described as being 5ft 10ins tall with dark brown hair, grey eyes and a fair complexion.

So ended Thomas' long and varied army career. His regiment had fought in many conflicts but his period of service had seen no major campaign to merit the award of battle honours. Thomas had joined up just too late for Waterloo and left too early to see the Battle of Maharajpore and the Indian Mutiny. Although he received no campaign medals he would have received a long service medal.

With his honourable discharge and long length of service Thomas was entitled to an army pension. He must have received it but no record of any actual payment has yet been found. On discharge his reckonable service was recorded as 24 years and 46 days, which is longer than the number of years from his 18$^{th}$ birthday in January 1817 (as assumed by the army) to his discharge in July 1838. The reason for this is that time spent in India was counted at time-and-a-half owing to the harshness of the conditions. According to the army records, Thomas' reckonable service was credited to him as follows.

| Period | Years | Days | Comments |
|---|---|---|---|
| 4 Jan 1816 – 3 Jan 1917 | - | - | under age |
| 4 Jan 1817 – 30 Nov 1837 | 20 | 331 | length of service from age 18 |
| 10 Oct 1832 – 30 Nov 1837 | 2 | 208 | 50% extra for service in India |
| 1 Dec 1837 – 25 Jul 1838 | - | 237 | additional time between certification of unfitness and final discharge |
| Total | 24 | 46 | |

Thomas returned to Somerset having been away for more than 20 years. It must have been quite a home-coming. He was now a widower and had to adjust and settle down to a new life. His children Ann, Elizabeth and William by his deceased wife Susannah definitely returned to England because they appear in the 1841 census, but when and how they returned from India is not known. Presumably they returned with their father.

There also remains a small inconsistency about their country of birth. Whilst there is little doubt that Ann, Elizabeth and William were born in Ireland, Australia and India, respectively, several subsequent censuses report them as being born in Somerset. Perhaps there was a stigma attached to any association with Australia and its convicts. One exception is the 1851 census which states that Elizabeth was born in "East Indies" [3], the collective name then given to the whole trading area of India, SE Asia and SW Australia. In a later census Elizabeth gives her place of birth as "South Wales", which the census coordinator marked as "unknown".

On 29 July 1839 Thomas took a new wife by marrying Mary Hines (sometimes spelt Hine) in the parish of Norton Fitzwarren, an agricultural area just two miles west of Taunton. Mary was born on 27 March 1808 at nearby Kingston St Mary, so it is possible that Thomas had known her as a child before his army career.

Thomas' 4th child, also to be christened Thomas, and his first child by Mary, was born in Norton Fitzwarren on 5 May 1840. Thomas' father Jacob lived just long enough to see this grandchild. Jacob died in Kingston St Mary on 31 October 1840.

By 1841 the family had returned to Kingston St Mary, the village of Thomas and Mary's birth, where they lived in the sub-district of Fulford. Thomas was working as an agricultural labourer. The family would have lived in a humble workers' cottage built mainly of cob and thatch, probably shared with other families. Most of these dwellings have disappeared. The few that remain have been totally transformed and extended by their current owners.

In 1841 the family comprised Thomas aged 44, his wife Mary aged 33, her step-children Ann, Elizabeth and William aged 16, 10 and 8, and her own child Thomas aged 2. Also living with them was Mary's father James Hines aged 70, an agricultural labourer.

---

[3] In 1600 a London-based trading company was established, known as the East India Company. It became the main vehicle for British commercial and imperial expansion in Asia, dominating both trade and Empire until its eventual demise after the Indian Mutiny (1857-59). Protected by Royal Charters, the Company obtained exclusive trading rights throughout Asia. Such was its influence and power that all the areas where it operated became known collectively as the "East Indies", which included New South Wales in Australia where Elizabeth was born.

Thomas' elder brother Jacob was then living close by at Middlebrooks. Jacob was also an agricultural labourer who married his wife Ann Hayes on 27 January 1817. In 1841 there were three children living with Jacob and Ann: William, Ann and Thomas aged 16, 13 and 7, respectively. Their eldest child James, born 1 June 1817, was no longer at home.

In 1851 Jacob and his family were still living at Kingston. They were lodgers with an agricultural labourer Mathew Pring to whom their daughter Ann was married. Jacob and his youngest son Thomas were both agricultural labourers whereas Jacob's wife Ann and son William are listed as paupers[4]. In 1861 Mathew Pring and his wife Ann were still living in Kingston, now with three young daughters Jane, Ellen and Sarah. Mathew died in 1866 aged 52.

There were also other Parsons living in Kingston, probably related. In 1841 Benjamin Parsons aged 38, publican of the Farmers Arms Inn, was living close to Thomas in the sub-district of Fulford with his wife Anna aged 30 and two children Emma Jane aged 6 and Eli aged 6 months. Benjamin had married Anna on 23 April 1834. Her maiden name was Pring, so she was no doubt related to the aforementioned Mathew Pring. Several others lived in the same household including a James Parsons, an agricultural labourer aged about 50. Benjamin died on 3 October 1848 whereupon Anna became the Inn Keeper. In 1851 she was living at the Inn with her two children Eli aged 11 and Emma Jane aged 16. The unknown James Parsons was still living with the family. He was now a pauper aged 70, and deaf and dumb. By 1861 Anna had become the licensee of the adjoining beer house and Eli had

---

[4] A pauper was someone receiving assistance from the Poor Law Guardians, either in the form of money or, after the Poor Law Amendment Act of 1834, entry into the local Union Workhouse. The Taunton Workhouse was erected in 1836-8 on what is now Trinity Road. Together with Kingston St Mary it served 28 parishes comprising a total population of about 33,000. Workhouses were introduced to help the destitute and to discourage idleness by their strict regime. It is not known whether any Parsons paupers entered the Workhouse.

become the Inn Keeper. His sister Emma Jane was then a grocer.

Just prior to 1843 Thomas and his family moved three miles to live in Taunton where their 5th child James, my great-grandfather, was born.

# James Parsons (1843 – 1888)

James, my great-grandfather, was born on 17 July 1843 in North Town, Taunton, in the sub-district of St James. His father Thomas was then working as a labourer. The sixth and last addition to the family was Charles, born in Taunton on 9 November 1849 (date unconfirmed). Thus, my great-grandfather James was the fifth of six surviving children named Ann, Elizabeth, William, Thomas, James and Charles.

By 1851 the family was living at 81 South Street, Taunton, where mother Mary worked as a mangler. A mangle was used for wringing out water from washed garments. Their terraced house must have been comfortable compared with their previous agricultural dwellings. For reasons unknown, Mary's husband Thomas was at this time a patient in nearby East Reach Taunton & Somerset Hospital, St Mary Magdalen. He was listed as a Chelsea Pensioner[5], a benefit he earned from his long service record and honourable army discharge. Thomas at this time was 54 years old. It is not clear whether he was still working. Perhaps his army pension and wife's earnings provided sufficient income.

In 1851 the eldest daughter Ann aged 26 and the eldest son William aged 18 were no longer at home.

By 1861 Thomas and his wife Mary had moved to Norman's Court, High Street, Taunton (between Nos 22 and 23 but no longer extant). They were aged 64 and 53, respectively. All their children had left home except Charles, the youngest, then

---

[5] A Chelsea Pensioner is a former soldier of the regular army who receives a pension after being disabled by wounds or accidents becoming unfit for service, or serving for 20 years or more. The term describes both 'In-' and 'Out-Pensioners'. An In-Pensioner surrenders his army pension in order to live in the Royal Hospital Chelsea. When the Hospital opened there were more Pensioners than places available. Those who were not given a place were termed Out-Pensioners. With the increasing size of the standing army the number of Out-Pensioners rose steadily, from 739 in 1708 to 14,700 in 1763, and 36,757 by 1815. Thomas was an Out-Pensioner. The Royal Hospital remained responsible for all army pensions until 1955.

aged 12. They had a lodger, a 69 year-old widower named Samuel Littlejohns who was a "cordwainer" (shoemaker). Thomas worked as a wood carver, possibly a skill he had learned in his youth from his father who was a carpenter, or maybe during his time in the army.

No record has been found of what happened to his wife Mary after 1861, but Thomas died on 21 April 1864 aged 67 of a chest infection. He was then living in Union Place, Taunton. His death was witnessed and reported by Eliza Williams, 38 years of age, living with her family at No 164 Union Place. It is not known what the relationship was between Thomas and Eliza Williams. But next-door at No 165 Union Place lived 67 year-old John Rossiter, a Chelsea Pensioner like Thomas. Possibly Thomas was living there at the time of his death and Eliza Williams was a helpful neighbour.

The life of my great-grandfather James will be detailed shortly. But what became of Thomas's other five children?

It is possible that the eldest daughter Ann married one Henry Davy in 1847 eventually settling in Bedminster, Bristol. But verification is needed that this is the right Ann Parsons. Whilst the marriage certificate gives her father as Thomas Parsons, subsequent censuses consistently give her birthplace as Stogumber, Somerset, and her year of birth as 1820, neither of which is consistent with other information about the "real" Ann Parsons.

The second eldest child Elizabeth married Thomas Baker on 15 May 1853 at the Parish Church of Taunton St James. Elizabeth aged 23 was a weaver by trade. Her husband aged 22 was a labourer. Elizabeth and Thomas lived for several years at Coal Orchard in Taunton near North Town Wharf. At some time between 1881 and 1891 they moved to 6 Prospect Terrace on Canal Road where they lived for the rest of their lives. They had six children named Mary, James, Thomas, Ann, Charles P, and Sarah. Elizabeth died in 1909, and Thomas in 1911.

The third child William was living in 1851, aged 18, in nearby Paul Street in Taunton with the family of Anthony Loynes, a farrier and postmaster. William is listed as a servant and driver. He has not been traceable after that date. Likewise, there is no trace of what happened to the fourth child, Thomas. Possibly they both joined the army.

There is much information about the sixth child, Charles, the youngest, of which more later.

With regard to the fifth child, James Parsons, my great-grandfather, there is a gap in information after 1851. He is not mentioned in either the 1861 or 1871 censuses. This is probably because he followed his father's footsteps by joining the army and may have been serving overseas. He was certainly in the army in 1871 and at the time of his marriage in 1873.

The eventual wife of my great-grandfather James was Harriet Andrew (occasionally spelt Andrews). She sometimes spelt her Christian name as Harriett. She was born in Loddington, Northamptonshire about 1837. Harriet's father was John Andrew, a master baker and shoemaker, born about 1783. Harriet's mother was Hannah born about 1795. Both came from the local area of Loddington and Kettering. They all lived at what is now called "The Old Bakehouse" at 12 Main Street, Loddington.

Harriet was the seventh of eight children whose names were, starting with the eldest, John, Henry, George, Hannah, Eliza, Fanny, Harriet and Robert. All the sons became involved in the bakery business, some of them opening their own shops in neighbouring villages.

In 1851 Harriet, aged 14, still lived in Loddington with her parents John and Hannah. On 16 March 1855 Harriet's father died, aged 72, of carbuncles and apoplexy (a stroke). In September of that year Harriet was in Birmingham, possibly working there or staying with relatives.

Her mother Hannah died aged 64 of jaundice on 16 June 1860.

In 1861 Harriet aged 24 was living in Loddington with her elder brother Henry and his family. Henry was then a baker aged 34, with a wife named Rebecca age 26 and a child named John aged 1.

In 1871 Harriet, aged 34, remained unmarried. She was then working as a Lady's Maid and domestic servant in the household of The Rev William F Hotham, Rector of Buckland in Surrey, located between Reigate and Dorking. The Rector had a wife and two children, Frances aged 13 and Frederick aged 8. There were five other domestic servants – a cook, nurse, housemaid, kitchen maid, and a footman.

The rectory at Buckland still exists (now called Glebe House). It is set in beautiful grounds deep in the Surrey countryside at the end of a narrow lane off the main Reigate to Dorking Road. Buckland village and church are equally idyllic.

In 2008 when passing by on my way to a business meeting in Kent, I summoned up the courage to knock at the door of the imposing rectory. The owner was a Mr Roger Daniels, a retired merchant banker who used the house to entertain friends at weekends. The house had been his family home since the 1930's when he lived there as a boy with his parents. He was most hospitable and invited me in for a cup of tea. He showed me round the ground floor. The house ceased to be a rectory well before 1930. It was not clear where the servants' quarters would have been.

To return to Reverend William Hotham, he was born in Rochester on 28 March 1819, the son of Reverend Frederick Hotham and Anne Elizabeth Hodges. His grandfather was Beaumont Hotham, 2nd Baron Hotham of South Dalton in Yorkshire. The Reverend William Hotham married Emma Carbonell in Reigate on 31 January 1855. He was Rector of Buckland from before 1861 to after 1881, probably until his death on 10 September 1883 at age 64. The son Frederick, aged 8 when Harriet was working there, eventually became the 6th Baron Hotham, the title passing by The Reverend William Hotham because he died too early. So Harriet worked in a grand environment.

It is not known how James and Harriet met, but by 1873 Harriet had left the Reverend Hotham's employment and at the time of her marriage to James was living in the coastal town of Sandgate, adjoining Folkestone. James was then a private in the 3rd Battalion of the 60th Regiment of Foot (The King's

Royal Rifle Corps), later to become the Royal Green Jackets, based at Shorncliffe Camp [6] overlooking Sandgate. James was discharged from the army in 1873. His marriage was probably planned to coincide with this.

The following is an extract that appeared in the Bombay Guardian in 1890. It is an account of Sandgate and its relationship with Shorncliffe army camp. If the article is to be believed, Sandgate seems to have been on the brink of total moral disintegration during the 1850's, long before Harriet was there:

*A typical English garrison. One of the military stations in England to which troops from India come and go is Shorncliffe, a breezy upland on the coast of Kent. At its foot is the pretty little watering place of Sandgate. The regular occupation of Shorncliffe Camp by the military changed the character of Sandgate. Troops returning from the Crimean war, and leaving for the scenes of the Indian Mutiny, were expected to indulge their animal, nay beastly, propensities to the full. Sandgate at that time reeked with the "trap-doors to hell". Low beerhouses abounded on every hand. Miscreants, mostly drunks, were regularly locked up in the old Sandgate Castle. [The article went on to deliver news of the renaissance... ]. Twenty five years have made great changes. The low beerhouses have been swept away, the number of liquor houses of every kind are reduced by fully one third. The 'Balaclava', the 'Sebastopol' and the 'Inkerman' are again peaceful cottage homes, while other dens of infamy have entirely disappeared. Rows of modern shops and handsome Queen Anne houses have taken their place and quite altered Sandgate.*

---

[6] Shorncliffe is the birthplace of modern infantry tactics. The barracks is currently the home of the Gurkha Rifles, but the fort (Shorncliffe Redoubt) is now derelict and overgrown with an uncertain future. The fort was the subject of a Channel 4 Time Team programme screened in Feb 2007.

An insight into James' possible army career prior to his posting to Shorncliffe can be gleaned from the published movements of his battalion, extracted below. It is not known when James joined the army, but the fact that he fails to appear in the 1861 Census suggests he joined up prior to that date. So he could well have served in the Far East, like his father. Army records show that in 1871 he was serving as a private (Service No 2093) in the 3$^{rd}$ Battalion of The King's Royal Rifle Corps stationed in Bellary, one of the chief military stations in Southern India. Bellary is just 200 miles north of Bangalore where his father had served some 35 years earlier.

| 18 | 61 | 62 | 63 | 64 | 65 | 66 | 67 | 68 | 69 | 70 | 71 | 72 | 73 |
|---|---|---|---|---|---|---|---|---|---|---|---|---|---|
| Jackatalla, India | X | X | | | | | | | | | | | |
| Thayet-Myo, Burma | | X | | | | | | | | | | | |
| Tonghoo, Burma | | X | X | | | | | | | | | | |
| Rangoon, Burma | | | X | X | X | | | | | | | | |
| Andaman Islands | | | | X | | | | | | | | | |
| Madras, India | | | | X | X | X | X | | | | | | |
| Bellary, India | | | | | | X | X | X | X | X | | | |
| Bangalore, India | | | | | | | | X | | | | | |
| Winchester Depot | | | | | | | | X | | | | | |
| Aden, Yemen | | | | | | | | | | | X | X | |
| Shorncliffe, Engand | | | | | | | | | | | | X | X |

*Locations of 60th Rifles, 3$^{rd}$ Battalion, in the period 1861-1874.*

According to the annals of his regiment, the 3$^{rd}$ Battalion embarked from India on 29 November 1872 and arrived in England on 24 December 1872 to be stationed at Shorncliffe. So it appears likely that nearby Sandgate is where James and Harriet met.

They were married on Thursday, 9 October 1873 at the Registry Office in the District of Elham. At that time the Elham registration district covered several towns and villages including Folkestone, so it is likely that the marriage actually took place near Shorncliffe Camp. James was then aged 30. Harriet also gave her age as 30, although she was actually 36. Possibly she had deceived James about her age. The witnesses at the ceremony were Francis and Mary White who lived at 2 Harvey Place in the centre of Folkestone, about one mile from the Shorncliffe Camp. Their relationship to James and Harriet is unknown.

It so happened that James' younger brother Charles was also in the army. In 1871 he too was based at Shorncliffe Camp as a private in the 6th Inniskilling Dragoons. (A dragoon is a soldier who travelled on horseback but fought on foot.) So it is possible that James' and Charles' postings to the Shorncliffe Camp overlapped. It is unlikely that Charles was still there in 1873 because he would surely have attended James' marriage as a witness. In fact, it appears from army records that Charles had been transferred to Aldershot Barracks.

Charles himself was married on 26 December 1876 to Eliza Sarah Nicholls at the church of St John the Baptist in Margate. Eliza was then 19 years old and had been living at 218 High Street, Margate. Her father James was a woodman. Charles was still in the dragoons and based at the Canterbury barracks where one of his duties was to give new recruits tuition in horse-riding. Probably he met Eliza while based in Canterbury because that is where she was born.

James left the army soon after his marriage and moved with Harriet to live near Minehead in Somerset. Their first child was Florence (Auntie Florrie), born in the parish of Dunster, in the village of Alcombe, near Minehead, on 12 November 1874. At that time James was working as a clerk.

It seems likely that James had relatives in Alcombe. There were certainly two families named Parsons in Alcombe between 1871 and 1881. The first was headed by Henry Parsons, an agricultural labourer married to Elizabeth, with two daughters Edith and Henrietta. The second was headed by George Parsons, also an agricultural labourer, married to Mary with four sons Frederick, William, Henry and John. Henry Parsons and family lived in Lower Staunton. George Parsons and family lived close by between Lower Staunton and Grove Place. If they were related to James it is not clear how. Also it is not known whether James and Harriet lodged with either of these families.

In about 1876 James and family moved from Alcombe to Loddington to live with, or near, Harriet's relatives. Florence's two brothers Henry James (Uncle Phil) and Thomas, my grandfather, were born in Loddington on 29 June 1877 and 20 October 1879, respectively.

# Thomas Parsons (1879 – 1967)

Thomas is my grandfather. At the time of his birth, on 20 October 1879, his father James was working in Loddington in the bakery at 12 Main Street owned by Harriett's elder brother George. But James and Harriett did not stay long in Loddington. Possibly there were too many people for the bakery business to support.

By 1881 James and family had moved to a terraced house at 40 Larches Street, Aston, Warwickshire (now Birmingham). Maybe there was a connection between this location and Harriett's earlier visit to Birmingham in 1855. James was working as a warehouseman and packer. The three children Florence, Henry and Thomas were then aged 6, 3 and 1, respectively. In the 1980s all the Victorian houses in Larches Street were pulled down to make way for a new housing estate.

By 1888 the family had moved to 259 Bradford Street, just half a mile from Larches Street and close to Birmingham's new city centre. The area still has many Victorian factory buildings and warehouses that look run down and ripe for redevelopment. James was now a Works Manager. It was about this time that James and Harriett had photographic portraits taken of themselves (Plate 6). The house where they lived no longer exists.

On 5 January 1888 James Parsons died prematurely aged 44 at 259 Bradford Street. The cause of death was pneumonia from which he had been suffering for 12 days. James' younger brother Charles was present at the death. Charles must have travelled the 70 odd miles from Buckingham barracks where he was then stationed.

After her husband's death Harriett did the sensible thing by returning to her roots in the Loddington area. In 1891 she and her young family were living at 5 Bowling Green Avenue in Kettering. Henry aged 13 was a printers apprentice and Thomas aged 11 was still at school. There were two lodgers, both tailors, who provided some income for Harriett.

The eldest child, Florence, often stayed nearby in Loddington with her uncle and aunt, George and Hannah Andrew, both unmarried and aged 62 and 60, respectively, and still living at what is now called "The Old Bakehouse" at 12 Main Street. George was still a self-employed baker. Florence aged 16 was a dressmaker. My grandfather Thomas stayed frequently in the 1890s both with his Uncle George in Loddington and with his Uncle John in nearby Aldwinkle St Peter.

By 1901 Harriett, then aged 63, had moved with her children to Leicester, to live at 18 Barclay Street. Florence aged 26 was still working as a dressmaker. Henry aged 23 was a print machine minder, and Thomas aged 21 was a shoe clicker, that is, a worker who cuts leather into the pieces that are later sewn together to make the shoe upper.

On 29 August 1902 Harriett died of emphysema and heart failure aged 64. She was then living round the corner from Barclay Street at 18 Browning Street. She had been working as a clerk in an umbrella factory up to her death.

Meanwhile, recall that James' brother Charles was a private in the 6th Inniskilling Dragoons. He had joined up in 1867, aged 18, to become a career soldier like his father. He married Eliza Sarah Nicholls in 1876 when his regiment was based in Canterbury. Their first child Florence Webster was born in Edinburgh on 2 March 1879 while Charles was stationed there. By 1881 Charles' army career had temporarily separated him from his wife and two year old daughter, who were then living with Eliza's uncle Octavus Webster, a wealthy tailor, and his wife Sarah at 40 London Road, 1 Auchers Villas, Canterbury, the city of her birth. That year Charles transferred to the 15$^{th}$ Hussars.

By 1882 Charles had been posted to Aldershot in Hampshire where his wife and young daughter joined him. A second daughter Ethel Maria was born in Aldershot in 1882. Two more children followed: Charles George born in Aldershot in 1883, and James Gordon born in Canterbury in 1885. Both children died young, in 1885 and 1888, respectively.

On 15 February 1886 Charles transferred to the Bucks Yeomanry Cavalry, later to become the Royal Bucks Hussars. By 1888, the year that his brother James died, Charles and his family were based at the barracks in Buckingham where Charles was now a Sergeant Major. There followed five more children all born in Buckingham: Katharine Eva born 1888, Victor William born 1889, Thomas Clarence born 10 November 1892, Maurice Reginald born 15 November 1893, and Jack Percival born 1896.

Charles' son Maurice Reginald had a daughter Dorothy Margaret born 15 October 1924 in Surbiton and living for many years in Cobham, Surrey. She is my second cousin once removed. She remembers Charles, the brother of my great-grandfather James, and has therefore been a valuable link to that generation. However, her side of the family had no knowledge of our side, and vice versa.

Charles retired from the army on 30 November 1894 after 26 years service. From December 1894 to 1900 Charles ran the Old Angel public house, probably located at North End near Buckingham.

On 20 February 1900 Charles, aged 51, reenlisted with the XVth Battalion of the 57th Company of Imperial Yeomanry (No 10375) to support the South Africa campaign. On 16 March 1900 he embarked for South Africa where he saw action in the Second Boer War. Charles returned to England on 6 October 1900 and was eventually discharged from the army on 18 April 1901. At this time the family, including seven children, were living in Buckingham at 14 Church Street.

On 16 April 1909 Eliza died aged 52 of heart disease culminating in a cerebral embolism (stroke) at 1 Bristle Hill, Buckingham. Her husband Charles, then an army pensioner, was present at her death. Charles died many years later on 23 November 1937 in Amersham Hospital, aged 88.

Cousin Dorothy recalled how, in his old age, her grandfather Charles was regularly collected by her father Maurice in his car to join them for Sunday lunch over their grocers shop in Surbiton. She recounted with humour that he was deaf and bad-tempered, and spent a lot of time in hospital during his army career as a result of often falling off his horse. Although Charles never spoke of James to Dorothy, the two brothers must have been close. They each had army careers that took them to the same locations, they each probably met and certainly married their wives while stationed in Kent, and Charles had made the journey to Birmingham to be present at James' death.

To return to my great-grandfather James and his wife Harriett, when she died in Leicester in 1902 her three children Florence, Henry and Thomas were aged 27, 25 and 22, respectively.

Florence (Auntie Florrie) married rather late in life, aged 43, to William Laney (Uncle Billy) who became a bank manager. Their wedding was on 11 June 1918. The reception was held at Hurtwood, Farnborough Park, Hampshire. The couple settled in a bungalow in Farnborough at 7 Brookwood Road, and never moved. They had no children. Florence died on 6 December 1968. Her husband Billy died much later on 8 October 1985.

Thomas' brother Henry became a music hall artist. It is a mystery how he chose this career because there is no prior history of show business in the family. Henry worked as a comedian and singer under the stage name "Phil Parsons". His career spanned the period from the early 1900s to about 1923. Although he never achieved top billing he became a successful supporting artist performing all around the UK, even doing tours of Ireland in 1910 and 1920. He was rarely out of work. His popularity can be measured by the large number of "hits" (175) in the Music Hall archives (www.archive.thestage.co.uk). Here are some of his reviews:

> *Phil Parsons, dandily attired in a green frockcoat and trousers, sings effectively "Choose her in the morning".* - Hackney Empire, Dec 1907.
>
> *Phil Parsons is a humorous entertainer; his sketches and monologues gain much applause.* - Newport Empire, Jan 1908.

*Phil Parsons delivers much sage advice in his song on the question of choosing a wife, and the breezy manner in which he sings adds greatly to the effect.* - Richmond Theatre, Apr 1908.

*Phil Parsons was among those who contributed to the programme, and was highly amusing in his song of the 'Pious Young Curate'.* - Pavilion, Torquay, Jul 1915.

*"Phil Parsons does well with two popular songs, as well as with an imitation of a chairman at a meeting of an international exhibition, which causes hearty laughter."* - Palace Theatre, Reading, Feb 1916.

Henry had a large repertoire of songs some of which he bought as signature songs. Examples of the songs he sang are: "I don't care what becomes of me" (1911), "Hello, who's your lady friend" (1914), "Good luck, Tommy Atkins; Good luck, Jack Tar" (1914), "Tommy's learning French" (1915), "Down Texas Way" (1917), "Tails Up" (1919), "Mr Pussyfoot" (1919), "Our Menagerie" (1920), and "I know where the flies go" (1920). One of his last songs was "Lil' old one pound notes" (1923).

The third child Thomas, my grandfather, worked in his teens as a shoe clicker. He was a fine sportsman. In the early 1900s he played football for the reserve team of Leicester City Fosse (now Leicester City). It is also possible he played for Oldham Athletic. My brother Nick has his boots and shin-pads.

More important to my grandfather were his running activities. He was a good runner and ran with Leicester Harriers. He often absconded from work, making credible excuses, in order to attend race meetings around the country. As part of the subterfuge he would run under an assumed name, which unfortunately makes his achievements difficult to trace. He won many prizes, including a medal for the Midland Counties AAA 880 yards flat in 1907, and a short case Vienna regulator clock made about 1890 that I inherited.

This clock was a poor timekeeper and stopped frequently for no apparent reason. My Aunt Margaret said that it became unreliable after being shaken by bombs in WW2 when they were living at 40 Henley Crescent. I eventually gave it to our daughter Isabelle, who reported in 2011 that after an overhaul it was working perfectly (until recently).

Thomas trialled to represent England in the 1908 Olympics. He failed to make the grade but received a medal for his efforts. Two of his running medals are in the possession of my brother Nick. One has MCAAA (Midland Counties Amateur Athletics Association) 880 YARDS FLAT 1907 written on the box and 880 YARDS FLAT CHAMP 1907 engraved on the medal; the other has AAA OLYMPIC GAMES TRIAL RACES 1908 on the box and AAA 1908 T. PARSONS engraved on the medal. He won other running medals including a gold one that he sold for two gold sovereigns when he was out of work shortly after his marriage in 1908.

The two brothers, Henry and Thomas, got on well together. Before they each married, Henry would accompany Thomas to racing events and Thomas would accompany Henry on his music hall rounds, often entailing three different venues on the same night. They apparently had a good time together.

Violet Blanche Bradsell, who was to be Thomas' wife and my grandmother, was born on 7 January 1893 at St Johns Wood. Her parents were Thomas Bradsell and Hannah Rosetta Pond Bradsell, known as Rose, maiden name Brewer. Thomas Bradsell was then working as a bailiff's clerk. Rose and Thomas went on to work in the licensing trade at various places including the Railway Hotel, Billericay, Essex.

Violet was the eldest of three children. Her brother Herbert Thomas (Uncle Bert) was born on 26 April 1894. He married Daisy Kemsley in Willesden in 1913 and died in 1983. Violet had a sister Doris May born on 30 December 1897, later to emigrate to Australia with her mother. A fourth child Ernest Leslie born on 28 June 1896 died an infant in hospital of a fever on 26 May 1897.

On Saturday 15 August 1908 my grandfather, Thomas Parsons, married Violet Bradsell in St Michaels Church at

Buslingthorpe, Leeds.  It may seem strange that Violet travelled all the way from her home in St Johns Wood to get married in Leeds.  A likely explanation is found in a publicity photo of Phil (Henry) Parsons supplied by Maureen Hudson, and taken by professional photographers J Bacon & Sons of Leeds.  So Leeds is clearly where Henry, and probably Thomas Parsons were based at that time.  Violet was just 15 when she married Thomas, who was 28.

It is of interest to note after 1837 the legal age of marriage was as low as 12 for a girl and 14 for a boy, although the consent of parents was required for both parties up to the age of 21.  It was not until 1929 that the legal age of marriage for both parties was raised to 16, but parental consent was still required up to the age of 21.  Now, the age at which people can marry is still 16 but the age for consent has been lowered to 18.

On 31 December 1909 Thomas and Violet Parsons had their first child, Frank Philip James, born in Walthamstow at 119 Winns Avenue.  Frank was baptised at St Luke's Church Walthamstow on 27 February 1910 and later confirmed on 22 March 1925 at St Mary's Church Prittlewell.  In October 1910 the family was living at 20 Roman Road, Walthamstow.

By April 1911 the family had moved to 237 Edgware Road, Broadway, Cricklewood.  Violet was still only 18 years old and had been married for more than two years.  Thomas' elder brother Henry was also living at the same address.  This was the site of the cycle shop that Henry had bought as an investment with his earnings on the stage.  Thomas was employed by brother Henry to manage the shop and do the cycle repairs. It appears to have been a profitable business.  It is likely that Henry rarely worked in the shop owing to his many music hall commitments.

It seems that everyone now knew Henry by his stage name Phil, which even appears on the 1911 Census, so from here on I shall refer to him as Phil, or Uncle Phil.  In the 1911 Census he is listed as unmarried with the occupation of motor dealer.  His music hall career is not mentioned.  Thomas is listed as a cycle maker and dealer.

On 23 July 1911 a second child, Florence Harriett Violet (to be known as Cissie or Chris) was born to Thomas and Violet. Cissie was baptised at St Michael's Church Cricklewood on 20 August 1911 and later confirmed, together with her brother Frank, on 22 March 1925 at St Mary's Church Prittlewell.

Some time between 1912 and 1915 Phil Parsons sold the cycle shop, possibly to raise cash for his impending marriage, of which more later. The sale left Thomas without a job and the family had to move house. There was possibly bad feeling about this.

On 18 October 1915 a third child, Thomas, my father, was born.

# Thomas Parsons (1915 - 2001)

On 18 October 1915 my father, Thomas, was born at 84 Audley Road, Hendon NW14. He was always known as Tom, or Tommy. He was baptised at St John's Church, Hendon, on 10 September 1916 and later confirmed on 14 March 1926 at St Mary's Church, Prittlewell.

Nine days after my father's birth his uncle, Phil Parsons, married Harriett Mary Josephine Brown. The following is the remarkable story of their marriage.

Harriett Brown was from Camberwell, South London, born in 1891. She was a Chorus Girl who performed as a dancer in London theatres. In 1910 she had become pregnant by a solicitor who was much interested in the theatre. Her baby was born illegitimate on 11 December 1910 at her parents' home at 267 Camberwell New Road, South London. The child was a girl whom she named Haidée Millicent. The mother re-spelt her surname as Browne on the birth certificate.

At some point between 1911 and 1915 Harriett Browne met Phil Parsons, no doubt in the theatre context. They became partners. On 27 October 1915 they married at the Register Office in Camberwell. Uncle Phil was then aged 38 and Harriett was 25. Uncle Phil accepted young Haidée as his own daughter. She was brought up to believe that Uncle Phil was her real father and that she was two years younger than she really was. This subterfuge was probably to let Haidée believe she was born in wedlock, because at that time there was great stigma attached to children born to unmarried mothers and as far as possible such things were kept quiet.

At the time of his marriage Uncle Phil was living with my grandparents, Thomas and Violet Parsons, at 84 Audley Road, Hendon, together with their children Frank and Cissie, and my father Thomas, born nine days before Uncle Phil's marriage.

It is not clear where Uncle Phil and his wife and daughter lived immediately after their marriage. They may have lived with my grandparents, but more likely they lived elsewhere. In any event they certainly lived close by in Hendon because it is known that Haidée and Cissie grew up closely together.

Whereas Haidée's parents were rather liberal, being theatrical and bohemian people, Cissie's parents were more strict, especially Grandma Parsons. Grandma got to know Haidée well and throughout their lives they remained close. Possibly Grandma Parsons looked after Haidée when her father and mother were working away.

Now jump forward several years. In early 1933 Haidée had a baby girl whom she named Yvonne. The father was William (Billy) Peters, a divorcee. On 7 June 1934 Haidée aged 22 (but in truth 24) married Billy Peters aged 32, at the Register Office in Hendon. Billy Peters was then working as a builders' merchant while living together with Haidée at 8a Devon Parade, Kenton. Haidée's mother Harriett was a witness at the marriage.

Shortly after their marriage, on 3 August 1934, Haidée's mother Harriett died aged 43. What began as a minor infection following a visit to the dentist developed into blood poisoning that turned out to be fatal. She died at the Hospital for Epilepsy and Paralysis in Marylebone. This loss must have further cemented the bond between Haidée and Grandma Parsons that had begun to develop when Haidée was young. Curiously, on her death certificate Harriett is referred to as Haidée. So it appears that Harriett had at some point renamed herself Haidée. Certainly, I recall her being referred to as "Big Haidée" to distinguish her from her daughter.

At the time of Harriett's death her address and that of Uncle Phil was 190 The Broadway, West Hendon, where Uncle Phil was once again running a business as a cycle agent, having been retired some 10 years from show business.

At the end of his life Uncle Phil lived at the house of his daughter Haidée and Uncle Billy at 20 Ilmington Road, Kenton, where Haidée nursed him through his last weeks. Uncle Phil died on 8 August 1954.

Although it is hard to believe, Haidée (the younger) knew nothing of her origins until 1972 when she applied for her old-age pension, believing she had reached the age of 60. The Pensions Office was unable to reconcile the dates in their records, leading to extensive searches by Haidée at Somerset

House. Eventually she discovered that she was 62 and that her biological father was not Uncle Phil. The Pensions Office dutifully back-paid the two years of pension to which she was entitled. Haidée coped well with this bombshell of news about her birth. It helped that she had been much loved by Uncle Phil as well as by her mother. Haidée died in 1994, aged 83, of heart failure. She outlived her husband Billy who died in 1985, also at the age of 83.

To return to my father, soon after his birth in 1915 the family moved round the corner from Audley Road in Hendon to 42 Bertram Road. It was the middle of WW1. Thomas Parsons senior, then aged 36, registered under the National Registration Act of 1915 as a Fitter (Sheet Metal). In 1917 he was working in the Army Reserve as an Engine Fitter at the Aircraft Manufacturing Co Ltd in West Hendon. This was the site of an important Military Airfield, now an RAF Military Museum.

After the war, circa 1919, the family left Uncle Phil and his family behind in London and moved to Southend closer to Violet's parents. It is not known where they moved to in Southend, but in the period 1925/6 to 1928 they were living at 15 Gainsborough Drive. Thomas had several jobs in the 1920s, working for at least two years as a fitter at Arthur N Sims, Cycle Manufacturer, of Ronald Hill Grove, Leigh-on-Sea.

Grandfather Thomas was not a religious man, but his wife Violet was a keen member of the Church of England. Early on in Southend she joined St Mary's Church in Prittlewell where she became a Sunday School Teacher and an ardent admirer of the vicar, Canon Ellis Gowing, who had been at St Mary's since 1917 (and died in post in 1960).

Circa 1924 my father Thomas joined the choir at St Mary's at the age of eight. He was delighted to be given time off school to attend practices. The organist and choirmaster was the renowned Mr Penny who, like the vicar, gave long service at St Mary's. Tom eventually became head chorister and soloist. Throughout his life he remembered in detail all the music he had learned as a boy. For example, in the year before he died he could still remember the different singing parts of Hubert Parry's anthem "I Was Glad" for double choir. He always spoke with great pride of his time in St Mary's choir.

My father had views about church choirs, for example, that the trebles should comprise boys only. There was no prejudice, simply the opinion that a boy's voice had greater purity than a girl's. Another opinion was that there should be a crescendo but no unnecessary slowing up at the end of the last verse of a hymn.

Whereas my father was undoubtedly a keen and talented choirboy he seems to have been less enthusiastic about other church activities. In 1928 at the age of 13 he received a certificate from the Church Missionary Society for passing the examination on "Talks on Friends in Africa" in the 3$^{rd}$ Class, which sounds like a euphemism for failure. The certificate was signed E W Gedge, Secretary of the Young People's Union.

In 1926 (possibly 1928) Violet Parsons' sister Doris emigrated to Australia to settle in New South Wales. To everyone's surprise her mother Rose decided to accompany her daughter, leaving behind her husband Thomas in Rochford. Rose died in Gosford NSW on 12 August 1955, outliving her husband who died in Rochford on 9 December 1943. In the 1980s my Uncle Frank made the long journey to Australia to visit Doris shortly before she died.

On 4 October 1928 grandfather Thomas was interviewed at 6 p.m. (prompt) for the job of caretaker at Southend Central Library in Victoria Avenue. Of the 119 candidates, 7 were selected for interview. Thomas was finally selected by a sub-committee. There was good reason for celebration because the job was both secure and pensionable.

This success may have triggered their subsequent house move towards the end of 1928 to "Miquette", Wells Avenue, Rochford, where on 25 November 1930 Thomas and Violet had a fourth child, Margaret Christine. Margaret was baptised on 19 December 1930 at St Mary's Church, Prittlewell.

At this time the other children Frank, Cissie and Thomas were aged 20, 19 and 15, respectively. Frank was learning his trade as a butcher, Cissie was training to be a nurse, and my father Thomas had already left school at the age of 14 to become a Post Office Messenger delivering telegrams. In the mid-30s he was promoted to postman, a job he greatly enjoyed. In later years he would recount amusing anecdotes of his time as a postman. He learned to drive with the Post Office and was given a licence without ever having to take a test.

In 1931 grandfather Thomas was promoted at Southend Library from $3^{rd}$ to $2^{nd}$ porter. In September 1931, despite glowing references from the Borough Librarian and Rev Ellis Gowing of St Mary's Church, Thomas failed to secure the job of Caretaker at Southend High School for Girls. Thomas continued working at the Library until his retirement in April 1946 aged 66½.

In 1933 their daughter Cissie married Donald Hudson, a police officer. They had three children: Donald, Paul and Douglas. Cissie died in 1989, and her husband Donald in 1996.

Soon after Cissie's marriage, probably in early 1934, the Parsons family moved with all the children, except Cissie, to rented accommodation at 40 Henley Crescent. They named the house "Loddington" after Thomas' birth place.

At that time people had to pay for medical treatment, there being no National Health Service until 1948. Throughout the 1920s and 1930s Thomas Parsons was a member of the Southend-on-Sea Hospital Provident Fund, paying 3d (about £0.90) per week, collected monthly. This entitled him to free basic hospital treatment on the recommendation of a GP. Violet continued to be a keen and active member of St Mary's Church, attending services, helping with Sunday School teaching, and helping with the various charitable events. Her passion was not shared by her husband Thomas.

Violet was a closet poet of some talent. In later life she had much of her work published in various poetry anthologies. Her poems are about unfulfilled love. It appears that the family was uneasy and disapproving of the sentiments expressed in her poetry and did their best to ignore it. Below are three examples of her work.

| DESTINY | PETITION |
|---|---|
| An aching heart that will not rest,<br>A sigh for a face<br>A desire for one's best<br>A longing to touch the object of this<br>A thanks to my God for so wondrous a Gift.<br><br>If a Thousand years hence<br>He should but glance my way<br>In one swift second of time I'll know<br>For my heart will say<br>He was my love of yesterday. | I do not ask for Castles or for gains<br>Neither a wish for beauty to attract<br>And make me vain,<br>And for the things I may achieve<br>Not to covet fame.<br><br>But give me Lord the deep deep love<br>For one good man<br>Whose power can fan the spark within<br>Into a flaming fire<br>That nothing, no not circumstance or age<br>Will ever dim so pure a flame. |

## HUMILITY

Tis the twilight hour
When I sit and muse
And make believe he visits me
We would sit and talk of mundane things
So happy I be.

And sometimes he would bring me flowers
Whose fragrance lingers on
Long long after that he is gone
And comforts me

And often as I pass his chair
My fingers would ruffle thro his hair
And he'd look up and smile at me
Such happiness for me.

And when its time for him to leave
He'd take my hand
The pressure of his fingers makes me glow
Like the setting of desert sun upon the silver sand.

And having said goodbye
I'd gently close the door
And in a supplicating mood,
Kneel upon the place where he'd just stood.

Dear Lord let me keep my twilight hour
From Sunset to rise let me dream
Bless me on waking to greet a new day
And Bless him dear Lord. Bless him too.

My wife, Lucienne, developed a close relationship with Grandma Parsons soon after we became engaged in 1964. Grandma told Lucienne that she had been unhappy in her marriage, which she said she entered into as an escape from the tedium of family chores imposed by her mother, in particular, all the washing-up. Her poetry was an escape into a world of what might have been with a certain man who could have loved her. At the time I was totally ignorant of all this. I certainly always had the impression that my grandparents were happy together.

Violet Parsons died on 4 October 1984 in Southend. The funeral was on 12 October at St Mary's Church, Prittlewell. She outlived her husband Thomas who had died 17 years earlier on 7 August 1967, just one month after the marriage of their youngest child Margaret to Malcolm Edwards.

Many years later in 2002, soon after Frank's death, I visited my grandparents' bungalow in Hobleythick Lane with Donald and Maureen Hudson to prepare it for auction. It evoked memories of times past but otherwise felt cold and empty. I found all Violet's poetry anthologies in the bookcase turned with their spines facing inwards thus hiding the titles from view.

Meanwhile, Dorothy Besent (my mother, born in Westcliff on 21 October 1911) was living in the early 30s with her parents and elder brother Harold. Their upbringing was middle class in the strict Methodist tradition. For example, they both took a solemn vow to abstain from alcohol. Harold kept to his vow throughout his life. My mother occasionally lapsed to be sociable. She always opted for cinzano and lemonade, which she did not really enjoy.

Both Harold and Dorothy worked in their parents' shop in the 1930s. They went on several holidays together, usually in association with their church or tennis club. Both Harold and my mother were keen ramblers.

My cousin Jean discovered a fascinating account of one of these holidays. It was written by Harold in 1933 about their walking holiday in North Devon. With the help of an old photograph album that had belonged to my mother, I was able to piece together the walks they had done and the places they had visited. In May 2018 I spent a memorable holiday with my brother Nick and cousin John in North Devon, where we trod the paths and enjoyed some of the beauty spots that our parents had visited all those years ago. I wrote up the experience, based on Harold's account, in an article that was published in the 2018 September/October issue of the "Devonshire Magazine", currently available on-line at http://devonshiremagazine.co.uk/.

In 1936 Harold married Jennie Irving in Gateshead. Soon after their marriage they bought 11 Henley Crescent and called it "Willersley", after Willersley Castle in Derbyshire where they first met in 1934 at a Methodist Convention holiday. Dorothy, my mother, had also attended this Convention. In June 1939 Harold and Jennie had their first baby, named June, who sadly died within a day or so of birth.

In her youth my mother was a keen tennis player. She joined St Mary's Lawn Tennis Club based in Priory Park. She must have been a good player because in 1938 she won the Ladies Singles Championship for which she was awarded a small silver cup, now in my possession. My father also became a member of St Mary's Lawn Tennis Club, as did their life-long friend Rene. It was here in 1938 that Tom met and began courting Dorothy. My father at that time was still working as a postman.

September 1939 saw the outbreak of war with Germany, the start of WW2. Soon after, on 6 March 1940, my father Thomas enlisted in Southend to join the army. He decided to enlist rather than await conscription[7] because he wanted the option of choosing his regiment. He chose the Grenadier Guards, probably because it was the most senior of the infantry regiments, and also because he met the exacting 6' 2" minimum height requirement (since reduced to 5' 10"). He was given the service number 2618314.

After training at the Guards Depot at Caterham, Thomas was posted on 11 May 1940 to the Training Battalion at Windsor Barracks. Soon after, on 5 August 1940 while stationed at Victoria Barracks in Windsor, he was promoted to Lance Corporal. He announced this achievement in an undated letter to his parents. He describes in his letter trips to Oxford and Brighton to collect "aliens" that he took to Kempton Park. The race course at Kempton Park had been

---

[7] Conscripted National Service in the UK existed for two periods in modern times, the first from 1916 to 1919 and the second from 27 Apr 1939 to 31 Dec 1960, the last conscripted soldiers leaving the service in 1963.

transformed into a holding camp where prisoners were interrogated before being despatched to other camps.

Two further undated letters from my father to his parents are in my possession. The first must have been sent soon after 11 December 1940 because he refers to bombings in London Road, Westcliff that occurred on that date. He mentions his sprained wrist that excused him from army duties for a few days, and he expresses his view that the war would surely be over by the following Christmas. The second letter must have been much closer to Christmas because he says he does not really need any Christmas present from his parents, who perhaps were short of money.

In the summer of 1940 Dorothy moved to Maidenhead, near Windsor, to be closer to Tom. She lived in digs at 21 St Leonards Avenue. In August 1940 she started work in a nearby fruiterers shop called International Stores. The move to Maidenhead must have required initiative on the part of my mother because no doubt she received little encouragement from her parents. My mother kept several passionate letters from my father, now in my possession.

Harold was at this time working at his parents' shop in Westcliff, but the business could no longer support both his parents and his own family. In July 1940 Harold began looking for work elsewhere in the retail fruit and vegetable trade. He wanted an opening with the potential to buy into a suitable business after proving his worth and saving the necessary funds. He found a job with a firm called Eastwell's as manager of a fruit and vegetable shop in Barnet, North London. Shortly before the start of the Blitz in September 1940 Jennie left Henley Crescent to join Harold in Barnet. At this point 11 Henley Crescent became empty. Harold and Jennie lived in rented accommodation in Bosworth Road, Barnet. Harold was soon appointed Managing Director of Eastwell's when it became a limited company. The business thrived, and they had at least one other shop in Russell Lane, Whetstone, until about the 1970s.

My parents were married in wartime by the Rev Ellis Gowing on Wednesday, 29 January 1941, at St Mary's

Church, Prittlewell, where my father had been a chorister. My mother was dressed rather austerely in an outfit she had made herself (Plate 15). It is a mystery why, as a dressmaker, she did not wear a white wedding dress like other wartime brides whose photos appear on the Internet. The reason could not have been clothes rationing, which did not begin until 1 June 1941.

My father is dressed in his army uniform. It will be noted (Plate 15) that the sleeves of his uniform bear two chevron stripes denoting the rank of Corporal, whereas his army records show he was never promoted beyond Lance Corporal. This anomaly was resolved by Guards Archivist Lt Col Conway Seymour who replied to my enquiry as follows: "Lance Corporals in the Guards Division wear two chevrons. Corporals are appointed Lance Sergeants and wear three. This is reputedly because Queen Victoria had thought that one chevron looked unseemly on her Household Troops."

Unfortunately there is no group wedding photo, but both sets of families must have been present because, as noted in my father's army record, the marriage ceremony was witnessed by the couple's two fathers. My father was still in the army stationed at Windsor, so presumably they began married life living apart.

Before his marriage my father's army record shows his next-of-kin to be his parents, living at 40 Henley Crescent. After his marriage his next-of-kin is amended to his wife Dorothy. Her address is also given as 40 Henley Crescent. Possibly she was living with her in-laws while Tom was away in the army.

Soon after the marriage my father became ill and was diagnosed with a stomach ulcer. This resulted in his discharge from the Grenadier Guards on 8 April 1941, less than three months after his marriage. His army record states that he was "permanently unfit for any form of Military Service .... character good". He returned from Windsor to work again at the Post Office in Southend. It is possible that he and Dorothy then rented 11 Henley Crescent from Harold and Jennie for a short period.

In July 1942 Harold and Jennie, together with Jean who was born in Barnet in September 1941, moved from their rented accommodation in Barnet to a house they bought at 90 Lynton Mead, Totteridge [8]. This is when my parents, Tom and Dorothy, bought 11 Henley Crescent from Harold and Jennie. My cousin John was born on 6 May 1944 when his parents and daughter Jean were living at Lynton Mead.

On Sunday 6 June 1943 I was born in the front room of 11 Henley Crescent.

---

[8] My cousin Jean has thoroughly researched the genealogy of the Besents, my mother's side of the family. Much of her work is written up in "Though Dynasties Pass" by Caroline Ingram & Jean McCarthy, printed by Fineline Print & Copy Service, Perth, W Australia (2004), ISBN 0 646 43675 9.

# BOOK TWO
# Childhood (1943 – 1954)

# Early Memories

My parents were married in wartime on Wednesday, 29 January 1941. My father was serving in the Grenadier Guards at Windsor Barracks having been granted special leave. The marriage was at St Mary's Church, Prittlewell, Essex, where my father had been head chorister.

I was born on Sunday, 6 June 1943 in the front room of 11 Henley Crescent, Westcliff-on-Sea, Essex, inside a Morrison air-raid shelter that resembled a reinforced rabbit hutch. My mother later recounted that she had never felt better than when she was carrying me, but the birth itself was difficult with only a midwife present. My younger brother Nick recalls that my mother vowed to have no more children after me, but she clearly relented. My father often recalled how a kind neighbour, Mr Ernie Barnet, knocked on our door on the night of my birth with a bottle of whisky to celebrate. This was generous because during the war such luxuries were scarce.

Unlike many local people we did not evacuate to safer havens. There were just a few of us left in Henley Crescent. My parents occasionally recalled how the German V-1 rockets, better known as doodlebugs or buzz-bombs (early examples of cruise missiles) could be heard over our house. The fearful moment was when the buzz from the motor stopped, indicating that the bomb was about to fall and explode. Such a bomb destroyed the manor house close by at the top of Hobleythick Lane. Another did significant damage to the school hall and organ at nearby Southend High School where I was later a pupil. These are examples of the few episodes that my parents told me about "The War". Generally, it was not a subject they discussed.

I was christened Alan Thomas at St Mary's Church, Prittlewell, where my parents had married. Thomas was my father's Christian name and is common to many ancestors on both sides of my family.

My godfather was Uncle Billy (William Laney) who lived in Farnborough with his wife Auntie Florrie (Florence), the sister of my paternal grandfather. I never saw much of Uncle Billy or

Auntie Florrie. My godmother was Rene (Irene) Cabuche, a close friend of my parents. She lived nearby in Southend and I saw her frequently. Rene was an attentive godmother and always remembered my birthday. She never failed to give me interesting presents, unlike many of my relatives.

Shortly before my birth, Rene's younger brother Alan was killed during the war with Germany. His full name was Alan Henry Cabuche, born in 1923 in Rochford, son of Eugene and Edith Cabuche of Southend-on-Sea. At the outbreak of WW2 he joined the Royal Artillery Unit (70 Anti-Tank Regiment) as a Gunner, No.1137510. He died on 19 May 1943 at Maidstone, Kent, and is buried at Sutton Cemetery, Southend, Plot O, Grave 9879. I believe his death was the result of injuries incurred from falling off the back of a lorry. His death was a great shock to Rene's family and also to my parents. The significance of this is that I was named after him.

The following account of my christening celebrations is according to my Auntie Margaret, my father's sister. After the service my maternal grandparents invited everyone back to their house. We assembled in the room over their shop at 314 London Road, Westcliff, together with my father's family. Uncle Frank, my father's elder brother, produced a bottle of sherry for the occasion, at that time an expensive offering. In my experience Frank was a man who throughout his life was always most generous and showed no malice to anyone. But the offering did not go down well with my mother's mother who was a Methodist and strictly teetotal. Her two children Harold (my uncle) and Dorothy (my mother) had been raised in the same tradition. My father's mother was offended by the rejection of Frank's sherry.

I cannot recall ever seeing my two sets of grandparents together either socially or at festive celebrations, but such family divisions are not unusual. I had a happy relationship with both sets of grandparents. It is possible that any coolness between them was because my maternal mother thought her daughter had married below her station. A more likely explanation is that my mother's mother was a devout Methodist whereas my father's mother was an equally devout

Anglican. They were both strong-willed women who dominated their husbands with personalities that no doubt clashed.

My maternal grandfather, Thomas Besent, known as Fred after his middle name, had a most generous and cheerful disposition. He was a skilled horticulturalist and a passionate gardener having trained in the profession at Mentmore Towers in Buckinghamshire where his father Thomas had been farm bailiff. His ancestors were all prosperous Dorset farmers[9]. His wife Gertrude was a handsome woman. Her father had been a bank manager in Marlow, Buckinghamshire, and a well-respected member of local society.

In 1909, while on their honeymoon in Southend (yes, it was viewed as a romantic venue), my maternal grandparents had seen a greengrocers shop for sale at 80 Leigh Road East, Westcliff-on-Sea, which they promptly bought. A few years later they moved from Leigh Road East to the aforementioned shop at 314 London Road, Westcliff-on-Sea. They lived over the shop and expanded the business selling fruit, vegetables and flowers.

I have a vivid memory of standing in the centre of the shop surrounded by displays of produce. My grandparents employed a female shop assistant who was always kind to me and remained close to the family, but her name eludes me.

In the early days my grandfather grew large quantities of flowers and vegetables for the shop in their back garden and on his three allotments. He must have worked hard. My mother and her brother Harold both worked in the shop. My grandmother was the business woman whereas my grandfather was too generous and trusting, allowing customers to run up large debts that were often never repaid.

---

[9] The father of my grandfather was also named Thomas. If this latter Thomas had not been disinherited for marrying the woodcutter's daughter I might now own a large area of Dorset farmland. Sadly, Thomas' young wife died shortly after their marriage, after which he remarried. The fact of Thomas' first marriage was not discovered until the mid 2000s, before which the reason for his disinheritance had been a mystery.

But the business clearly prospered because as early as 1928 they had a telephone (Southend 4180), but they never owned a car.

A related memory is of visiting the back garden of a neighbouring shop called Cramphorns that sold hardware, seeds and grain. The owners kept a nanny goat. I was offered some of its milk which I disliked. On another occasion I found a nest of woodlice under a box in their back garden. I was fascinated by the way they rolled into little balls when touched. Apparently I called them "woodies".

On leaving my grandparents' shop my mother and I often called in at a neighbouring fishmonger just up the road near the junction with Hamlet Court Road. I enjoyed the artistic displays of fish laid out neatly on white tiled slabs. My mother often bought herrings or herring roe. I particularly liked the creamy texture of the latter. Fish was cheap because it was in abundant supply. Herrings were probably the cheapest fish of all and the roe was almost given away.

These memories must all relate to a period when I was less than 6 or 7 years old because my maternal grandparents retired to 15 Highfield Grove in 1949/50 after selling their business. My grandfather and grandmother were then about 73 and 68, respectively.

A vivid memory is when I first wore spectacles. My poor eyesight was identified when I was only 3 or 4 years old. I recall being pushed by my mother in my push-chair down Hobleythick Lane. We were on our way home from the optician's shop in Westcliff and I was wearing my new glasses for the first time. They felt weird. As we passed the spinney on the corner of Carlingford Drive I looked up at the tall trees and was amazed that I could see all the twigs in the treetops. This was a new experience for me. It was probably autumn because dusk was falling and the trees had lost their leaves, leaving the bare branches in stark contrast against the darkening sky.

From then on I never took off my glasses except when I went to bed. Without them everything was one big blur. I did not even take them off to go swimming. This brings back an early memory, the one and only time that I did take off my glasses to go in the water, when I became frantic because I could not locate my mother amongst the crowds on the Westcliff beach when I left the water. It was as if I was blind. This occurred soon after I had my glasses so I must have been about 5 years old. The experience made such an impression on me that even in my teens I kept my glasses on when I went swimming with my friends in the public pool or in the sea. They were the old-fashioned National Health design that hooked round the ears and so stayed on even if I dived into the water.

I had my eyes examined regularly by a doctor at the top of Crowstone Road, Westcliff. He was an Indian, one of the few coloured people at that time in Southend. I regarded him as a strange curiosity. When I was six he referred me to Moorfields Eye Hospital in London to have my eyes examined more closely. I remember the visit and the lengthy tests but only now do I understand the reason and the purpose of it all.

The other fear that was comparable with losing my glasses was going to the dentist. I dreaded these visits that were frequent because of the poor state of my teeth. I often had toothache. I do not know why this was because I did not eat many sweets and I cleaned my teeth regularly. When I was very young my mother actually took me to a private dentist in Chalkwell Avenue, a rather affluent area of Westcliff. Maybe she thought it would be less traumatic for me. In the immediate lead-up to the appointment she would sit me in our hallowed front room with a Rupert book to calm me down. If I was having an extraction (always with gas) we went there by bus and returned by taxi. All this must have been very expensive for her.

Perhaps this is why my mother later returned to NHS dentists. After a bad experience with the schools dentist I went to Mr Cutler at the bottom of Southend High Street. This was also not without its trauma. I must have been about 11. Mr Cutler said that two of my second teeth, two lower incisors, had grown out of the gum in juxtaposition. His partner in the upstairs surgery was called down to examine me and confirm the opinion. They both said they had never seen anything like that before. An extraction was necessary. The dentist injected my jaw with a local anaesthetic and sent me back to the waiting room. In those days it took about 15 minutes to take effect, like waiting for an execution. I had the extraction and on leaving the surgery, which was on the first floor, I passed out and fell down the stairs. I never ceased to be terrified of visits to the dentist. The countdown would start a week in advance. The fear was not helped by my father's often-told anecdote about the pain he experienced on having all his teeth extracted. The feeling of horror of dentists has never really left me.

All this may sound a bit depressing and give the impression that my early life was a series of unpleasant experiences. This is far from the truth. It is just that the traumas stick in the memory.

When I was very young my mother took me regularly to the Westcliff beach in summer months. It was only a short walk from my grandparents' shop. The route took us down Hamlet Court Road, past Woolworths, over the railway bridge and down to the seafront, usually to the paddling pool. This was an ingenious facility. The tide on that part of the Thames Estuary retreats more than a mile over the mud flats so at low tide there is no water to swim in. Westcliff paddling pool is a three-sided enclosure made of concrete and wood, about 60 x 30m, built on the shelving part of the beach to trap the water with each high tide to a depth of about 1m in the deepest part. The water would slowly leak through the retaining walls but there was always enough left for swimming before it was replenished by the next high tide. This is where I taught myself to swim, doing the doggy paddle.

Another memory is of Bucknells the butchers on Earls Hall Parade where my mother bought all our meat. The precise memory is of sitting in my push-chair inside the shop and surveying the sawdust scattered over the floor. This was to soak up the blood on which people might slip, an early example of health and safety. Apparently I was watching the butcher skin a rabbit. This was cheap meat highly sought after in those times of food shortage because rabbit, like other game meat, was not rationed. My mother prepared rabbit regularly, usually in tasty casseroles. They must have been wild rabbits because the meat had a strong gamey taste. This was before the dreadful plague of myxomatosis that put people off rabbit for more than a generation.

This experience of butchering put me off eating meat, so my mother registered me as a vegetarian. This enabled her to exchange my allocation of meat ration coupons for extra cheese coupons. I was fond of cheese. Of course it was Cheddar that we bought with the coupons. It was always fine quality in those days and was the only English cheese that did not crumble when cut and served in the shop, so producing less wastage.

Most foods were rationed both during and after the War. The weekly allowance for an adult included 2oz of cheese, 1 egg, 4oz of bacon or ham, and other meat equivalent to two chops. Consequently, we ate more vegetables, particularly root vegetables. Rationing did not completely end until 4 July 1954 by which time I was a carnivore again. Interestingly, because of the abundance of fish the staple fare of fish-and-chips was never rationed.

Apparently I had deep dimples in my cheeks. Neighbours and relatives were always making remarks like, "Hasn't he got big dimples". This was supposed to be complimentary but I found it embarrassing. It made such an impression on me that when at age 18 I applied for my first passport and encountered the question about distinguishing features, I wrote "dimples".

Several other memories relate to my push-chair. One memory is of being pushed home by my mother in bright afternoon sunshine having just collected my treasured weekly copy of Enid Blyton's magazine "Sunny Stories" from the local sweet shop, run by Eileen and her mother Mrs Shannon. My mother often read to me and I was a good reader from the age of four. I was a great fan of Enid Blyton and remain so.

A neighbouring shop that I often visited in my push-chair with my mother was Jepps the grocers where we bought all our eggs, flour, sugar, biscuits, ham, bacon etc, all sold loose.

If you bought a dozen eggs they always gave you 13 because more than likely one of them would be broken or rotten. When I went into this shop in winter my glasses instantly steamed up making it impossible to see. There must have been a high humidity in that shop. The optician once gave me a small booklet of special paper tissues that you tore off and wiped on your lenses to stop the condensation forming. It worked quite well. I have never since seen tissues like this (until recently at Campbells, the opticians in Wantage).

These memories of the push-chair must all relate to a period before I was six years old because after that I was able to propel myself on a red tricycle that my parents bought me. I would ride it alongside my mother to visit her parents in Highfield Grove. I think we visited them several afternoons each week, usually after I came out of Earls Hall primary school in Carlton Avenue. My mother met me after school bringing my tricycle with her. This must have been quite an effort.

I learned to ride my tricycle at breakneck speed round the narrow rectangular concrete path in our back garden that my father had built to provide an edging for his central vegetable plot. At the two bottom corners of the plot my father had planted a pear tree and an apple tree with a small herb garden in between. In front of this he built a greenhouse. The large plot nearest the house was for flowers and vegetables. When I rode my bike round the path I took the corners at high speed on just two wheels. I was told off for doing this because I was damaging the crops bordering the path.

Before I went to school I often played on the doorstep with my bricks. I had a remarkable book called something like "Things to Do". It was simple and beautifully illustrated with black-and-white photos. It explained how to make printing blocks out of potatoes, sculptures from bars of soap, invisible writing with lemon juice, mobile toy vehicles made from matchsticks and cotton reels powered with elastic bands, communication systems made with two tin cans and a taut length of string, miniature divers in milk bottles, periscopes to look covertly over high walls, self-propelled boats made from balsa wood and a length of bent metal tubing filled with water that was heated by a candle, and many other interesting activities. I do not know what happened to this treasured book. I liked making these things on the front porch, or more likely I was put there so as not to make a mess indoors. I spent a lot of time happily playing by myself when I was young.

There were two milk rounds. One was a horse-drawn cart from Cow & Gate, but our milkman was from the Coop. He used a relatively modern hand-pulled pedestrian-controlled float assisted by an electric motor. I somehow saw the horse and cart as being more up-market than the electric float because it was big and impressive, and the milkman could ride high up on the cart. I and all the other local children were eager to have a ride on the horse-drawn cart but the offer was never made. One benefit of the horse and cart was that it provided manure which the horse had the habit of dropping in the same place each day just outside our house. If he was not beaten to it, my father would collect the manure in a bucket when he came home from work and use it round the garden. Coop customers collected points towards a dividend and each customer had his own identification number. When the milkman knocked at our door every Friday to settle the bill he would ask my mother for her number. Hearing it so many times it is engraved in my memory – 394262.

There were other frequent pedlars doing their rounds. There was the rag-and-bone man with his horse-drawn cart who bought old clothes that he weighed with a spring balance. I was disappointed never to see any bones. Gypsy women often came round selling posies of lavender and useless charms. If my mother turned them away they would curse her.

When it came to discipline my mother was a soft touch. I could twist her round my little finger. I was not a particularly naughty boy but if I did do something wrong, rather than punish me herself she would say, "I will tell your father when he gets home". I soon learned that if I pleaded with her she did not carry out the threat. I think this was because she was fearful my father might beat me. My father was much stricter than my mother, but although he believed in corporal punishment he was certainly not a violent person. He beat me with a cane no more than once or twice, and that was for damaging his precious plants in the garden.

The afternoon visits to my mother's parents were quite eventful. My grandfather was a patient and kindly man. In my memory he was always in the garden, never in the house. He loved being outside and in the sunshine even though, like me, he had a fair complexion with gingerish hair and a freckled skin sensitive to the sun. One afternoon he arranged for me to paint a kitchen chair. I did this in the back garden just below the steps outside their kitchen door. I am sure I splattered paint everywhere but he was most appreciative and complimentary of my efforts. I know I was proud to have painted this chair. I was about seven at the time.

My maternal grandmother always smelt of tea. She also suffered badly from arthritis. Once in her back room she showed me exercises her doctor had given her. They involved making her patella (knee bone) move up and down by flexing the muscles in her knee joint. It is a bit like trying to make your ears wiggle. My grandmother was unable to do the exercise, but I remember having no difficulty myself.

A vivid memory is of lying on my back under the table in my grandmother's kitchen trying to see up my mother's skirt. My grandmother told me off for doing this. She was kind but did not stand for any nonsense.

Very occasionally I was allowed into the front room of my grandparents' house where everything was meticulously clean and tidy. It was a sanctuary reserved for special occasions or important visitors. On one occasion, a neighbour came to tea and the best china came out with embroidered doilies under the plates on little side tables. Strangely, everyone used to have a front room like this for special occasions. The room was a bit like a museum housing the best furniture, a piano, lace curtains and trinkets. (They still do this in Holland, where they never draw the curtains at night so that passers-by can see inside and marvel). The front room remained cold and uninhabited for most of the year until the next honoured guest or festive celebration. Our own front room at Henley Crescent was just like this. It was used at Christmas when a coal fire was lit in the grate, and occasionally at Easter. Central heating saw an end to these restrictions.

Our front room was also used as a fitting room for my mother's dress-making customers. Although my mother never went out to work she did earn money from dress-making. She must have learned this skill when she was young. She was certainly competent and over the years built up a small business of several private clients. These were mostly women of her own age recommended to her by word-of-mouth.

My mother had some massive books of catalogues, one of which was Vogue. Her customers would select a dress from the book and then supply the requisite length of material. My mother then bought the design and laid the whole thing out on the dining room table, with both extensions. There were many sheets of tissue paper of peculiar shape punched with circles, squares and triangles, and adorned with lines and squiggles in various directions. She always complained that it was impossible to understand the instructions. Then came the moment of truth. She was terrified of committing herself and cutting the material wrongly, which of course would incur great expense. I clearly remember these times of tension. But she must have been good at her job because her customers always returned. I am sure she was far too cheap.

When a customer came for a fitting my mother drew the curtains in the front room and excluded me while the customer undressed. If the door was left ajar I could sometimes peep through the crack and see what was going on. Some clients had children of their own whom I played with while their mother was occupied. One of these was Helen, a girl of my own age who would eat earth from our back garden, including worms. I was horrified but rather impressed by this feat.

All my mother's customers were women, with one rare exception when she made a suit for a man. She did not like the experience and it was not repeated.

I have a vague recollection of our cat called Moggy. He was a tom-cat and often disappeared for days at a time. Once he returned home with a dead rabbit. Another time he returned with a broken leg, probably the result of being caught in a trap. We never had another cat after that. As for dogs, my father hated them. But I did have several pets including mice, budgerigars, and goldfish, all short-lived, and I soon tired of cleaning out their habitats.

My father had many hobbies over the years including: growing prize sweet peas and carnations (he won several national cups in the 1950s); growing fruit and vegetables in the greenhouse and garden, and later on allotments in Carlton Avenue and off Rochford Road near Southend Airport that he rented from the council; type setting and printing, (he bought an old professional printing press resembling a giant mangle with which he made Christmas cards); oil painting with all the paraphernalia; photography including developing and printing; marquetry; beer, slow gin and ginger wine making; ginger beer which exploded in the pantry on the first attempt; draughts and chess; learning the piano; collecting LPs of popular classics for which he bought a stereogram, and many other interests. The list was endless.

He pursued each hobby with great enthusiasm but they were mostly transitory. There were however some exceptions. One was Vernons football pools. Every Saturday evening in winter we had to be totally silent at around 6 p.m. while my father sat by the fire listening to the football results on the radio. The top prize for correctly predicting eight draws was £75,000. This was a fortune, equivalent today to winning £10 million on the National Lottery. The most my father ever won was a few hundred pounds. I am sure that on balance he was a loser, just like his betting on the horses.

My father's most enduring passion was the Masons. He was a member of Thundersley Lodge. I do not know why he never joined the local Southend Lodge. The Masons was always top of his agenda. He studied a lot, religiously attended all the meetings and dinners and held a variety of offices, to become Worshipful Master in 1965. He made many close friends that cut right across the hierarchy of Southend society. He mixed with shopkeepers, estate agents, solicitors,

57

doctors, police chiefs, teachers and headmasters. He was invited and attended functions at other Lodges, often far afield. He enjoyed all this because he was a sociable person. Possibly it compensated for the fact that, lacking a good education, he never reached his full potential in his Civil Service career.

Over the years my father gave a great deal of money to charitable Masonic causes. Our family could ill-afford this. My father's freedom with money was a cause of tension between my parents. He was not good at saving and tended to spend anything he had in his pocket. This was quite the opposite to my mother who was very frugal. I did not appreciate these points until many years later when my brother Nick pointed them out.

I benefited from my father's many passing interests because some of them rubbed off onto me. When I was about 8 years old I became interested in his hobby of the moment – wickerwork and basketry. I learned how to make modest trays with wicker borders. This was quite advanced for me and I remember the great sense of achievement it gave me. My father would cut out a circular or rectangular piece of plywood and I drilled the holes round the edge about every centimetre using a mechanical hand-drill. The next job was to paint the tray with two coats of garish pink paint that seemed to be available in abundance in our house. The wickerwork entailed cutting the uprights out of large gauge wicker, immersing them in a bowl of water to make them pliable, inserting them in the holes and then securing them with a special locking weave underneath. I then used a narrower gauge wicker to build up the walls of the tray to a height of an inch or so. Finally, I wove the thicker uprights into a matted interleaved corn-effect finish. I chose this finish from a book on wickerwork that my father lent me. The really tricky part was tying in the pattern at the end to make it look continuous. I was delighted to master this technique and can remember how to do it to this day.

I produced many trays to this design for which I had to find an outlet. Fortunately generous grown-ups were at hand, the main ones being my grandparents, a neighbour of my mother's parents called Mrs Cleverly, and Mrs Lotte Hart next door. I think they actually paid me a shilling (5p) for each tray, quite a lot of money. I never saw these trays again.

Another hobby my father taught me was marquetry. I learned all the basic skills including how to select the appropriate woods, how to use the razor-sharp marquetry knife for template cutting, wax polishing and French polishing. The latter was by far the more difficult method of polishing but gave stunning results when done properly. I became proficient in the various techniques and produced several creditable pictures.

Yet another of my father's passing hobbies was oil painting. I liked the smell of the paints and the turpentine. I had a go at oil painting but without great success. In later life I took up watercolour painting which was more to my liking because of its spontaneity.

My favourite special treat was Heinz Sandwich Spread. This was quite expensive at the time. For elevens's my mother sometimes made me a plate of Sandwich Spread fingers from a slice of bread. I would put the plate under the lid of my desk in my bedroom. I sat there doing drawing or painting, occasionally opening the desk to reward myself with a bite. This made it last a long time.

I also liked Shipham's Sardine & Tomato paste, in preference to meat pastes. Marmite was another favourite. Whilst I preferred Bovril, it was more expensive than Marmite and rarely available in our house.

I had all the usual childhood illnesses when I was young, including measles, mumps, chicken pox and scarlet fever. In fact, it was considered to be a good thing because by having them young you became immune to them in later life when they could be more serious. My bout of measles confined me to my bed and a dark room for 10 days on the instructions of our family doctor, Dr Rayne-Davis. I remember that when the curtains in my bedroom were finally drawn back to reveal the light of day, the piece of carpet on the floor next to my bed looked brown instead of the usual red and green. I never thought any more about this until as a young adult I did some sight tests and discovered that I was partially colour-blind.

Nobody had television when I was young. The radio was the main source of entertainment. One programme I liked was "Dick Barton, Special Agent", a sort of junior Dick Tracy. This series was broadcast for ten minutes every morning around 11 a.m. I remember on one occasion rushing home from playing in the fields in Midhurst Avenue to hear the latest episode. Apparently, 711 episodes were broadcast between 1946 and 1951, suggesting that I was less than 8 years old at the time. I am not even sure if Dick Barton was meant for children.

Another radio programme that was a weekly highlight was "Journey Into Space", a science fiction drama that ran for some 60 episodes starting in 1953. Each half-hour episode ended with a dramatic cliff-hanger to increase the incentive to tune in the following week. I think my mother enjoyed "Mrs Dale's Diary", a serial drama that ran for many years.

In the winter months we all sat listening to the radio (called the wireless) in the back room huddled round the coal fire, which was the focus of the room there being no television, my father prodding chestnuts on the hearth that were cooking and spitting in the heat.

Our use of the coal fire in the back room required the chimney to be swept regularly. The chimney sweep accessed the chimney flue with his brushes from the grate, and from an access behind a metal plate high up above the picture rail on the white-washed wall near the ceiling. My mother knew there would be a mess and covered all the furniture with sheets.

My father had a sweet tooth. There were periods when he regularly brought sweets home to eat in the evening. I think these periods coincided with his efforts to give up smoking. His favourites were Needler's Fruit Drops, Murray Mints, and American Hard Gums. He could have chosen to buy them from across the road at Shannons, but he preferred to go further afield to Jennings on the Bell Parade because Mr Jennings was a mason.

I remember my nightly ritual of arranging my dolls each side of me across the pillow. Closest to me was Baa-Lamb who was always my favourite. Baa-Lamb was given to me by my godmother Rene when I was a baby. Then there was Pussy-Cat, Golliwog and various other lesser animals. I never had a teddy bear. After arranging the dolls I said my prayers under the bedclothes: "God bless Mummy, God bless Daddy", then all my relatives and friends being sure not to leave anyone out. I still have Baa-Lamb and Pussy-Cat, both threadbare. They live in a bottom drawer next to my bed.

I don't know what happened to Golliwog. Sadly, like the Robertson's Jam logo, golliwogs are no longer PC. I should say that I never associated my golliwog with black people. It was just another doll. I made the connection only after golliwogs were banned. In September 2008 a woman in Stockport was arrested, finger-printed, and DNA-sampled by the police for keeping a "golly doll" in her shop window. The PC brigade now monitors everything we do and say.

# Life at Henley Crescent

The road where I lived was unmade with a rough surface comprising a bed of compacted stone that over the years had disintegrated into numerous potholes. I think it would originally have been called a "metalled" road. I knew the location of the potholes intimately from negotiating them on my bike. In the rain large puddles formed that were connected by rivulets of water running down the shallow gradient of the road towards Midhurst Avenue. In winter the puddles froze over. When it snowed there were only about two places where the road was sufficiently smooth to support a decent slide, unless the snow lay very thick. Broken kerbstones were present bordering rough footpaths and untended grass verges. The general effect was quite dilapidated and rural, unlike most of the roads in the surrounding area.

I liked our road because it was a fruitful play area. Being a crescent it led nowhere, so the only people walking down it were those who lived there. It was safe because there were few cars, and those there were had to drive slowly over the rutted surface. Our road remained in this condition until I left for university when it was finally "made up".

Our house in Henley Crescent was No 11, built in 1936. Curiously, the first owners were my mother's brother Harold and his wife Jennie. They bought the house at the time of their marriage and lived there for about four years. They called the house "Willersely", after Willersely Castle in Derbyshire where, in 1934, Harold and Jennie first met at a Methodist Convention holiday also attended by my mother.

During their time in Henley Crescent Harold worked at his parents' shop in Westcliff. At the outbreak of war, it no doubt became apparent that the business could not support both families. So Harold and Jennie moved to North London where Harold found work in a greengrocer's business in New Barnet. They left Henley Crescent at some point between July 1940 and the start of the Blitz in September 1940. Jennie was at that time heavily pregnant with Jean. They moved to rented accommodation in Barnet before buying their house in

Totteridge. The greengrocer's business in Barnet was to become very successful, largely through the hard work and business acumen of Harold.

Meanwhile, 11 Henley Crescent must have been empty for some time until, after their marriage in January 1941, my parents bought it from Harold and Jennie. The house was a modest semi with a small front garden, a larger back garden, and a sideway that was just a bit too narrow for a garage. Downstairs there was a front hallway and stairs, a front lounge, a back room that was the main living room also used as a dining room, and a kitchen with a small pantry. Between the back room and the kitchen was a hatchway through which dishes could be passed. Upstairs was my parents' bedroom at the front, my bedroom at the back, a small corner room at the front that was my father's hobby room before becoming Nick's bedroom, a bathroom, and a separate toilet. The main rooms had a central carpet with linoleum (lino) round the edge. There was also a small loft with difficult access requiring step ladders.

We had no central heating when I was young. In winter we had a coal fire in the back room and a coke boiler in the kitchen supplying the hot water. On special occasions we had a coal fire in the front room.

Upstairs there was no heating except in the bathroom that contained the hot water tank. My mother gave me a bath on Wednesday and Saturday evenings before I went to bed. We were still in the post-war era of saving hot water. I sat at the plughole end to make it easier to wash my hair under the tap. In those days nobody had a shower. That was considered a form of torment that only hardened rugby players suffered after a match.

My bedroom and my parents' bedroom each had a small electric heater built into the wall, but they were rarely used because they were too expensive to run. I recall that the electric fire in my bedroom was used only once when I was ill, and the one in my parents' bedroom only early on Christmas morning when I opened my presents from Father Christmas.

The bathroom and kitchen were the warmest rooms in the house. In winter it was bitterly cold going to bed with the sheets so freezing that I shivered. Sometimes I had a hot-water bottle to help me get warm. My mother told me not to rest my feet on the hot-water bottle because it would give me chilblains. In the morning my bedroom was so cold that I often went downstairs to dress in the warm kitchen, first warming my clothes in front of the coke boiler. Men and women wore vests under their shirts and blouses in those days. Vests are now almost non-existent, like left- and right-handed socks. To help me combat winter viruses my mother gave me regular doses of cod liver oil, a supplement that I actually enjoyed.

Outside there was a coal shed and a coke bunker, both built by my father. Fuel was delivered a ton at a time, requiring the coalman to carry in and empty 20 x 1 cwt sacks (1 cwt = 1 hundredweight = 112 lbs = 51 kg). This took some time. We counted each bag as it came in because it was well-known that coalmen made money on the side by delivering one bag short. A neighbour once challenged the coalman that he had been delivered short. The weights and measures people had to come to rebag and weigh all the coal. I do not recall the outcome.

In retrospect our house was perfectly located. It was off the main road and so had some seclusion. It was less than 10 minutes walk to both sets of grandparents, 5 minutes walk to the primary and secondary schools that I was to attend, 5 minutes walk to the best recreational areas that any boy could wish for, close to an array of local shops, 10 minutes bus ride to Southend town centre and the beach, and 10 minutes walk to Southend General Hospital if the need arose.

When my parents bought our house, my father's parents Thomas Parsons (known as Tom) and Violet (known as Vi) were also living in Henley Crescent at No 40 with their two unmarried children Frank and Margaret. Families in those days were close-knit. My grandparents' house was named "Loddington" after Thomas' birthplace near Kettering. In 1944 they paid a weekly rent for No 40 of 18s 6d.

My grandfather Thomas had a working class background. He was only eight years old when his father died in 1888. His mother had to work hard to raise Thomas and his elder brother and sister. Thomas was a great sportsman and was particularly good at football and running events. He had many jobs throughout his married life and the family moved house several times. His last job was caretaker at Southend Central Library where his average weekly wage was £3-17s. This was his job when they were living at 40 Henley Crescent.

His wife Violet was 14 years younger and came from London. She was only 15 when she married in 1908. She had a strong personality and dominated her husband Thomas. I was to discover much later that she was a closet poet of some talent and wrote about unconsummated love.

While living in Henley Crescent, Violet worked as manageress in the chemist shop of Williams & Lane (Druggists) at 661 London Road, Westcliff. Every week, usually on a Monday, she took the shop's cash takings to Barclays Bank, amounting to between £50 and £70. In Tax Year 1945/46 she earned £260-13-6 and paid £48-11-00 in income tax. (The average annual wage at that time was about £350). She also received a quarterly bonus, typically between £10 and £30 before tax. This was calculated as 10% of the amount that the total quarterly takings exceeded £500. She usually gave the bonus to St Mary's Church in Prittlewell where she had been a Sunday School Teacher. For example, in November 1947 she gave £10 towards the fund for St Stephen's Church Hall in Manners Way. In return she received grateful letters of thanks from the Rev. Ellis Gowing. He was the Vicar, later Canon, at St Mary's from 1917 until his death in 1960.

On 14 October 1944 Thomas and Violet ceased to be tenants and became owners when they bought 40 Henley Crescent from Frank Oscar Wardell with a mortgage of £575 from the Abbey National building society. In April 1946 Thomas retired from full-time work at the age of 66½. He began receiving a pension of £1-2-9 per week. On 20 August 1946 the family moved to rented accommodation at 16a St

George's Park Avenue, Westcliff-on-Sea. This move was to clear the way for the sale of 40 Henley Crescent in September 1946 to a Mr Bocquet for £1700, giving Thomas and Violet a handsome profit. To put all these figures in perspective, the cost of living rose by a factor well in excess of 60 between 1944 and 2016. House prices rose by a much larger factor. So the profit of £1125 they made on the sale of 40 Henley Crescent would be worth about £75,000 in today's money. Their rented accommodation in St George's Park Avenue was a staging post while they waited for the completion of a bungalow they were having built on a vacant plot of land they had previously bought in Hobleythick Lane.

Thomas was not a particularly religious man, but my grandmother Vi always remained a keen member of St Mary's Church where she developed an ardent admiration for the vicar, Canon Ellis Gowing. (I was to discover her emotional attachment to the vicar only much later). My father had been a choirboy at St Mary's under the direction of the choirmaster and organist Mr Penny. Both Cannon Gowing and Mr Penny served at the church for many years and I recall them well.

I always had the impression that my grandfather was mainly interested in watching football while my grandmother made all the household decisions. He was easy-going whereas she had a strong will. My early memories of these grandparents are not too distinct. This is probably because my mother did not take me to see them as often as to her own parents. I think she found her mother-in-law rather overpowering and disapproving.

Whereas two of my paternal grandparents' four children were married (Cissie and my father) their other two children, Frank and Margaret, remained single and continued to live with their parents.

Margaret started work in a bank when she left school. At one point she wanted to join the armed forces (Women's Royal Naval Service) but Frank advised her against this saying it was foolhardy and needed thinking through. This advice was set out in a letter he wrote from London in November 1951. In the event Margaret did not join up. Soon after this she had a boyfriend whom she almost married, but I believe Grandma Parsons disapproved and put him off. I think Margaret remained sad about this for the rest of her life. She continued to live with and support her parents until 1967 when, aged 36, she married Malcolm Edwards.

Frank never left home except for a few extended periods when working as a butcher in London. He trained as a meat cutter with Dewhurst and so knew a lot about the butchering trade. He was skilled at making sausages and drawing poultry. He often practised these skills on the Christmas turkey. Frank loved the hustle and bustle of working and living in London and had digs at various places. Between March 1950 and November 1951 he was living at 259 Camden Road, London N7. In August 1954 he was living at 60 Eade Road, London N4. Frank's work required him to be in and out of the butcher's chill room all day long. This was bad for his health and he was twice taken ill with pneumonia. These bouts forced him to give up butchering in the 1950s when he returned to Southend to live again with his parents.

In the 1950s grandfather Thomas earned extra money doing occasional house-to-house canvassing work for the Register of Electors for Southend Town Clerk's Office.

In July/August 1949 my paternal grandparents moved into their new bungalow in Hobleythick Lane with their two unmarried children, Frank and Margaret. The bungalow was just two minutes walk from our house in Henley Crescent. At this time my grandparents had a car, a great luxury just after the war and also a rather reckless purchase.

My grandparents had bought the car the previous year, in February 1948, while living in St Georges Avenue. It was a dark blue 1935 Morris Type 10/4, 10 HP Saloon, registration AMU 874 (engine no 42100, chassis no 46892). My father had a third share in this car. The previous owner from Benfleet had kept chickens in it. The purchase price was £106.10.00, but my grandparents immediately had to spend another £25 to make it driveable.

My grandmother had just learned to drive at the age of 55, quite an achievement. Her husband Thomas never learned to drive, nor did Frank. In September 1948 Margaret began a course of six driving lessons with L.G.Collings of 853 London Road, Westcliff, at a cost of £3-17-6, but it appears she did not pass her test because she never drove subsequently.

It was in the summer of 1949 when I was six that my father borrowed the car to take my mother and me on a short holiday to stay with relatives in Parkstone, near Bournemouth. It was probably Uncle Ernest and Aunt Ethel, although it could have been Uncle Percy and Aunt Irene, both families living in the Parkstone area. Earnest and Percy were brothers of my maternal grandmother. I remember the tree-lined street outside the big house where we stayed. Our hosts had a musical box that fascinated me. Whenever you opened the lid it began to play. This holiday was an exception. Generally, we did not go away on holiday when I was young.

I have another early recollection of this car. The incident in question must date to when I was six years old because the car was sold in March 1950. We all went out in the car one afternoon for a drive in the country. My memory is of being sandwiched between two adults on the deep, leathery back seat when the car broke down on Ashingdon Hill. There was much panic and confusion in trying to get the car up the hill. People shouted that it was rolling backwards and we all had to get out and push. My grandmother was the driver. Somehow we managed to get the car home.

This car was the only one my grandparents ever owned. It did not stay in their possession for long. It was always breaking down and was a drain on their limited resources. A list of some of the repair bills is as follows: £3-12-7 on 18 Feb 48, £4-2-7 on 29 Mar 48, £3-4-8 on 23 Apr 48, £0-4-2 on 12 May 48, £2-3-0 on 18 May 48, £0-9-3 on 26 May 48, £1-9-9 on 23 Sep 48, £11-3-10 on 26 Nov 48. They sold the car in March 1950 having owned it for just two years.

The bungalow in Hobleythick Lane developed serious structural problems in the roof area soon after my grandparents moved in. It was 1949. The bungalow was brand new having been built specially for them by a local builder. Possibly he was incompetent or used poor quality materials after the war. I did not know the details but I was aware that arguments with the builder were causing my grandmother great consternation. The issue dragged on for some years before being resolved at great expense.

I have clear memories of my paternal grandparents at Christmas time. This was a magic time for me. The Christmas ritual was always the same. My father put up the Christmas tree in the front room, and the decorations which had to be red, green and white. He bought the turkey on Christmas Eve, usually from Garons in Southend High Street. In the evening of Christmas Eve the family would meet at our house in Henley Crescent. After a late meal of salads, home-cooked ham, homemade pork pie and pickles, followed by nuts, fruit and dates, some of the adults would find the energy to go to midnight mass at St Mary's Church.

Christmas morning was always exciting with presents. I would wake up early to find a pillow-case full of presents beside my bed. When I was very young I opened my presents in my parents' bed. It was only on Christmas morning that I ever entered my parents' bedroom. My presents always included Dandy and Rupert Annuals. I particularly liked the sweet smell of the fresh paper, something that is impossible to experience with e-books. While I was opening my presents I was vaguely aware of activity downstairs in the kitchen. It was my father preparing the turkey. I can remember hearing him pronounce with a sense of achievement and relief, "It's in the oven".

Christmas was the one time of the year when the whole house was warm with fires in all the grates. The main festive meal was at lunchtime, alternating each year between our house and my father's parents' house in Hobleythick Lane. The core family members always present were myself, (not Nick early on – his birth was not until 1951), my mother and father, my paternal grandparents, and Uncle Frank and Auntie Margaret.

Some years we were joined by Haidée, her husband Billy and their daughter Jane, all from Kenton, North London. Haidée was the only child of my grandfather's brother-in-law

Phil whom I never remember seeing in Southend, although he apparently visited frequently. Jane was younger than me and a bit of a tom-boy. She had a sister Yvonne who was much older. Some Christmases Yvonne made a brief appearance before she eventually moved to work in Africa with her husband. Yvonne divorced some years later and has since been living in Kashmir.

Haidée was an out-going person and a great friend to everyone. Her husband Billy was gruff and entertaining. After Christmas dinner he usually drove by himself at high speed all the 50 miles back to Kenton to feed the cats, returning to us in time for tea. Billy ran a successful roofing business. He had two serious roof falls each of which almost killed him.

My father's other sister Cissie and her family had their own Christmas celebrations. They lived nearby in Beeleigh Close. Cissie was married to Don Hudson. They had three sons, Donald, Paul, and Douglas who was the youngest and a year younger than me. Occasionally I played with Douglas and once took him to Priory Park, but Cissie discouraged this because she thought I got him too excited and made his stammer worse. After that I never played with him again.

We sometimes visited the Hudsons at their immaculate bungalow for sherry on Christmas morning. Paul always managed to say something outspoken and outrageous that greatly annoyed my father. Otherwise we had little contact with the Hudsons.

Cissie and Don Hudson usually attended midnight mass at St Mary's Church. They always sat at the very back of the church where, according to my parents, they could survey everyone else. Basically, Cissie was in a higher social class because her husband Don was a detective in the police force, but I am not sure whether he was a mason. Curiously, my father only came close to Cissie and Don many years later, when he played a counselling role in saving their marriage after Don's infidelity.

I do not recall that the Christmas company ever included my mother's parents. I am not even sure what they did at Christmas time.

To return to the Christmas meal, this was always late because we had to wait for grandfather Tom to return home from the Southend United football match where he had a permanent seat in the Director's box. I believe he earned this privilege because of his long support of Southend United and because he had worked as a trainer for the club in the 1920s.

If the meal was at our house Margaret always said "Can I just have the white meat, Tom", to the annoyance of my father while he was carving, and Frank always said, "Nice bird Tom" which was greeted with noises of agreement. Then my mother would say, "There's plenty more gravy if anyone wants some". We pulled our crackers and put on the silly hats. My father would remark on what good value the crackers were. If the meal was at Grandma Parsons' house the routine was basically the same. Frank prepared the turkey and Margaret did most of the cooking. I enjoyed all this ritual immensely and loved the Christmas meal.

When the adults had recovered from the meal we played games until late into the evening. The games included charades, "The Tray" memory game, "The Wand is Passing", and various other amusements and card tricks which never lost their appeal for me from repetition each year. Supper was always cheese and celery followed by homemade Christmas cake. The celery in those days was a variety now known as "dirty celery" but no longer readily available. It was then the regular celery and had to be washed thoroughly. It had an intense, sweet flavour and was not stringy like its modern equivalent.

There were plenty of sweets available at Christmas. Favourites were Newberry Fruits and Quality Street. I liked the green chocolate triangles best. Sometimes my father bought an enormous box of special Newmans Chocolates (he pronounced it "choclits"), twice the size of the standard chocolate with exotic fillings. He bought them from a small shop in The Aldwych in London, one of only two or three exclusive outlets, and long since closed.

At some point over Christmas my parents' great friend Rene Cabuche, later to become Mrs Rene Isaac, usually visited us. Rene was my attentive godmother and I always enjoyed her company. She had a cheeky sense of humour that I understood. She often came with her elderly mother Mrs Cabuche. I was fascinated by the fox stole that Mrs Cabuche always seemed to have draped round her neck, the fox's head dangling with its glassy eyes staring at me.

On Boxing Day the whole performance of the Christmas meal was repeated at the other house. Sometimes, if I was tired, I slept over at my grandparents' house sharing Margaret's bed. She was a kind and attentive aunt. The festive period was always over much too quickly for me. On twelfth night the Christmas Tree was removed, all the decorations came down and the cards on display were discarded. The house felt sad and bare for a short time. I never wanted Christmas to end.

People those days were much more neighbourly than today. There was a sense of community and we knew our neighbours well, for better or worse. There was less mobility with people living in the same place for a long time. This was probably because they tended to stay in the same job and did not divorce. There were no unmarried couples living together and certainly no gays. Anything like that was unthinkable.

In our road only the men worked. The women may have once had jobs but after marrying they stayed at home to keep house and bring up a family. Even if they had no family they still stayed at home. This may have been unfair on the women, but the family unit and family values were all-important. In any case, it really was a full-time job being a wife and mother because there were no refrigerators, freezers, dishwashers or washing machines, no disposable nappies, ready-meals, supermarkets or internet deliveries, and most people were without a car. All meals were prepared by mothers from scratch with the individual ingredients bought daily from a range of local shops. Having said this, I believe that WW2, like WW1, made an important contribution to the cause of women's liberation, simply because they had to do the jobs that the men traditionally did.

Almost all the houses in Henley Crescent were semi-detached. The neighbours adjoining us at No 9 were Mr & Mrs Marsh, an elderly couple who kept very much to themselves. They were friendly and kind to me. My parents always said they had been good neighbours during the war years. The Marshes were one of the few couples that I remember moved house to live elsewhere. They moved when I was about ten to a small bungalow in Southend Road, close to the airport.

The family that replaced them at No 9 was Mr and Mrs Sid Gibbons and their daughter Maureen. They came from London's "East End", reputedly a rough part of London where crooks and thugs lived. Most Southend people had never been there, but plenty of East Enders came to Southend as day-trippers. There were "special" trains for them at weekends during the summer, and they poured in to enjoy the many bars, cockle and jellied eel stalls, amusement arcades on Southend sea front, and of course the beach. They were a rough lot and often got involved in drunken fights on the sea front. So my parents waited to see how our new neighbours would fit in. In fact, they were nice people and my father even managed to enrol Sid as a mason. I was never sure about Sid's job. It was something to do with the wholesale supply of newspapers and magazines. He used to travel to work in Southend every day, even at weekends. I believe he also owned a sweet shop in Basildon.

Their only child was a girl called Maureen. She was about my age. I was interested in her as the "girl next door", and she became the subject of innocent fantasies. I was never courageous enough to talk to her, and in any case she never showed the least interest and barely acknowledged me.

Our bedrooms were at the back of the house and adjacent to each other. It was exciting to reflect that our beds were separated only by a thin wall. In warm weather Maureen often left her bedroom window open. When looking out of my window at dusk I discovered that her open window became partially reflecting, and if the angle was just right I could make out a dim shape in her bedroom, but that was about all. Maureen remained elusive.

Occasionally, a nephew of Sid's travelled down from London to visit for the day. He was older than me and was called Cliff. He would arrive on a big motorbike clad in all his leathers. Maureen fawned over Cliff and rode on the back of his motorbike, making me very jealous.

Sid was always generous and gave me two shillings each Sunday morning to wash, wax and polish his car and hoover the inside. This took me about 2 hours. Maybe it was this experience that deterred me years later from ever cleaning my own car. Cars in those days used to rust easily, especially the chrome, so car polishing was essential and quite hard work. Later, I started doing the car of Sid's affluent friends, Mr & Mrs Harvey, when they came down on day trips from London. I also occasionally cleaned Mr Barber's car at No 4. This was all useful pocket money.

Next door to Mr & Mrs Gibbons at No 7 was Mr & Mrs Garrard. They were an elderly couple with whom I had little interaction. At some stage they either moved out or died, and were replaced by Sid's aforementioned London friends, Mr & Mrs Harvey, whose car I had started cleaning on his prior visits.

Next door to the Garrards at No 5 lived Mr and Mrs Jack Perrett. I rarely saw Mrs Perrett, but Mr Perrett could be seen regularly walking his faithful and mangy dog Jip round the block. Mr Perrett was a dour, stooping character who was either retired or just about to retire from gents hairdressing. He had a shop somewhere in Westcliff that I occasionally visited. Like many men in his profession he was severely round-shouldered after a life-time of leaning over people.

Next door to us on the other side at No 15 (there was no No 13) was Mr Laurie Hart and his wife Lotte. They had a son, Gerald, who was two years older than me. I always knew that the Harts were a rather unusual family unit but I never understood why until I was much older. Apparently, Gerald was Laurie's son by a previous marriage. When war broke out Laurie was living alone with his son except for visits from their regular housekeeper who was Lotte. I have no idea what happened to Gerald's mother and nobody ever talked about it.

Anyway, Laurie and Lotte grew close and it seemed a good idea for them to marry, which they did. I think it was a marriage of convenience. So Lotte became Gerald's stepmother.

Lotte Hart was a kind and gentle woman, and one of my mother's closest friends. She never had children of her own. Laurie Hart on the other hand was a distant, short-tempered man. He became quite irate if my ball went over the fence into his garden. He reputedly was making lots of money in various shady commercial ventures at Southend dog racing stadium. I somehow saw him as a bit of a spiv and conman. According to my mother he was mean with Lotte and vetted every penny she spent as though it was one of his business ventures. He was one of the first people in the road to own a car, a small Triumph Herald that he drove every day to work. He kept very much to himself and did not fraternise with anyone in the Henley Crescent community. It was perhaps because of my wariness of Mr Hart that I never really made friends with Gerald, although he was a very approachable boy. He never came into our house and I rarely went into his. Our distance was not helped by the fact that he went to Westcliff High School, whereas I was to go to Southend High School.

It was many years later, working as a dustman during a university vacation, when fellow dustmen on the Henley Crescent round to which I was assigned at the time (yes, I emptied our own dustbin) remarked upon the virtues of this neighbour. They said that at Christmas and other times during the year they received generous tips from Mr Hart at Southend dogs stadium. They told me that he was manager of the catering and bar facilities and they held him in the highest esteem as a real gentleman. This made me look at Laurie in a new light. He was a chain smoker and eventually died of cancer.

I have two particular memories of Lotte. The first was when she slipped and fell with all her shopping on an ice slide that I and Trevor Williams had made in the road one snowy winter's day. She was upset and bruised but soon forgave me. The second was when my father took me to one side and instructed me that when I met any of our neighbours like Mrs Hart in the street I was to doff my cap and say "Good afternoon Mrs Hart". He made me practise this several times. I was taught to do a similar thing with my cap if I saw a funeral hearse pass by. It was common practice at such times for people to stop in the street and bow their heads as a mark of respect. My father impressed upon me several other rules of common courtesy such as, offering up my seat to a lady on public transport and saying "Please may I" instead of "Please can I". I cannot imagine much of this happening today.

Next to the Harts at No 17 were Mr and Mrs Ernie Barnet of whisky fame at my birth. They were another elderly couple who remained in the Crescent during the war years and who had been most helpful to my parents resulting in a lasting friendship. They had a son and a daughter whose families often came to visit the Crescent.

Opposite our house at No 12 were Mr Glyn and Mrs Enid Williams, with their son Trevor and younger daughter Barbara. Glyn was one of my father's closest friends, and while I was at Junior School Trevor was my closest friend. Trevor was two years older than me but we got on well together. I will say more later about my friendship with Trevor.

I sometimes stayed to tea with the Williams. I was surprised that Trevor and his sister Barbara were trained to do the washing-up, which they did without any question. This was something my mother never asked me to do, although I did other chores. I also never helped Lucienne do the washing-up. In fact, she positively did not want me to because she thought it was demeaning. So at the end of my life I shall be able to say that I never did any washing-up. However, I later discovered there was a good reason that Trevor and Barbara helped their mother. It was because Mrs Williams could not hold a plate between her fingers. She suffered from chronic arthritis which doctors and faith healers failed to cure. Over the years it became progressively worse affecting her whole body and as a result of which she sadly died while still relatively young. Mr Williams was later to die of a heart attack.

Trevor's father, Mr Williams, was a mason which was one reason why he and my father were so close. Mr Williams was a teacher and deputy headmaster at Eastwood School. This was a Secondary School as opposed to a Grammar School. The headmaster was Mr Hepple, also a mason and a good friend of my father. His wife, Mrs Hepple, was a teacher at my junior school of which more later. So it was a small world with a network of masonic connections.

Moving down the road from the Williams to No 10 were Mr Frank and Mrs Olive Mandry and their son Robert. Mr Mandry was some sort of salesman, possibly insurance. He certainly looked the part. He was stocky with dark hair sleeked back either side of a central parting. He wore thick tortoise-shell glasses. If he was a mason he was a closet one. His wife was slim and had probably once been attractive, and maybe still was. She always gave me the impression of being lonely. Their only child Robert was a few years younger than me. The problem for Robert was that his mother had always wanted a girl, so Robert was brought up as if he were a girl. This possibly had a lasting effect on Robert. He later became friendly with my brother Nick who reported that he was quite normal and that his main interest was photography. Mr Mandry died relatively young of a stroke. Robert went on to become a school teacher. I do not know what became of Mrs Mandry.

Next door to the Mandrys at No 8 lived Mr Les and Mrs Peggy Harris. They had money because Mr Harris was a bookmaker. His betting shop was round the corner on the Bell Parade. When I was young I did not know the meaning of "Bookmaker" or "Turf Accountant" but I sensed there was something shady and not quite respectable about the job. This was confirmed to me by his flashy cars, his short moustache and greased hair style, his dapper outfits and the fancy clothes that he and Mrs Harris wore. Mr Harris was definitely a spiv. His wife had a stunning set of false teeth in alabaster white. Mr and Mrs Harris were the Henley Crescent "nouveaux riches". They had no children and so could be self-indulgent. They sometimes went "abroad" on holiday, which was most unusual at the time. They actually used to eat in restaurants, one of their favourites being the new restaurant at Southend Airport. You had to be well off to do all this. They hosted lavish parties with large quantities of alcohol. I think my mother felt uncomfortable with these parties but my father lapped it up. I know that Les was also a mason.

At weekends my father often spent long evenings into the early hours at Les's house. Apparently they played poker in a smoky, alcoholic atmosphere. Mr Williams also attended. It is quite possible that large sums of money changed hands. I am sure my mother disapproved of these goings-on.

Over several years my father and Mr Williams worked for Les on Saturdays at his nearby betting shop. This obviously earned my father extra money. Later, my father told me that Les was not a good Turf Accountant because he failed to manage his risks by laying off large bets. This meant that he sometimes incurred heavy losses.

For Lucienne, the defining incident with the Harris's was when we were married in 1965. Mrs Harris gave us some towels saying, "They always come in useful", adding to the 50 towels we already had. Mr and Mrs Harris were chain smokers and both eventually died of lung cancer.

No doubt Les Harris' influence was behind some of the horse-racing scams that my father invested in. The one I recall was a syndicate that only backed the first and second favourites, or something like that. The scheme was supposed to be a sure winner, but only it seems for the organisers. I cannot recall the details but I know it cost my father a lot of money. Again, my mother was not pleased.

Moving down to No 6, there lived a family whose names I have forgotten and who were generally regarded as rough people. The house was a mess and the garden overgrown with a nasty dog that fouled everywhere. They must have been tenants rather than owners. My mother did not approve and we did not talk to them.

The end house at No 4 belonged to the Barbers whose car I occasionally cleaned. They had a young son called Paul who became a close friend of my brother Nick, and a daughter called Sally who was either very clever or mentally retarded. It is sometimes difficult to tell the difference with such people. Mr Barber had a good job as an accountant. They eventually moved out of the Crescent to a smart house in Thorpe Bay. My mother kept in touch with Mrs Barber. Mr Barber died young of a heart attack or stroke. Paul was intelligent and did well at school. On leaving he took up employment as a bus conductor, a job that he continued for many years.

Moving up the road from the Williams to No 14 there lived an elderly couple, Mr and Mrs Bateman, and their two spinster daughters in their early forties who used to ride to work on mopeds. There were many middle-aged single women in those days owing to the shortage of men after two World Wars. The main claim to fame for Mr Bateman was that he must have been one of the first in the road to own a television. It looked like a Dalek. Once a year he would generously invite all the boys in the road to watch the Oxford v. Cambridge boat race. This was tremendously exciting. The TV set was massive but the screen itself was tiny in the shape of an oblate spheroid. It took about 10 minutes to warm up and you had to draw the curtains to see anything. It was all in black and white and rather blurred. This was entertainment at its finest.

Next door to the Batemans at No 16 were Mr and Mrs Yates and their son Brian, plus a younger daughter. I think that Mr Yates was some sort of builder. Their son Brian was of similar age to me, and I sometimes played with him. But my friendship terminated after the flying rake incident that occurred when I was about 7 years old. At the side of Brian's house was a flat roof garage. One day we were having fun climbing onto the garage roof and jumping off to land on a pile of soft compost and grass cuttings. After doing this for some time Brian decided to make it more interesting by carrying objects onto the roof and throwing them off. One such object was a garden rake. I jumped off the roof and Brian threw the rake after me. The rake struck me prong-first on the head. I was stunned, there was a great deal of blood and I managed to stagger home. The next thing I remember was being treated in the out-patients at the local hospital. The injury was not as bad as it looked but it left two nasty scabs on my head that took some time to heal. The benefit of this experience was that I was able to boast that I had been given an anti-tetanus injection, which sounded quite important. On the downside I developed a fear of going to the hairdresser in case it opened up the wounds.

I should say that men went to the hairdressers in those days almost as frequently as women go now. Possibly it was the only place that a man could readily buy certain things. My father went at least every fortnight, usually on a Saturday morning, to Ravens in Southend's Victoria Circus. Ravens was a large men's outfitters that included a prosperous hairdressing business with six chairs. I often went there with my father. Fortunately my fear of hairdressers was short-lived, unlike that of dentists.

The other people I remember on that side of the road were Mr and Mrs Hall at No 24 with their son Dave and a younger daughter Marion. Mr Hall was an optician who sold spectacles in a shop in Southend High Street. Mrs Hall was a petite, attractive brunette who my father secretly admired. Dave was to become a good friend of mine, of which more later.

At the top of the Crescent at No 32 lived Mr and Mrs Swotton, an elderly and bad-tempered couple. They were always hovering behind their curtains to rush out and scold us if we dared play on the grass outside their house. Strangely, Grandma Parsons, from her time living at No 40, said what a kind and generous couple they were during the war years.

Just round the corner at the top of the Crescent at No 21 lived Mr and Mrs Absalom with their son Terry and daughter Moira. Mr Absalom was an administrator at Southend Fire Station. Terry told me with pride that during the war it had been his father's job to defend Southend Pier. Mrs Absalom had been a nurse and was now a housewife. Terry's sister Moira was some years older and attractive. She had large doe eyes with hooded lids, rather like the French actress Stéphane Audran. Terry was my age and another close friend. Again, more of Terry later.

Still further round at No 42 were Mr and Mrs Eaves. Their youngest child Nigel was my age. I sometimes played with Nigel but he was a bit rougher than me. He was a boy who could look after himself. He always wore heavy hobnail ankle boots that generated sparks when he scuffed them on the pavement. This greatly impressed me. My parents often remarked how well-mannered Nigel was in spite of his tough upbringing. He had an elder sister who scared me with her sharp tongue, and a much older brother who was a sailor in the Navy. On the few occasions that I met him he had a great aura of worldliness. Nigel failed the 11-plus and went to Fairfax Secondary School. I then lost contact with him. He left school to work nearby at Jepps the grocers, first as a delivery boy and then behind the counter. My mother always said what a pleasure it was to be served by Nigel. He later left for a more secure job in the Fire Service.

At the bottom of our road in Midhurst Avenue lived Tony Watts whom I occasionally played with. Tony lived alone with his widowed mother. His father had died of beri-beri as a prisoner-of-war at the hands of the Japanese. My mother always expressed sadness about Tony and the fate of his father. I did not appreciate the significance of The War. I just accepted what little I was told without emotion.

I must say that nobody spoke outright about the horrors of the recent war, at least not to me, although memories must have been fresh. I had little knowledge of these terrible events. The closest I got was overhearing a conversation about the buzz-bombs and my family playing Monopoly throughout the night in the air-raid shelters.

Also in Midhurst Avenue lived another boy called Doug Harfield who was two years older than me. His parents were regarded as working class with pretensions above their station. They had built various extensions to their semi-detached house that the neighbourhood considered to be unsightly and out-of-character. I did not get on with Doug Harfield because he was a bully. He used to make joking remarks about my father and the way he looked with his handlebar moustache. This shocked me, not for what he said, but because I was unable to see my father as a figure of ridicule.

Midhurst Avenue, at the bottom of our road, was "made up" many years before Henley Crescent. It was a long job and a source of great interest to us boys. All the hard work to prepare the road surface and dig trenches for the essential supplies was done by Irish navvies using just pick-axes and shovels. They were a rough bunch who introduced me to many new swear words. One of them, called Mick, always called me "ginger". When a group of us once came too close to the work in progress Mick shouted that he would "tan our arses" if we got any closer. I committed this phrase to memory. Every evening a coke stove was lit for the night watchman who sat in a small wooden hut guarding the tools.

The colour of my hair often prompted remarks like "ginger" or "carrot-tops". Nowadays these would be considered terms of abuse but I can't say I was ever bothered. It helped distinguish me from the rest. People are nowadays much too easily offended. My mother always insisted that my hair was golden.

We knew pretty well everyone in and around our street. I don't think there were many secrets. It was an age quite different from now. Doors were left unlocked and bicycles left unattended without fear of theft. In the 1950s crime was low and falling, people looked out for each other and children looked to adults for guidance and assistance. We could play in the streets, climb trees, and get into fights without fear of being restrained by health and safety or political correctness. You ate what you were given with no fuss, and chocolate bars were a special treat rather than part of lunch. We respected our parents and felt close to them, but never hugged them. It was a society of primness and a certain amount of repression. I think this was generally healthy because freedom without restraint is not a good thing.

There was respect for authority, including adults, policemen, school teachers, and other public servants. Most children knew their place and accepted that if they stepped out of line they would be punished. This was part of life and not a problem. For us children there was little distinction between punishment from parents, teachers, or any other adult in authority. We were minors and did not have the same rights as adults. The term "child abuse" did not exist, although there were certainly instances of children being beaten. These were the days when parents supported the decisions of teachers. We were not inclined to tell our parents about any punishment received at school for fear they would be angry and punish us again. Teachers were to be obeyed by us children, and even feared. They were certainly not our friends, like today. Life had strict order and rules that were understood without question.

There are those who claim that the environment in which we then lived was just as dangerous as it is now, arguing that we were less aware in those days. There is some truth in this, yet the facts speak for themselves. The period immediately after WW2 was a period of low crime-rate. Notifiable offences recorded by police in 1957 were 10 per 1000 of the population. By 1997, 40 years later, the figure had risen to 77 per 1000 of the population. The corresponding figures for violent crimes against the person were 0.2 in 1957 rising to 4.3 in 1997. The figures have since risen further, particularly with regard to fraud and violent crime. I lived my childhood in an age of relative innocence compared with the present day.

# Early Friends and Interests

When I was young I spent a lot of time with my best friend Trevor Williams who was two years older than me. He lived opposite at No 12 Henley Crescent.

A favourite place to play was Priory Park, just 10 minutes walk from home. The park was like an adventure playground covering some 200 acres and containing everything a small boy could wish for. There were extensive playing fields for football or flying kites, large areas of undergrowth and woodland for exploration, tree climbing, hiding and stealing birds' eggs, a complex network of paths and secret places, fast running streams for jumping across and messing about in, two large lakes for swans and for sailing toy boats, plenty of conker trees, steep slopes for snow sledging in winter, an excellent maze made from box hedge, a café for buying ice cream, a playground with roundabouts and swings, and much more. Priory Park was an idyllic place for any child. A great deal of my childhood was happily spent there with no parental supervision. We learned to take risks and to understand that every action had its consequences.

Priory Park also had a large, secluded walled garden called the Old World Garden. This was out of bounds to unaccompanied children and therefore of special fascination, particularly because it had ponds with enormous goldfish. There were park-keepers who chased us if we broke the park rules, like cycling on the footpaths or throwing sticks up trees to dislodge conkers. For the older generation there was a museum, a bowling green, a bandstand, tennis courts where my parents met and where I later played, and extensive displays of exotic shrubs and formal flower gardens. For small boys it was a haven. The maintenance costs of the park must have been enormous.

One hot afternoon my father walked with Trevor and me to Priory Park. I think he came to help fly a new kite. We were lying on the grass in the sunshine looking for four-leaf clovers when we found two silver half-crowns sitting side-by-side in the grass (2 x 12½p in today's money). There was no sign of any possible owner so my father suggested we spend the money in the sweet shop on Earls Hall Parade. We all particularly liked the new variety of milk lolly that was creamy and fruit-flavoured on a stick. They were called Milko Lollies. My favourite flavour was banana. The money was enough to buy nine lollies to share between the three of us. I was rather surprised that my father encouraged this extravagance. We spent all the money we had found.

One of the park-keepers was a bad-tempered man with a wooden leg. I never questioned why he had only one leg, but on reflection he must have lost it during the war, rather like the man with only one arm selling newspapers on the pavement at Victoria Circus in Southend. Nobody ever talked about these things. Peg-leg was at a severe disadvantage when chasing us for transgression of the park rules, but he made up for it with his imaginative invective.

My Uncle Frank was caretaker and later warden of Priory Park Museum, the site of a medieval priory of the Cluniac Order. His position allowed me and my friends privileged access to the museum, which was normally strictly out of bounds to unaccompanied children. The museum had a strange musty smell and filled us with a sense of mystery and adventure. We often visited to look at the artefacts, the most interesting being a macabre collection of thumb-screws, man-traps used long ago to catch poachers, and a range of rusty Victorian bicycles including a "bone-shaker" and a penny-farthing. Also of some fascination was the Hoy Collection of Stuffed Birds. Uncle Frank hosted us in his warden's uniform on these visits and recounted interesting stories of ghosts in the Old Priory. I think he was happy in his job.

It is curious how so many of our activities involved throwing things: throwing balls to each other over ever-increasing distances, throwing heavy sticks high up into chestnut trees to dislodge conkers, throwing stones over water to make them skim and bounce, throwing stones at tin cans and at wooden boats in streams, throwing snowballs. Some boys (not friends of mine) would throw stones at the swans and geese in Priory Park with occasional fatalities, but I thought this was cruel and did not take part. As a result of all this throwing I developed a good technique achieving great range and accuracy.

Every boy had a pen-knife, or in a few cases a sheath knife. A smart pen-knife conferred a certain status on its owner. Pen-knives were used to cut sticks, to make bows and arrows or prise open sea shells. They also had more erudite uses such as for making musical flutes out of the tubular stems of cow parsley, a skill now largely lost. A particular skill to be acquired with a pen-knife was the ability to throw it into the ground or a tree so that it stuck point first.

Unfortunately, the pen-knife my father bought me had an unconventional curved blade making it difficult to throw. It always seemed to me that my parents bought me presents that were not quite in the style I wanted, like the roller skates that had a strange lace-up design and my first bicycle that had peculiar-shaped handlebars. I was not comfortable with these things because they were not the same as what my friends had. I was very ungrateful.

To return to the pen-knives, we knew you could cut yourself if you were not careful but it never occurred to us that we could use them to injure other people. I am thinking of how different it is today with our knife culture.

I grew up in an era when there were no mobile phones, no iPlayers, no Internet, no computers, no PlayStations, no social media, hardly any television, and no such thing as "celebrities". New technology for me was my birthday present of a levered fountain pen. It might sound that life was limited, but we had complete freedom. We were rarely bored because we learned to use our imagination.

At home, Trevor Williams and I invented a sort of shove-ha'penny football game called "Crickenny", so named because the players used two old **cr**own coins (belonging to my father), because you had to be s**ick** to play the game, and because the football was a p**enny** - hence cr-ick-enny. The football pitch was a large piece of hardboard that I stored behind the sofa in our front room. The goal posts were made of short Meccano rods bolted to the hardboard. You took turns using a ruler as a cue to strike your crown to hit the penny. The aim was to knock the penny into the opponent's goal. This was a wonderful game for a rainy day and we used to play for hours. My father enjoyed it too. It was much more exciting than the later well-known table-top football game Subbuteo.

When I was nine years old my parents gave me a Hornby Dublo electric train set for Christmas. The engine was the Duchess of Atholl with a coal tender and two carriages. There was quite a lot of track that I added to over the years. I had no permanent place for my train set. Each time I wanted to play I had to set it up on the floor of the front room, usually on the big piece of hardboard used for my Crickenny game. Both my father and I enjoyed playing with the train set. It was eventually inherited by my brother Nick. I hope he still has it.

I also liked making things with my Meccano, which taught me about the hazards of cross-threading nuts and bolts. My interest was greatly enhanced when I acquired a battery-powered dynamo motor for driving automated models like cars and windmills. Some of my friends had Trix, but this never caught on with me.

But my favourite construction set was Bayko for building model houses. Bayko was a forerunner of Lego, but required more skill to plan and assemble the buildings with an end result that was totally realistic.

At about this time, when I was nine years old, my father had a serious illness. It was December 1952. I remember being woken up by a commotion in the middle of the night to witness my father being carried downstairs on a stretcher to a waiting ambulance. He was groaning and in obvious pain. My mother was very upset and so was I. It turned out that my father had a burst stomach ulcer requiring urgent surgery.

I remember visiting him in hospital after the operation. This was not in Southend but in London. I did not question the reason for this and I have no idea how he got there. I now realise that he had the operation at the Royal Masonic Hospital in Hammersmith, located just behind Queen Charlotte's Hospital where our daughter Caroline was later born. It must have been organised by his GP, Dr Rayne-

Davis, who was also a mason. My father spent Christmas and New Year in hospital. He sent me a card that I still have.

My father made a good recovery with only minor relapses. He told people how two thirds of his stomach had been cut away, but this never seemed to affect his appetite. My mother told me that stomach ulcers were brought on by worry. It was common knowledge that our neighbour Mr Williams also had an ulcer, but the duodenal sort. I wondered what these men worried about so much. It is now known that stomach ulcers are usually the result of a bacterial infection and nothing to do with worry.

On reflection, I cannot recall my mother ever being ill. She suffered from the occasional sick headache that she cured by lying down for a couple of hours. Otherwise she was in good health. She did have regular work done at the dentist, but I don't remember her ever going to the doctor except perhaps in later life. My father, on the other hand, was always at the doctors for his various ailments that disqualified him from ever obtaining life assurance. My mother once told me that she would hate to have to go into hospital. As it turned out, the only night she ever spent in hospital was the night before she died.

For a few summers we shared a beach hut at Thorpe Bay with the Williams family for a week. Not having a car we travelled there and back each day by bus. When it was sunny we swam and had great fun with an inflatable army dinghy that my father had acquired. Seeing my father in his swimming trunks I was shocked to see the enormous 10 inch scar up the middle of his abdomen from his recent operation.

We dug enormous holes in the sand surrounded by battlements to try and hold back the tide, rather like King Canute. When it rained we huddled inside the beach hut round a small spirit stove used to boil water for tea. I recall the sickly smell of the methylated spirits. Picnics in the beach hut gave me my first pleasurable taste of what was supposed to be French mustard, out of a tube.

I spent a lot of time with my friends on the beach and went swimming at all times of the year, even in the dead of winter.

At low tide we could walk out on the mud flats for over a mile searching for cockles, mussels and fat lugworms. When the tide turned it would race in faster than we could walk. We had to be careful not be stranded by any of the deep channels.

When my friend Trevor joined the choir at St Peter's Church I soon followed him. I was about nine when I joined. This was the nearest "Church of England" church and was in Eastbourne Grove, about 15 minutes walk from my house. The church building was actually a large, rather austere hall. It was meant to be a temporary church while money was raised to build a proper church on the adjacent plot of land. This piece of land was overgrown and densely wooded, and known by everyone as The Wilderness. It was a place of great adventure on dark nights. The "temporary" church building must have been used for 50 years. It was not until long after I left home that the new church was finally built on the adjacent Wilderness site.

St Peter's was a popular church with a large congregation and many supporting activities. The vicar was Mr Lawrence Reading, a tall athletic man with a full black beard that made him look a bit Jewish. For this reason we called him "The Shonk", but not to his face. It was rumoured that the beard was to hide an ugly scar earned in the Navy during the war. Mr Reading had a daughter Anne about my age and a younger son called Timothy. Anne was to become notorious amongst the teenage youth as a girl of easy virtue.

When I joined St Peter's choir I was presented with a copy of "The Chorister's Pocket Book", still in my possession. This small volume gave important information about the church ritual, church composers, how to read music, and how choirboys should behave. I regret to say that I never opened this book until later life.

Our choir comprised 8 to 10 boys, plus about 6 men and 4 women. The choirmaster and organist was Mr Herbert whose skill and dedication I never appreciated at the time. We each wore a black cassock underneath a white surplus and neck ruff. We had choir practice on Friday evenings and had to attend at least two of the three Sunday Services, which were 8 o'clock Holy Communion, 11 o'clock Matins, and 6:30 Evensong. I enjoyed singing in the choir and particularly liked the anthems that we sang on special occasions in the church year. My favourite anthem was "From the Rising of the Sun" which we regularly sang at Epiphany.

To give the choir boys an extra incentive the church paid us a small sum of money for our services, something like one penny for each service attended. But for a wedding we were each paid the enormous sum of 2/6 (2 shillings and 6 pence). It was the groom of course who had to pay. We collected our pay every quarter when each boy typically received about 5 shillings (25p), quite a lot of money.

In the week before Christmas Mr Herbert would organise an evening of carol singing round the streets, to which we boys applied ourselves with gusto. I do not recall that we collected money for our carol singing. I think it was more for the simple pleasure.

Every year there was a competition held for local church choirs at nearby Leigh-on-Sea. We must have been a good choir because we always did well and actually came first one year. We also took part in various festivals including the annual Choirs Festival at Chelmsford Cathedral. One year it was my turn to be appointed Head Chorister. This meant that I could wear a special medallion round my neck on a length of purple ribbon.

The service I enjoyed least was early morning Holy Communion at 8 a.m. We were not allowed to eat any breakfast before this service, which entailed long periods of kneeling and standing. On at least two occasions I almost fainted and had to sit down or stagger into the vestry to get some fresh air. I became in regular fear of this happening.

I was quite musical and was able to make up my own harmonies even though I had no knowledge of the theory of music. I could do this spontaneously, inventing a credible bass, tenor, alto or descant in real time as I was singing. I could hear these different parts in my head all at once. So instead of singing the treble line in the choir I would sometimes sing alto because it was more interesting and fitted the range of my voice. This got me into frequent trouble with Mr Herbert. I never appreciated that I had this talent until I realised much later that few other people could do it.

I also liked organ music from an early age. I had an old wind-up gramophone with doors at the front that controlled the volume like a swell box. I kept it in my bedroom. My records were all old 78 r.p.m. shellac discs. Some belonged to my father including Grenadier Guards marches and 1930's American song hits like "Loaded Pistols and Loaded Dice", "The Opossum Song" and "Woodman, Spare that Tree" by Phil Harris, that my father loved to play. More interesting were records of my own such as that well-known number "Teddy Bears Picnic" with "Dicky Bird Hop" on the B side, and also "Sparky and the Talking Train", an American story of a small boy who saved a big ship from sinking in a storm with the help of a talking steam engine, complete with songs and sound effects. I would listen to these records over and over.

But the record that really excited me had organ music on it. I don't know where this record came from. It was LP size and required me to slow down the speed of the turntable to 33 r.p.m. On one side was Liszt's Fantasia and Fugue on B.A.C.H. and on the other was Julius Reubke's Sonata on the 94th Psalm. I had never before heard music as exciting as this. I lay on the floor with my head stuck between the doors of the gramophone in order to maximise the volume. The virtuosity and power were mind-blowing. It was many years before I was able to develop this interest positively.

I never graduated to the modern style of record player that played the new type of unbreakable plastic vinyl records, even though just about every one of my friends had one. I don't think pop music interested me enough. I also discovered that I did not like listening to music as a background to doing something else. If I liked the music I had to give it my full attention to the exclusion of all else. If I did not like it then it was not worth listening to anyway. I realised that there were two types of people: those who like music and those who just like the noise it makes.

To return to the choir, although the choirboys worked hard we were not angels. In fact we were rather naughty. The ringleader was usually a boy called Dave Hopkins. In fact it was he who encouraged me to smoke my first cigarette.

A favourite prank was performed after choir practice. The boys would arrive for practice at 7 p.m. and were joined by the adults at 8 p.m. for a joint practice. The boys then left at 9 p.m. leaving the adults to finish practising. Instead of going straight home we would wait outside the church for an opportune moment to turn off all the power in the church. This could be done by throwing the mains switch located in a small store room just inside the entrance door to the church. For maximum effect we waited until the adult choir was in full flow and the organ playing loudly. One of us then threw the switch. All the lights went out and the organ ground to a halt making a moaning noise as the last breath of air escaped from the wind chest. This caused great confusion in the church on a dark night and great merriment to us choirboys who would run away at high speed. Of course, the real excitement was accepting the dare of the other boys to be the one to perform the deed. This took some courage, especially when it was done the second or even the third time on the same evening.

I am ashamed to say that we once played this prank at a Sunday evening church service. A group of us choir boys had been excused from attending the evening service because we had sung at both Communion and Matins. We nevertheless met up in the evening. It was winter, so it was dark with all the lights on in the church. Everyone was singing loudly and the organ playing. We crept into the store room just inside the church entrance and threw the power switch. This made the church pitch dark and stopped the organ. There was much commotion from the worshippers inside. The Scout Master, known as Hawkeye, who was on sidesman duty that day must have spotted us because he was hot on our heels as we fled from the church. Fortunately we were able to hide in the dark undergrowth of The Wilderness close by.

A similar prank was called "knock-down-ginger" which entailed knocking at someone's door and running away.

There were certain preferred houses for this prank, either because we did not like the person who lived there (perhaps they had confiscated a ball or something) or because the front path was particularly long which increased the exposure and therefore made it a bigger dare. Again, the biggest dare of all was to repeat the prank at the same house in quick succession.

Once I accepted the dare to do the third knock-down within ten minutes at the same house. This was a house in Carlton Avenue. It was a dark winter's evening. I crept up to the front door and knocked but he was waiting for me. The door was flung open and the owner, in plimsolls I remember, tried to grab me. I escaped and ran down the path and onto the pavement with the man in hot pursuit. He ran fast and caught me, shouting and shaking me. I was so terrified I wet myself. The man must have realised and let go, whereupon I ran off to join my friends trying to show bravado. I could not go home until my trousers had dried out. I told my parents that we had a late choir practice that night.

Sometimes Trevor and I played truant from the Sunday evening service. On one occasion on a fine summer's evening we were tempted to go for a walk in the country instead of going to church. We walked over the fields at the back of Manners Way and stood watching the trains go past, hauled by steam engines in those days. Somehow we got the idea of putting pennies on the line to see if the heavy engine would squash them. It did.

Another time, on a cold night in winter, Trevor and I were too late for the evening service. So we messed around outside peering through the windows and making faces at our chorister friends inside. Before the service finished we had the brilliant idea of climbing into the small roof space in the porch (since demolished) that was built over the exit gate to the church premises. We found there was just enough room for us in this small dark space, sitting one at each end of the cross-beam and completely hidden from the people passing below. Sometimes groups of people would stop and carry on a conversation directly beneath us. We could reach out and touch their hats and place pieces of paper on them. It was difficult for us not to burst out laughing. This became a favourite hiding place.

Towards the end of my time in St Peter's choir we were commissioned to perform a special service at the laying of the Foundation Stone for a new church to be called St Cedd's. It was being built about a mile away from St Peter's in an area bordering open fields where there is now a housing estate. It was a windy day for this open-air service. I can pin-point the date from the foundation stone to 30 October 1954. One year later our choir assisted at a special service led by the Bishop of Chelmsford to commission the finished church.

Many years later, in September 2009, during a visit to Southend to visit my Aunt Margaret (the last time I was to see her) and her husband Malcolm, I called in at St Cedd's church out of curiosity. I was shocked to find it had been divided into two to provide a community hall at one end where a tea-dance was in full swing. I was invited in and took part in the dancing.

At home in Henley Crescent we had occasional visits from Uncle Harold, my mother's brother, his wife Jennie and their children Jean and John. They lived in Totteridge, North London. Their visits were always on a Sunday because Harold was busy in his greengrocers shop in Barnet all other days of the week. Neither Harold nor Jennie ever learned to drive, but the business owned a car. This was a smart Bentley driven by Joe, who worked for Harold fetching fruit and vegetables from Covent Garden Market every morning. Joe would drop the family off at our house in the morning then drive into Southend to enjoy the pleasures of the seafront before collecting the family again late afternoon. On leaving, Uncle Harold usually pressed a half-crown into my palm, a welcome gift.

Guy Fawkes Day on November 5 was always a great event. For the two or three weeks before the big night there were fireworks available that we bought across the road at Shannons. There was no ban on children buying fireworks. It seemed to be at the discretion of the shopkeeper. We could not afford fancy fireworks and were not really interested in them anyway. What we wanted was penny "bangers". Pains fireworks were good but Brocks made a bigger bang.

On Saturday mornings leading up to the big day we would play in the fields at the bottom of the road (now a housing estate) devising new purposes for these explosive devices. There were so many things you could do with a penny banger. You could light it, let it fizz for a few seconds, then throw it like a hand-grenade. The object was to time the throw so it exploded in mid-air. You could put them in bottles and explode them. You could bury them in ants nests or, if you were lucky enough to find one, in a wasps nest and cause mayhem. You could drop them down rabbit holes. You could sometimes get them to explode underwater with interesting results. It is amazing that we did all these things without injury. I was no more than 10 or 11 years old. I only once had an accident when a banger went off in my hand causing some numbness for a few hours.

In order to fund these activities we once did something rather dishonest. In those days empty glass lemonade bottles and beer bottles were reused by the brewery. They had a refund value, typically one penny. You could often find discarded bottles on waste sites and make a few pennies by taking them to an off-licence. The nearest off-licence to my house was by the side of the Bell Hotel. The yard behind the off-licence was used for storing empties awaiting collection from the brewery. We once climbed over the fence, topped with barbed wire, to steal bottles from the yard and reclaim the refunds from the off-licence. This was bad and I am slightly ashamed. But it did enable us to buy a few extra bangers that year. It is a pity they stopped giving refunds on empty bottles. It would stop a lot of the litter you see today.

Firework night itself was always exciting. First of all my father usually had a small display in our back garden to which Trevor and friends were invited. This was enjoyable but tame. My friends and I then went out on the streets armed with pocketfuls of bangers. I probably told my parents I was going to a friend's house. The first port of call was always the Francis' Firework Display. Mr Francis had a butcher's shop at the end of the parade of shops by the Bell Hotel, just five minutes walk from my house. Mr Francis was always a bit of a wild man, but on bonfire night he became totally irresponsible.

He and his two sons, one of whom was my age and called Graham, collected material for a huge bonfire in the weeks leading up to firework night. The bonfire was built on the vacant field next to his shop. On firework night they lit the fire and entertained all the local families gathered in the field to a massive display with the most expensive set pieces. His supply of fireworks seemed to be endless. After about half an hour Mr Francis would get carried away and start firing roman candles and star busters into the crowds. He held these big fireworks in gloved hands, aiming them at people. This livened things up considerably. People in the crowd retaliated and a free-for-all usually developed. It is amazing nobody was ever hurt.

After this excitement we roamed the streets looking for other bonfire parties to join. Over the course of the evening we had to carefully ration the use of our precious bangers. Thinking back, it is strange how we never put lighted bangers through people's letter boxes or in post boxes. Is it because we had not thought of doing that sort of thing, or did we realise that this was beyond the bounds of acceptable behaviour?

A weekly treat was Dandy, a comic delivered on Tuesdays with the daily newspaper (Daily Herald) at about 6:30 a.m. If it arrived early enough my father would read it before going to work. He then brought it upstairs to me in bed. We both liked Desperate Dan who used to eat cow pie, but our favourite was Rusty on the back page.

I followed Trevor into the Cubs at St Peter's Church at about the same time as I joined St Peter's Choir. We met once a week in the church hall. We played games, had inspections and pledged our loyalty to God and King. We chanted various things like DOB DOB DOB which stood for "Do Our Best", the wolf cub's motto.

All this was fine, but what I disliked was the part of the evening devoted to earning badges. One of the most basic badges was the fitness badge that required you to do a head-over-heels. I was unable to do this properly. I think I was scared of damaging my glasses. My father tried to teach me and so did the cub master, but every time I collapsed sideways. This turned into a nightmare for me that went on for weeks totally over-shadowing the other badge work like learning knots and recognising birds. So I did not greatly enjoy these weekly meetings. I have never been a club type of person.

However, I did enjoy Bob-a-Job week which came round once a year at Easter time. A "Bob" was one shilling. Each cub worked on his own and had to earn as much money as possible by doing odd jobs for people. After the obvious candidates like parents and relatives, I visited the neighbouring streets dressed in my cub uniform, knocking at people's doors to ask if they had a Bob-a-Job for me. The responses were varied but usually generous. They would ask me to do some shopping or polish the brass. Some mean people would ask for their whole lawn to be mown. You were not allowed to turn a job down. A few people paid a shilling just to get rid of me. By the end of the week I usually had about £1 in earnings which I thought was pretty good. I gave this money to the cub master. I don't know what he did with it. Bob-a-Job has now been banned because it is not politically correct and because it is considered too dangerous for unaccompanied little boys to knock on the doors of strangers.

There were occasional camping expeditions which I managed to avoid. Even at that age I was not attracted to eating and sleeping outside. Singing round a camp fire held little attraction for me. The whole thing seemed very Neanderthal.

One year there was a cub and scout show held in the circus arena at Southend Kursaal, a large entertainment complex on Southend Seafront. All the packs in the area were to take part. The show was to be watched by a large audience including parents. The particular contribution of our pack was a Red Indian War Dance. My mother made me a Red Indian outfit complete with head-dress. My father helped me make a tomahawk from a piece of broom handle topped with a thick cardboard axe head painted silver, which I daubed with red paint to look like blood. I was proud of my outfit.

We had special rehearsals for our display. The War Dance required us to be in pairs. My partner was Terry Absalom, my friend who lived up the road at No 21. Everything went well until the dress rehearsal when they told me that Red Indians did not wear glasses. So when it came to the real thing at the Kursaal, horror of horrors, I had to do the whole thing without my glasses. I could not see a thing, let alone the audience. I managed to get through the routine by staying close to Terry and following the boy in front. To crown it, Terry became carried away with his role and after the show started laying into me with his tomahawk in the dressing room. This was before I had time to get my glasses back on. I was unable to defend myself and got badly beaten. Altogether this was a bad experience. I left the Cubs when it was time to graduate to the Boy Scouts.

I continued playing with Trevor until I was at least 12 or 13 years old. I can be sure of this for two reasons. Firstly, I remember being in Trevor's back room while he was playing repeatedly a certain record called "The Rock Island Line" by Lonnie Donegan, which came out in 1955. Secondly, I remember one year when we were sharing the beach hut at Thorpe Bay hearing with some concern the man selling newspapers shouting "There's going to be another war". He was referring to what was to become known as the 1956 Suez Crisis.

However, my friendship with Trevor was beginning to wane because our two-year age gap was increasingly significant. Trevor began to have new friends and when I joined them I found I was the odd one out. Being the youngest I was more vulnerable and was even bullied by his friends. On one occasion we were all playing in a large field off Bridgewater Drive (now a housing estate). There was Trevor, Dave Hopkins and one of his friends whose father was high up in the police force. They tied me to a tree and lit a fire at my feet. I was terrified. I managed to escape but the fire got out of control. The grass in the field was bone dry and the fire swept across it towards the houses. The other boys tried to put the fire out. A worried house owner called the fire brigade but by this time I was on my way home.

In the summer of 1951 when I was eight years old my father took me on a day trip to London to visit the Festival of Britain. It was located on the South Bank close to Waterloo Station. The aim of this national exhibition was to give Britons a feeling of recovery and progress after the desolation of British towns and cities in the war. The Festival also celebrated the centenary of the Great Exhibition of 1851. There was not a great deal there to interest me but I do recall seeing the Skylon and the Dome. The thing that impressed me most was the moving escalators and the London Underground.

My father bought me several "First Day Cover" letters which I posted to myself with Festival of Britain stamps. Apparently, stamps are sometimes more valuable if franked. I still have them, worth about £10 - £15 each. I was interested in stamp collecting at the time and had a large collection of worthless stamps from many countries.

At about this time I entered a competition run by Cadbury's Chocolate. We were given a basic shape which you had to incorporate imaginatively into a complete picture in the form of a pencil drawing. The basic shape and the drawing I made are roughly reproduced in Plate 17. In case it is not intelligible, my drawing is supposed to show the back view of a man in tails waving an umbrella. Amazingly I won first prize. Every month for a year I received through the post a large package containing an assortment of Cadburys chocolate boxes and bars. I made myself sick of Cadbury's chocolate and no doubt did much damage to my teeth.

When I was young my parents never went away for summer holidays. They could not afford it. But late in the summer of 1951 I did go away with my father to visit distant relatives, an Auntie Sylvia and her husband. They lived at 14 Boys Lane, Cadley, near Preston in Lancashire. I know the precise year because it was just before the birth of my brother Nicholas. I was unaware of this imminent event at the time, but later I was told that our Preston trip was to give my mother a break before the birth.

I have several vivid memories of this holiday with my father. We travelled to Lancashire by coach, probably joining it at London, Victoria. It was a long journey because there were no motorways and it rained all the way. The coach had an accident trying to avoid a car pulling out of a side road. I remember the coach going into a long skid on the wet road and finishing up at right angles to the road in someone's front garden. I don't think anyone was hurt but we were shaken and delayed for a bit. We stopped in Wigan for a short break. I recall that the streets were filthy and covered in orange peel. Perhaps it had been market day. Ever since, Wigan and discarded orange peel have always been synonymous.

Our relatives collected us from Preston bus station in their open-top vintage car. They took us at high speed through country lanes to their house. They lived in a rural setting. Opposite their house was a field full of ducks. On arrival we went to feed them the remains of a large pork pie left over from our coach journey. To our astonishment one duck appeared from nowhere, picked up the pork pie in its beak and ran off round the field at high speed with its prize, all the other ducks chasing it. We collapsed with laughter.

I had recently learned from my Rupert Annual how to make an inflatable paper kettle. I spent a long time in our relatives' front room making bigger and bigger kettles from ever-larger sheets of newspaper.

During this holiday my father took me on two day trips, one to Blackpool where we climbed the famous Tower and another on a train to the Lake District via Windermere and Ambleside. We picnicked overlooking a lake which must have been Rydal Water near Grasmere. This was to be my first of many trips to Lakeland.

My brother Nicholas Andrew was born on 7 December 1951. According to Nick he owes his existence to £500 that my father had won the previous year from gambling on the horses. This was a lot of money, probably close to his annual salary. He used it to persuade my mother to overcome her earlier vow to have no other child after me by promising to pay for her confinement in a private nursing home. She obviously agreed to this. Nick was born in Hayesleigh Nursing Home, Westcliff, where he and my mother stayed for over a week, as was normal in those days. I was staying with my grandparents in Hobleythick Lane. Soon after we celebrated Nick's christening.

At about this time I began staying for short summer holidays with relatives. I had several holidays with my Auntie Haidée and Uncle Billy at 20 Ilmington Road, Kenton, in North London. I liked Haidée because she was generous, young at heart, and had time for everyone.

On reflection, I think that Haidée was unusually tolerant and liberal in her views. There were always visitors staying at her house, people of all nationalities and backgrounds with whom she had somehow developed a connection. Sometimes they stayed for two days, sometimes for weeks because she was so hospitable. Life at her house was always busy. Uncle Billy remained calm, just sitting in his armchair with a pack of cigarettes by his side and watching television. Haidée often asked me what I would like for dinner and I usually said Shepherds Pie. For breakfast she would give me two boiled eggs. I had never before had two eggs at once. I remember thinking that my mother would have considered this extravagant. Haidée always made a fuss of me. Once she took me into London and treated me to a *mille feuilles* at Lyons Cornerhouse.

Haidée's husband Billy was a kindly but gruff man with a dry sense of humour and a high-pitched voice. He ran a successful roofing business. He was a mason and often invited my father to dinners at his Lodge in Kenton. Both Haidée and Billy were chain smokers and invariably lit up at the dinner table between courses. Smoking was socially acceptable, but I was surprised by this breach of table manners.

Their younger daughter Jane was slightly younger than me. We played together but did not have much in common. Rather like Doreen Yarwood of Henley Crescent notoriety, she was an exhibitionist and would show me her private parts, in which I had a mild curiosity. It was all very innocent.

I recall meeting Haidée's father Uncle Phil, the retired music hall artist. He was elderly and frail and lived at Haidée's in his final years. On the few occasions I saw him he was always in his dressing gown and took little interest in me. Uncle Phil died on 8 August 1954.

I had several holidays with my cousin John Besent in Totteridge, North London. I always enjoyed staying with John. We had immense fun playing together and teasing his elder sister Jean. I introduced John to the game of battleships which we played for many hours. John's mother, Auntie Jennie, introduced me to porridge. This and her northern accent made me think she was Scottish. I learned only much later that she was from Gateshead.

One of the highlights of staying with John was visiting his father's fruit and vegetable shop in Barnet. There was a backyard where all the empty boxes and stale produce were temporarily stored. John and I had great fun clambering over the mountain of boxes and throwing rotten fruit at each other. Once Uncle Harold caught us and scolded us severely. This was one of the few times when I saw him lose his temper.

We often visited nearby Hadley Woods to explore and climb trees. Once John recounted how the woods were the site of an ancient battle fought during the Wars of the Roses (the Battle of Barnet in 1471 when the Yorkist King, Edward IV, defeated the Lancastrians led by the Earl of Warwick, who was killed in the battle). This was scary. The woods were dense and we began to see the shadowy ghosts of dead soldiers darting through the undergrowth. We scared each other more and more as our imaginations escalated until we ran out of the woods in terror to the safety of daylight. If we had our bikes with us, John borrowing Jean's, we would cycle to the railway bridge in the middle of the woods to watch the express trains emerging at high speed from the distant tunnel on their way to Kings Cross. Once we saw Mallard.

During John's return visits to Southend we always visited Southend Pier. This was an amazing attraction. Built in 1830, it is the longest pleasure pier in the world at 1⅓ miles (2.13km). There were two ways of getting to the Pier Head; you could either take the train or walk. Walking was quite exciting particularly in bad weather. Usually we would walk out and take the train back.

The Pier Head was built on several levels and offered a wealth of entertainment. There was the Life Boat to visit, a theatre and orchestra, a smart restaurant, various snack and ice cream stalls, a fun fair, games such as deck quoits, fishing for those inclined, speed boats that I only ever watched because they were too expensive to ride (5 shillings each for 10 minutes), illuminations at night, and plenty of deck chairs and sundecks for the less mobile. But the biggest attraction was the penny slot machines that included peep shows such as "What the Butler Saw". It was possible to spend a whole day at the end of the pier and never get bored.

Big passenger ships used to run from the Pier Head. One was a paddle steamer called the Medway Queen (since saved for posterity by the Medway Queen Preservation Society) that did day excursions to Margate, Herne Bay and Rochester. Her sister ship was the Crested Eagle. My father once took me on one of these day trips to Rochester where we visited the cathedral. There was also a regular service to Ostend, Calais and Boulogne on the Royal Sovereign and the Royal Daffodil. These were larger vessels based up the Thames in London, calling in at Southend pier on their way to and from France. I took advantage of this service when I was older, as did Lucienne.

The pier was later blighted with many disasters. In 1959 a fire destroyed the land-end pavilion which was rebuilt as a bowling alley. In 1976 a fire at the pier head destroyed most of the amenities. In 1986 a tanker crashed into the pier creating a 70 foot gap. In 1995 another fire at the land-end destroyed the bowling alley. In 2005 yet another fire at the pier head destroyed the remaining amenities. The light railway that ran the full length of the pier was powered by electricity and was twin track. This has been replaced by noisy diesels on a single track. The pier is now recognised as an icon and efforts are being made to reinstate it to its former glory. Jamie Oliver has used a café at the end of the pier as a base for one of his cooking series.

Next to the pier was an attraction of comparable interest to small boys. It was a full-size replica of the Golden Hind, the sailing ship used by Sir Frances Drake to sail round the world. You could explore the whole ship and visit the adjacent waxworks exhibition of medieval tortures.

I used to go regularly with my friends to the cinema, especially on wet Saturday afternoons. There was no shortage of cinemas to choose from. In Southend there was the Odeon in the High Street which was the poshest, the Ritz at the top of Pier Hill, the Rivoli in Alexandra Street, the Gaumont in Southchurch Road, the Strand reached via an alley from the High Street (later renamed the Essoldo), Garons at the back of the High Street, and in the arcade by Victoria Circus was a tiny cinema which had many incarnations and names including the Civic, New Vic and Continental. Lastly, in Tylers Avenue, off the High Street, was the Regal Theatre specialising in nude films, which I never went to.

Also in the neighbourhood was the Plaza at Southchurch, the Mascot and the Metropole (later renamed the Essoldo) in Westcliff, and the Corona and Coliseum in Leigh. These last two were the furthest away but very conveniently served by the No 21 bus that started its journey from the bottom of our road and finished at the Coliseum.

When I was at junior school we often went to see comedy films like Dean Martin & Jerry Lewis or Norman Wisdom. These films had us in uncontrollable fits of laughter. Some of my friends went to "Saturday Morning Pictures" especially for children where they showed several cartoons followed by a film, usually something like Tarzan. This never caught on with me.

In my early teens we often tried to gain entry to "A" and "X" certificate films. "X" films were strictly for adults aged 18 and over, whereas "A" films were for adults or children over 16 accompanied by an adult. To gain entry to "A" films a common strategy was to ask a sympathetic-looking man in the queue if he would say we were with him. This seemed to work best if the man had his girlfriend with him, perhaps because it made him feel important. Otherwise, we simply pretended we were 18 years old, which sometimes worked and other times failed. In this way we saw many "X" rated films like Dracula and Frankenstein, and notably, we saw in 1958 "The Fly" (now a cult film) when I was 14. The excitement was not just that such films were scary, but that they were forbidden. I often found them rather disappointing. They were certainly tame compared with today's films. In those days even Doctor Who would have been "X" rated, possibly even banned.

Immediately after the evening showing of a film the cinema always played the National Anthem. Everyone was expected to stand to attention until it had finished. Most people accepted this. A few saw it as an inconvenience and escaped up the aisle while the credits were still rolling.

In 1954 my maternal grandmother gave me the money for myself and a friend to see The Robe, a Biblical epic film about the Roman military tribune who commanded the unit that crucified Jesus. I don't know why she was so keen. She must have thought it would be good for my religious education. She certainly had not seen the film herself. I went to see it at the Odeon cinema with Richard Coakes and found it quite disturbing in its realism.

So altogether there were 13 cinemas, each with its own programme that changed every week. They all advertised their films for the coming week in the Southend Standard, published every Thursday. Hardly any of these cinemas remain.

Altogether, I consider I was most fortunate to have lived in an environment with so much freedom and so many amenities. This was partly because Southend was a holiday resort. For an entertaining and nostalgic look at the many attractions of Southend circa 1958, see the following publicity film on YouTube: https://www.youtube.com/watch?v=0lDvCtvDkOQ

# Primary and Junior School (1949-1954)

I remember well my first day at Earls Hall Primary School. It was in Carlton Avenue just a few minutes walk from our house. My mother walked me round to the school and left me there in the reception class. I was quite scared because I had never been separated from my mother for such a long period but I quickly settled down and began to enjoy it. The teacher was a large motherly woman whom I trusted. Because my birthday was 6 June I started late in the school year when I was 5½ years old, which would have been after Christmas in January 1949. I do not remember much about this class except that it seemed to be all play.

I recall a debate between the teachers about whether I was ready after only six months in the reception class to go straight to the top class in the infants. I was taken several times to a private room for maths and reading tests. I don't think I had any problem. I was good at sums and was able to read before I went to school. My mother had taught me well, and I was also encouraged to read to my maternal grandmother on our afternoon visits. Sixty years later my cousin John told me that my early reading feats were a cause of great annoyance to him when our grandmother compared our performances.

I should say that there was no such condition as "dyslexia" when I was at school. Some children had reading difficulties or they were just illiterate. I do not recall any illiterate children leaving my junior school. Similarly, pupils did not have "behavioural problems", "chronic disobedience" or "conduct disorders". They were just badly behaved and punished accordingly.

So in September 1949 I moved up to the top class with a teacher called Miss Isaac. (Many years later Miss Isaac's brother, George Isaac, was to marry my parents' close friend and my godmother, Rene Cabuche). Miss Isaac was an excellent teacher and I think I made good progress in the 3 Rs.

I have two vivid memories of my short time in the infants. The first was my part in the annual Nativity play. I played the part of an angel dressed in a white sheet and did not have to say anything. There was another angel called Gabriel who did have something to say. My mother had a close lady-friend called Gabrielle for whom she did dress-making. As a result I believed for 40 years that the Angel Gabriel was a woman. My parents were in the audience of the Nativity play and I was very proud. It was Christmas 1949.

The second incident made a greater impression on me. A teacher overheard some boys in the playground using bad language. Apparently I was one of these boys although I cannot recall what offensive words I was supposed to have said. A group of us was sent immediately to the Headmistress, Miss Warman, a strict and respected lady. We were ushered into her office and stood in a line in front of her desk. She reprimanded us and finished by saying that if we were caught using these words again she would "wash out our mouths with soap". This seemed a strange punishment whose metaphorical significance was lost on me, but it had the desired effect. I was in great fear of Miss Warman.

The photo in Plate 19 shows me in my final year in the Infants. I am in the middle row, far left. It will be noted that I am standing at the end of the row and that I am the only child wearing glasses. This must have given me a complex. My mother also wore glasses, and in extant group photos that include my mother she is also usually at the end of the row. I think this must be significant. Similarly, in secondary school photos I am one of very few wearing glasses. Did children have better eyesight in those days, or did poor eyesight just go undetected and neglected?

In the summer of 1950 I transferred to the Junior School at Earls Hall. This was a simple move upstairs. The Infant and Junior schools were run as two separate schools but occupied the same building, with the Infants downstairs and the Juniors upstairs. There were separate entrances and playground areas for the two schools. I was to spend four happy years in Junior School.

We all had to wear the school uniform which included a grey blazer, white shirt, tie, and short grey trousers for the boys. A boy did not usually wear long trousers until he was 12 or 13, or when his voice broke.

My first class in the Juniors was taught by Miss Pollard. I think my mother knew Miss Pollard through dress-making work that she had done for her. I really liked Miss Pollard. She was a young, enthusiastic and caring teacher. She married that year but kept the use of her maiden name at school. So she became Mrs Pollard. She lived with her husband in a detached house in Hobleythick Lane next door to Paul Archard, a future friend.

As a result of some staff reorganisation Mrs Pollard was again my teacher in my second year at Junior School. Then Mrs Pollard got pregnant. She announced to the class that she would be leaving at Easter. I was sad about this. For the remainder of that year we had a supply teacher.

In 1952 I moved up into Miss Milliken's class. I remember little about this class although I think she must have been a good teacher. We called her Miss Milk-Can.

Usually I went home for lunch, or dinner as we used to call it. Dinner hour was from 12:30 to 1:30. My mother always prepared a delicious hot meal which we ate together at the kitchen table. This was my main meal of the day. It was also my mother's main meal. On Mondays we usually had shepherds pie made from the remains of the Sunday roast. On Fridays it was always fish, often in cheese sauce, sometimes fried with chips. On other days she cooked things like sausage and mash, toad-in-the-hole, cauliflower cheese, steak and kidney pudding, or my favourite spam (spiced ham) fritters. In summer we often had a salad with pilchards in tomato sauce. There were always vegetables of some sort and always a pudding, either hot or cold. I particularly liked treacle pudding and banana custard.

Throughout my school career until I was eighteen I usually went home for lunch. I never appreciated the care and trouble my mother took over all those years to prepare a tasty and healthy meal for me every day on a low budget. It must have been quite a commitment. She never once said she was too busy or was going out to see a friend and so please would I eat at school that day. She was always there and I was obviously her priority. I regret that I took all this so much for granted.

In the evenings I had something light to eat before my father arrived home from work for his main meal, which he ate alone usually at about 6:30 pm and which my mother prepared specially for him. He enjoyed such things as oxtail stew and tripe, the latter apparently being good for his stomach condition.

We all ate together only at the weekends, the most important meal being Sunday lunch (called dinner), an enjoyable ritual. One of my jobs was to whet the carving knife, a narrow blade that my father used for all meats from Sunday roasts to hams and Christmas turkey. I would pour some water on the concrete wall surrounding the outside drain and hone the knife until it was razor sharp. My father carved the roast and my mother served the vegetables, usually accompanied by Family Favourites on the wireless (radio) with requests from various armed forces BFPO stations (British Forces Post Office) from mysterious places overseas. After Sunday lunch my father invariably sat down in his armchair to listen to the radio comedy programme "Take It From Here" which ran from 1948 to 1958.

I did occasionally stay at school for dinner, usually when there was something on in the lunch hour leaving too little time to go home. Nobody admitted to liking school dinners. In fact they were not too bad. The school had its own kitchens in a separate building at the side of the playing fields, so the food was freshly cooked. We always had a robust, healthy hot meal of "meat and two veg" followed by a hot pudding. The menu was varied but there was no choice. Sometimes the meat was a bit rubbery and fatty. We were served our portions by the dinner ladies and we had to eat it all up. This was accepted and I do not recall it ever causing any problem or conflict. We were basically hungry and ate what was in front of us. The problem today is that children are not hungry enough.

Parents now say to their children, "Don't worry, leave what you can't eat". Such a statement would have been iniquitous when I was young and food was scarce. A more likely form of encouragement was, "Eat what you're given and don't leave the table until you've finished". My parents did not need to force me to eat my food. I was hungry and I knew that if I did not eat what was in front of me there would be nothing else. We were grateful for what we had.

There were one or two plump children at Junior school but fat children were rare. Some adults were portly but not really obese. It is now commonplace to see whole families of fatties wobbling down the street, some so fat they can hardly walk. Such monsters once existed only in comic books (Billy Bunter) or at funfairs where we paid a penny to stare at the fat lady.

It was during my second year in the Juniors that I started playing the recorder. We were taught by Mrs Hepple. She did this in addition to her regular teaching duties and her role as the school pianist, always playing at morning assemblies and other school functions. We met at lunch time every Tuesday from one to half-past. I really enjoyed these lessons. This was where I learned to read music which has been of great value to me ever since. Each week we learned to play a new note on the recorder which slowly increased the range of pieces available for us to play. The last notes to be learned were the high ones that required the thumb hole to be "pinched" with a slight increase in the flow rate of air. I quickly acquired these skills and was soon in the recorder band that played regularly at morning assembly. I became the best player and was chosen to do the descants and the occasional solo parts.

Our recorder group played at most school functions, notably, in support of the school play. In consecutive years we did "Alice in Wonderland" and "Alice Through the Looking Glass". These were quite ambitious productions involving many rehearsals, costumes and stage props. The major parts in the play always seemed to be taken by Marilyn Dawes, Sheila Stuart and goodie-goodie Alan West. I never had any desire to be on the stage and would have been terrified at the prospect. But I was quite happy and confident to play the recorder music in front of an audience.

We had an upright piano in the front room at Henley Crescent. As soon as I could read music from playing the recorder I tried to play piano pieces from my mother's old music books of Brahms, Schubert and Tchaikovsky. Someone told me that to play the base cleft you just had to shift everything up two notes. I tried hard but progress was slow.

My mother must have been a good pianist because in her teens she accompanied the singing at her local Methodist Sunday School. Strangely she never played at home and did not really provide me with any encouragement. She did later join the Southend Women's Guild choir and sang with them for several years. I remember that my father had piano lessons for a short time with a woman at the top of the Crescent, and my brother Nick also had lessons for several years which Nick says he paid for himself. In retrospect I wish I had been offered piano lessons. It is much more difficult to develop technique in later life. Probably my parents could not afford the cost of lessons for me.

My enthusiasm for music prompted my parents to buy me musical instruments as presents. One Christmas they gave me a silver flute and piccolo in a beautiful purple-padded case. They no doubt thought that with my prowess on the recorder I would soon master these new instruments, but they are actually quite difficult to play. I was keen and made an effort, but without lessons I was unable to make much progress. I eventually gave up. Another Christmas they bought me a harmonica (still extant), a proper one like Larry Adler's, but this went the same way as the flute and piccolo and it made my lips sore. It is difficult to learn any instrument without instruction from a teacher.

In 1953 I moved up to the top class taught by Mr Edwards, another good teacher probably in his 50s. This was the year we prepared for the 11-plus. There were about 24 children in the class. We sat at double desks. My partner was Terry Absalom who lived at the top of my road. I was friendly with Terry in spite of his nasty streak (recall the Red Indian incident). Terry took pleasure in jogging me while I was writing so that the ink smudged the page. He enjoyed this sort of prank.

Other friends I had in this class were Richard Coakes, Peter Lee, Nigel Eaves, Roger Herbert, and Roger Irving. There was also Richard Fuller, Douglas May, Michael Pratt, Alan West, Howard Dulwich and two inseparable boys, John Gooch and Ian Brown who sat at the back of the class and were always clowning. In later life they both turned out to be as queer as coots. Another boy was Brian Portsmouth whose claim to fame was that he could defecate without dropping his trousers. He performed this feat by pulling his trouser leg to one side (we all wore short trousers), and allowing the material to roll down his inner thigh. This was successful only because he produced particularly dry material.

A distant friend in my class was Graham Lockwood. Graham was hit by a stone from a catapult and sadly lost an eye. After this he wore a glass eye, which intrigued us.

Last but not least was "Bossy" Bartlett. He joined our class only in my last year at the school. He was rather chubby. His father was an important executive in the Council Offices. He was chief education officer for Southend as Bossy never ceased to remind us. He thought he was superior to the rest of us and tried to organise everyone, but without success. I do not know what happened to Bossy, but his name lives on.

I recall five girls: Sheila Stuart, Marilyn Dawes, Brenda Davis, Dawn Freeman and Madeleine Ford who were all bright in class. Marilyn Dawes in particular was an all-rounder. She was clever and a fast runner. She would also take her pants down behind the school kitchens to show the boys her private parts. She was clearly an exhibitionist. Doreen Yarwood who lived at the top of Henley Crescent at No 44 performed similar stunts on her front porch. This was about the limit of my experience of girls.

At the start of the year in the top class a head girl and boy were appointed. The head girl was Sheila Stuart and the head boy was goodie-goodie Alan West. I was disappointed not to be chosen, but also relieved because of my shyness about speaking in public. Nevertheless, I don't think they had to do very much in these roles except thank the guest speaker on Empire Day.

In all my classes, great attention was paid to mental arithmetic. We spent time every morning reciting the times tables and being tested out loud on multiplications and divisions, chosen at random by the teacher, until we knew the answers without thinking. It was understood what had been known for centuries, that concepts are acquired not taught, which is itself a concept that modern educationists have not acquired. You cannot build anything creative without the solid foundation of learning by rote. This was noted by Michael Gove when he was Secretary of State for Education. Once you have the ground-work the rest falls into place. We actually took pride and pleasure in knowing our tables off-by-heart. Of course, we had to learn up to the 12 times table because our currency was still based on the 12-penny shilling.

Handwriting and spelling were equally important. We were regularly tested on all aspects. My own handwriting was messy. I was never able to achieve free flow like some children. Our school was entered for a national handwriting competition for which the criterion was neatness, not content. My own effort was poor. Even I could see that.

We also had lessons in diction, something that would be politically incorrect nowadays. It was all about how to speak distinctly and give the proper emphasis to the different syllables in words. This was important if you were in a choir. The instruction was done largely through reciting poetry together as a class such that every word was intelligible, no matter whether we were shouting or whispering. There were even inter-school competitions to see who could do it best. In one of these competitions held at nearby Princes Avenue School our class had to recite the poem "Song of Hiawatha" by Longfellow. It started:

By the shore of Gitche Gumee,
By the shining Big-Sea-Water,
At the doorway of his wigwam,
In the pleasant Summer morning ......

This was total nonsense to me. I was never able to learn poetry off-by-heart. I don't think we won this competition.

Break times and lunch times in the playground were always enjoyable. We played all the usual games and there were many "crazes". One of the most popular and enduring was cigarette cards known as "fag cards". Cigarette Cards were given away in packets of cigarettes. Most were produced in the 1920s and 30s. In early 1940 the production of cards was greatly reduced because of wartime restrictions, and any hope of a return to the era of Cigarette Cards was dashed by the high cost of materials after the war. Nevertheless there were many thousands still in circulation when I was young.

The basic idea was to collect as many cards as possible, preferably in sets. A set usually comprised either 25 or 50 cards on a particular theme, for example, British Butterflies, Vintage Cars, Regimental Uniforms, Flags of the Empire, etc. I used to collect "Celebrities of British History" which included people like Lord Nelson, Clive of India, Oliver Cromwell, and Geoffrey Chaucer. It was a thrilling moment to find someone with a card you wanted. You then had to barter for it. I never actually completed my set.

Of course, at some expense it was possible to buy complete sets but this amounted to cheating. There was a dark and dingy Dickensian shop in Southend Arcade that sold many mysterious things including card sets, knives, pea-shooters and old comics. We often visited the shop to view, not to buy. We particularly wanted to view the woman who ran the shop. She was rumoured to be a "woman of the night". We did not know quite what this meant which added to the mystery.

Bartering was only one aspect of the fag-card hobby. The other way to increase your collection was to win them. This was of course a risky business because you could also lose them. There were basically two games we played in the school playground.

The first game was low risk. It involved two or more players who each flicked a single card from an agreed distance against a wall. The owner of the card that finished up nearest the wall won the other cards that had been flicked. By pre-agreement it was possible for a losing player to retain the card he had flicked and give instead some other card to the winner. This meant that if you had a card you particularly valued because it was good for flicking, you did not risk losing it.

The second game was high risk involving just two players. A card was stood up against the wall and each player took turns to flick a card from a set distance to try and knock down the standing card. The first player to succeed won all the cards on the ground. Sometimes there could be as many as 100 cards before somebody won. This was a game of great skill and high stakes, often attracting a crowd of onlookers as it got underway. Sometimes whole sets changed hands with this game. These entertaining games are no longer played. Cigarette packets ceased to include cigarette cards a long time ago.

Another playground game was "conkers", a seasonal game for the autumn term. You first made a hole in your conker with a skewer and threaded it onto a piece of string with a knot in the end to hold the conker in place. To play the game, one player held up his conker on the end of its string while the other player took careful aim and struck his opponent's conker with his own conker as hard as possible. The players took turns at this until one of the conkers shattered. Sometimes a poorly aimed shot or a small movement by the player holding up his conker would result in a mis-hit, with the two strings twisting round each other and becoming entangled. If this happened you shouted "strings" which entitled you to another go. The real purpose of this game was to be in possession of a champion conker that was responsible for the demise of as many other conkers as possible. The value of the conker was equal to the number of other conkers it had disposed of. So a conker that had smashed 6 other conkers was a "sixer". The count was cumulative, so a "three-er" that disposed of a "fiver" became a "niner" (3+5+1).

Before the start of combat each player inspected the other's conker to check it had not been tampered with. For example, it was well-known that baking your conker or pickling it in vinegar would increase its strength. Such treatments were considered to be cheating. Of course, you had to take someone's word on the pedigree of his conker and undoubtedly some boys told fibs. But there were legitimate ways of obtaining ownership of a high value conker. One way that I devised was to select 11 conkers, hang one up from the clothes line at home, and hit it with a second conker until one broke. By repeating this process the single surviving conker was automatically a tenner.

Occasionally when playing this game a conker would swing out of control and hit one of the players on the head. Because of such risks "conkers" is now viewed as a dangerous sport and is banned in school playgrounds.

Another playground game that was popular for a time was toy racing cars. In the early 1950s Dinky started producing a range of toy cars of much greater authenticity than was previously available. Racing models included Aston Martin, Alpha Romeo, Ferrari, Talbot, and Maserati. Of particular interest to young boys was the fact that the wheels and axles of these cars were built to a standard that made them run smoothly, enabling them to free-wheel over large distances. Each boy associated himself with a particular make of racing car, mine being a Talbot. The objective was to see who could roll their car the furthest over the playground. This simple game was most consuming and encouraged each boy to look after his toy car, changing the tyres regularly and keeping the axles well-oiled.

Yet another game was marbles. These were small balls made of coloured glass, about half an inch in diameter, with attractive swirly patterns inside. Some marbles had exotic designs that made them sought after and collectable. There were several games you could play, the simplest being to take turns at rolling your marble towards your opponent's marble in an effort to hit it. If you hit it soundly you won the marble. If you only struck it a glancing hit your opponent shouted "tipsies", and you then had a chance to drop your marble from eye-height on top of your opponent's marble. If you hit it, you won it.

A sedate game that could be played all year round was Five Stones. You could play with ordinary stones, but we usually played with purpose-made cubic stones that could be cheaply bought. They each measured about half an inch across. The first player throws the five stones into the air with one hand and tries to catch as many as possible on the back of the same hand. The stones that are caught are then thrown up again from where they came to rest on the back of the hand, and as many as possible are caught in the palm of the same hand. If no stones end up being caught, the player's turn is over. If, however, at least one stone is caught, the player prepares for the next throw by keeping one of the caught stones in the same hand and throwing all remaining stones on the ground. The player then tosses the single stone into the air, attempts to pick up one of the stones that was missed before catching the stone that was tossed, all with the same hand. The player repeats this until all the stones have been picked up. That done successfully, the player throws down four of the stones again, throws the single stone in the air, and attempts to pick up two stones with the same hand before catching the tossed stone. This is repeated to pick up the remaining two stones. The process is then repeated for three stones followed by one stone and finally, all four stones are picked up before catching the single tossed stone. The game can continue in an agreed way with further permutations and challenges.

There were other playground games which, strictly speaking, were not allowed but which teachers did not make any great effort to stop. One of these was called "King of the Castle". The side entrance to the Dining Hall led off the playground by way of three steps surmounted by a concrete platform. It was possible for about 10 boys to stand on this platform, at a squash. The game was for one gang of boys to attack and dislodge the other gang of boys from the platform. This game could become quite rough and I was always in fear of breaking my glasses.

In winter we always seemed to have snow, but never enough. Every morning when I woke up in winter the first thing I did was to look at the gap between the top of the drawn curtains and the window frame. If the narrow strip of ceiling above it was extra white I knew it had snowed and I would jump out of bed to pull back the curtains. It was so exciting to discover everything covered in a blanket of soft virgin snow. It seemed to deaden all sound.

The school playground was an ideal place for making long slides of compacted icy snow because it was on a slight incline. The more you slid on it the more the ice turned slippery and glassy enabling you to slide sometimes as far as 10 metres. Boys with rubber soles to their shoes were discouraged by other boys from using the slide because it made the slide slower. Hob-nailed boots also damaged slides. Leather soles were best. This is no doubt another game that would now be banned.

I do not recall that schools ever closed when there was snow. It was the duty of the school to remain open in all circumstances even if some of the children or teachers could not make it. "Health and Safety" had not yet been invented. It was the duty of everyone to manage as best they could. When there was snow we looked forward to break time when we could play snowballs and make slides in the playground. I only remember once being allowed to leave school half an hour early because there was a blizzard outside. Children were not taken to school in their parents' cars. They probably did not have a car anyway. Most children walked or cycled to school, and those who caught buses were delighted to walk in the snow if their transport did not turn up.

The Oxford versus Cambridge boat race always attracted interest, possibly more than the football Cup Final. In the days leading up to the race you had to decide which team you supported. A boy would walk up to you in the playground and ask if you were for Oxford or Cambridge. If the answer was wrong, he gave you a sharp push causing you to fall over his accomplice who had already knelt down discreetly behind you.

I own up to playing these pranks on other boys. One lunchtime going home from school I pushed a boy into a large bank of stinging nettles.  Unfortunately for me, his mother arrived to meet him at the same time and she was furious.  I ran away.

I should point out that the boys did not mix much with the girls.  Although our classes were mixed we had separate playground areas.  The girls' playground was between the infants' and boys' playground, and perhaps was designed this way as a buffer.  The girls seemed to spend their spare time playing hop-scotch and skipping, all totally alien to us boys.

It was in the school playground that I learned of the death of King George VI on 6 February 1952.  This was a shock because we all held the Royal Family in high regard.  Some time after this Queen Elizabeth II had her coronation.  It was on 2 June 1953, a day of great celebration.  There were lots of mementos of this occasion, some of which I still have including special edition stamps and decorated mugs.  The euphoria was heightened by the fact that Edmund Hillary had four days earlier claimed the record of being the first man to climb the world's highest peak, Mount Everest.  In those days Great Britain was really great.  The fact that Edmund Hillary was actually a New Zealander was glossed over because, after all, New Zealand was part of the Commonwealth.

Morning break always started with milk.  Each child was entitled to a small bottle containing one third of a pint.  It was declared that milk was good for you and we all enjoyed it.  You collected your bottle from crates in the cloakroom where it had to be drunk and the empty bottle returned to its crate before you could go into the playground.

In my final year I was appointed as one of three milk monitors responsible for arranging the milk crates, collecting the silver tops, and generally keeping everything tidy.  One day we hit upon a brilliant idea for using up the cat's milk, the name given to the dregs left in the bottom of an empty bottle.  We collected all the cat's milk from the empties to make up a full bottle, and carefully fitted a discarded silver top to make it look like new.  We then placed the rogue bottle amongst all the others and waited to see who would take it, hardly able to contain ourselves with glee.  A poor unsuspecting boy would eventually take the bottle and we watched with a mixture of amusement and disgust as he drank it all. This cruel activity became a regular sport until word got out and we had to stop.

The end of break was signalled by the teacher on playground duty ringing a hand-bell. We all had to line up in the playground, one line for each class. The teacher then inspected the lines to check we were all presentable, quiet, and standing to attention. The first class to achieve this state was instructed to lead in. There was one teacher, Mr Tubby, who never actually taught me, who had the nasty habit of creeping up behind any boy in the line who was talking, and flicking his ear with a fat nicotine-stained middle finger. He once did this to me and it was painful.

Not many teachers administered corporal punishment but some did, and not just the male teachers. The punishment for a boy would be a whack on the bottom with a cane, and for a girl a smack on the palm of the hand with the flat of a ruler. There was one teacher, a rather short-tempered Scot called Mr Baptie, who would beat boys regularly, even more than the Headmaster, Mr Woodvine. Whereas Mr Woodvine always used a cane Mr Baptie used a heavy, curved chair leg.

I was rarely punished for being naughty, partly because I was well-behaved and partly because I managed to avoid being caught. But I was once punished for something I hadn't done and the incident is ingrained in my memory. For some reason the children in the top class were occasionally placed for an hour or two in one of the lower classes. By "lower" I mean either the 'B' stream class or the dunces' class. This may have been because our regular teacher was otherwise engaged but I have the feeling it was to homogenise us a bit and help remove class barriers, perhaps an early example of social engineering. So one day I was removed from Mr Edwards' class to spend the afternoon in Mr Baptie's class. Whereas Mr Edwards allowed a certain amount of restrained talking in his class, Mr Baptie required absolute silence at all times. I was not used to this regime. In the event it was my neighbour who was talking but I who was hauled out to the front. I had the impression that Mr Baptie was making an example of me, trying to show his regular class that he made no distinction between the brighter children and his own lesser flock. The punishment was inflicted without delay. He took the dreaded chair leg from his desk drawer, bent me over, and delivered two punishing blows to my backside. My initial reaction was one of great shock from the impact followed by a sense of searing, prickly pain on my bottom. I was sent back to my desk where for the next half-hour I sat wriggling, trying to find a comfortable position. It was a relief to return to the friendly fold of Mr Edwards' class where for a short time I was able to languish in a celebrity status.

A lesser punishment was to write lines. Once Mr Edwards made me write out 300 times "I must listen to the teacher and talk less". I cheated by tying some pencils together so that I could write several lines at once. Carbon paper was too much of a give-away. The teacher must have recognised the multiple pencil trick, but he said nothing. Another time the punishment was to write out and learn the 13 and 14 times tables. This was actually an exercise that has been useful to this day.

Every week we had a lesson in the school hall called "Music with Movement". The teacher played us some evocative music such as Mendelssohn's "Fingals Cave" or Tchaikovsky's "Swan Lake", and we had to prance around in our socks trying to capture the mood. The exercise always seemed a bit pointless to me. One day the teacher was called away for a short period and we were left in the hall to continue the musical interpretation on our own. This developed into sliding across the polished floor at high speed and shouting loudly. The headmaster's study was next to the entrance to the school hall and he must have heard that something was wrong. He entered the hall with his cane, which he thwacked loudly against his trousers as he strutted up and down. I don't recall he said anything. His presence and the ominous sound of the cane were enough to produce instant silence.

I never excelled at school sport although I did enjoy football. For a short period I was in the school football team and played in a few matches against other junior schools in the area. These matches took place after school or on Saturday mornings. I recall that Roger Irving was a fine "dribbler" of the ball, as is often the case with knock-kneed people. I played in at least one away match against our old rivals Princes Avenue School. I think we lost. Although I did not excel at the game, I did play football regularly with friends in Priory Park and thoroughly enjoyed it. I think my problem was confidence. I could never head the ball or commit to a strong tackle for fear of breaking my glasses. I did enjoy looking after my football boots and loved the smell of fresh dubbing, second only to the smell of coal.

Once a year Earls Hall had a school sports day to which parents were invited. All the children were encouraged to take part. In addition to the standard track events like the 100 yards sprint and hurdle race (both always won by Marilyn Dawes and Roger Herbert), there were various other events aimed at giving everyone a chance. There was the slow cycle race, always won by Roger Irving because he was able to balance on his bicycle without moving an inch, and several other bizarre events like the egg-and-spoon race, the sack race, and the 3-legged race, none of which I ever won. There were no races for the parents because that would have been considered undignified.

Another annual event was the celebration of Empire Day on 24 May. This was all about Queen, Country, and our Dominions. We would assemble in the school hall and sing with great gusto hymns like "Jerusalem" and "I Vow to Thee My Country" to Gustav Holst's stirring tune. If I wasn't singing I would be on the recorder. There was always a local dignitary present, like a town councillor, who would make a short speech. The highlight was the expected request from the dignitary to the Headmaster to grant the school a half-day's holiday, to which the Headmaster always acceded. One year the dignitary did not make the request, to the shock and disgust of all the children. Perhaps he had been primed to forget it.

Political correctness finally won the day when in 1958 Empire Day was renamed British Commonwealth Day, and still later in 1966 when it became known simply as Commonwealth Day. The date of Commonwealth Day was then changed to 10 June, the official birthday of the present Queen Elizabeth II. By this time there no longer seemed any point in celebrating it.

There was one teacher who terrified me and made Mr Baptie seem like a Sunday School teacher. Her name was Miss Smith. She appeared at the school in my final year in some sort of part-time role. I and some of the children in my class had a weekly lesson with Miss Smith together with a selection of children from the lower classes. This may have been another example of social engineering. The lesson was immediately after lunch at 1:30 every Monday. I am not sure what we did. It may have been craft work. But Miss Smith would drag me out to the front of the class and ridicule me with her waspish comments. For some reason she seemed to pick on me specifically. I would cringe in my seat trying to be invisible hoping to avoid the humiliation. I would even be sick on Monday morning in an effort to miss school. I do not know how I survived the ordeal of Miss Smith. She must have set me back several years in my relations with women. My father told me some years later that Miss Smith was a cracking busty blonde of about 25, but at the time I was oblivious to this.

Throughout my time at Earls Hall I always walked to school, most days with Terry Absalom. He lived at No 21 at the top of Henley Crescent. I left home around 8:30 after a breakfast of cereals, usually Shredded Wheat or Wheetabix (I rarely had a hot breakfast except at weekends and in winter) and knocked on Terry's door. We would walk together round the Crescent into Midhurst Avenue, through a back gate into the school grounds and along a path that led to the playground via the playing fields. This route was quicker than walking the other way along Carlton Avenue. It was a short walk and we loitered along the route, stepping in puddles and throwing stones.

On dewy mornings in autumn a favourite diversion was making spiders' web mirrors. You took a pliable twig and bent it round to form a loop. You then used the loop to scoop up spiders' webs that would be glistening with dew. There were usually plenty of webs in the hedges along the route.

Otherwise you could use "cuckoo spit". When you had enough web material covering the loop you gently smeared it with your finger and miraculously it all coalesced into a shimmering transparent sheet. This pointless activity never lost its fascination. There are not enough spiders nowadays to do this.

On part of the walk to school we had to pass the bungalow of Mr & Mrs Mills. This was on the corner of Henley Crescent and Midhurst Avenue. They were another elderly and bad-tempered couple. Mr Mills was exceptionally tall and spindly whereas his wife was a midget. They looked rather strange together. Just walking past their house was enough to set Mr Mills running down his front path shouting at us. It was most unfair because we didn't even do anything.

After school I often went home to Terry Absalom's house for tea. A big attraction was their TV set. We watched the children's programmes that usually finished with a western, either "The Adventures of Rin-Tin-Tin", Gene Autrey, Roy Rogers, or the Lone Ranger and his faithful Indian friend Tonto of "qui mo sabe" fame. But these were all American productions. The programme I enjoyed most was the BBC's adaptation of E. Nesbit's "The Railway Children" which first ran in 1951 as a serial of eight 30-minute episodes. This was beautifully produced with a haunting theme tune. After watching television we would sit down to tea. Mrs Absalom often served dripping, always a favourite. Dripping was the name given to the left-over juices from the Sunday roast which separated on cooling into a rich jelly topped with white animal fat – really scrummy. Roast pork made the tastiest dripping. On one occasion at Terry's house I spread some on a slice of bread and sprinkled it with a generous helping of what I thought was salt. In fact it was sugar. The combination made me gag and I was almost sick, but I had to continue eating as if everything was normal. Terry's attractive elder sister Moira never seemed to be around. Perhaps she was doing her homework, or just didn't like dripping. She went to Westcliff High School for Girls.

Sometimes after school or on Saturday mornings we visited a local shop called Costins that sold many interesting and essential items including Spanish wood for chewing (sometimes known as Liquorice Wood), a range of chewing gums and gob stoppers, tattoo transfers, stink bombs (tiny glass phials containing liquid $H_2S$), itching powder, reels of percussion caps for our cowboy guns, and extra large percussion caps for "bombsies". We sometimes made our own itching powder from maple seeds collected from nearby trees on Prince Avenue. You put it down the back of people's necks and it was very difficult to get rid of. Costins also sold pea-shooters, but not the ammunition which we bought from a hardware shop on Victoria Avenue near the junction with Fairfax Drive. The preferred ammunition was pearl barley, much smaller and more effective than dried peas. You could fill your mouth with pearl barley and fire a continuous stinging barrage, remembering not to swallow.

There was one particular use of percussion caps which for me was the greatest fun. When I was about nine years old my parents gave me the remarkable present of a toy army field gun. It had a bank of 6 barrels arranged in 2 lines of 3 and fired metal shells about 1 inch long. Each shell could be primed with a percussion cap. When a shell was fired it made a loud bang with a satisfying flash accompanied by smoke, sending the shell at speed across the room. None of my friends had a gun like this.

I also had a large collection of toy bricks and wooden boards salvaged from Christmas date boxes, all of which were stored in a big tin box in the cupboard under the stairs. I became adept at building large complicated forts with several floors and galleries, some of them cantilevered with the aid of the boards and heavy bricks. I then manned the fort with my toy soldiers. The exciting part was to destroy the fort with my field gun. It was even possible to fire a full broadside with great effect. I could play these games for hours, often by myself. One day in an act of wanton destruction I used my toy field gun to destroy a Keil Kraft model aeroplane I had made out of balsa wood and which no longer flew properly. It was a model of a Beechcraft Bonanza.

At the start of my last year at Earls Hall Juniors, at the age of 10, my parents wanted to prepare me as best they could for the 11-plus examination for entry to the Grammar School. I think they were fairly confident I would pass, but since there was a teacher and family friend, Mr Williams, living opposite I began private lessons with him once a week. I shared the lesson with a boy called Howard Dulwich. Howard was a rather boring, pasty boy on the periphery of my circle of friends. He would have been a good role model for a Hubert Lane acolyte. I had previously been invited to one of Howard's birthday parties organised by his mother. We had played dreadfully boring games like musical chairs and trying to remember objects on a tray, and I hoped I would never be invited again. I remember that I had no difficulty with the coaching work, but Howard was not doing so well and started having extra lessons with Mr Williams.

I passed the 11-plus whereas Howard Dulwich failed. So Howard went to Fairfax High School instead of the more prestigious Southend High School where I went. Howard's mother began a long battle with the authorities. According to my mother, Mrs Dulwich wrote to the examination board, teachers and town councilors arguing that an injustice had been done and that her Howard was in the wrong school. Amazingly she succeeded and Howard moved to Southend High School having spent two years at Fairfax. He went straight into the C stream and stayed there. Howard never succeeded in his academic studies but perhaps his mother's pride had been satisfied.

Unknown to me, my father was also sitting exams at this time. He had been working at the Post Office in Southend, first of all as a postman then later behind the counter and in the offices as a clerical assistant. After the war, as an ex-serviceman, he had the opportunity of free training to become a teacher. This was an emergency scheme set up by the government to replenish the teacher pool. My father did not take up this option and possibly he later regretted this. I think he would have made a good teacher. Nevertheless, he was ambitious for a better job and more money so he took exams to join the Civil Service. He was successful and started work in London about 1954, commuting every weekday by train. He wore a dark suit, a bow-tie, a bowler hat, and carried a rolled black umbrella. This used to be the standard uniform of a "City Gent" that elevated him to a higher social class. I am sure he was proud of this. He started work within the Ministry of Defence in the Air Ministry. This put him in close contact with Air Force officers. It was no doubt through this association that he grew his beloved handlebar moustache.

My father would get up at 6 a.m., and after a full breakfast walk all the way to Westcliff station to catch the 7:30 train to Fenchurch Street. He returned home about 6:30 p.m. This routine did not change even in pouring rain, for which he wore galoshes. Except for a five year period in the early seventies when he worked in Southend, my father continued working in London until 1978 when he was 63 with various jobs in the Ministry of Defence, and finally in the Price Commission. I know that he thoroughly enjoyed the bustle and atmosphere of working in London.

In the second term of my last year at Earls Hall, Mr Edwards began preparing us for the "London Trip" that we would all be going on. This was an annual event for the top class. It involved doing research on the sights to be seen in the great capital. Since my father was now working in London he was able to obtain a lot of brochures and background material for me to study.

The trip took place on Friday 14 May 1954 for which we each paid 6s/6d (32.5p). We travelled up by coach to visit and view the sights, including the Tower of London, Tower Bridge, St Paul's Cathedral, Buckingham Palace, the Science Museum, and Trafalgar Square. One of the highlights of the trip was a short tour by boat round London's docks. I think we visited the Royal Docks possibly including the Royal Victoria, Royal Albert, and King George V docks. In the 1950s these docks were very busy with nests of cranes unloading enormous cargo ships, but none of it exists now. It was a long day and I did not return home until about 9 o'clock at night. My father lent me his camera for the occasion. I took some fine photos for a ten year old. I had a natural understanding of composition.

We had to produce a written account of our trip in the form of a hardcover book that we made ourselves. I put a lot of effort into this project and produced a fully illustrated account in the form of a guided tour of London (Plate 20). There was a competition for the best book, which I won. I was proud of this achievement and still have the original book.

So ended my time at Earls Hall School, altogether a happy experience.

# BOOK THREE
Youth (1954 – 1964)

# Secondary School to "O" Levels (1954-1959)

In September 1954 I began my seven years at Southend High School for Boys. This was a good school and remains so to this day, retaining its status as a Grammar School and achieving high numbers of university and Oxbridge entrants.

I was most fortunate that my secondary school, like my junior school, was just a few minutes walk from home. I had only to walk to the top of Henley Crescent, through the alley, up Hobleythick Lane to a second alley leading to Earls Hall Avenue then straight into the back entrance of the school. The journey took no more than five or six minutes.

The first few days at a new school are always daunting. My first impression was one of size. Everything was on a larger scale, much more serious and grown-up than Junior School with impressive laboratories, classrooms and sports facilities including 12 acres of playing fields. One immediate big difference was that all the teachers wore long black gowns that flew behind them as they strode down the corridors. A second difference was that the teachers addressed us by our surnames. I was proud to go to this school and quickly understood that loyalty to the school and its values was paramount.

I started in Class 1A. We were told we had been assigned randomly to our classes 1A, B, C and D but none of us really believed this. Children in schools are never in doubt about where they stand.

Our form master was Mr (Cliff) Wyatt who taught biology. He was an amiable man whose father was also a teacher at the school. On the first day he began by assigning us to "Houses". Any boy whose father or brother had been to the school was automatically assigned to his House. If we had no prior allegiance we were assigned randomly. I was placed in Athens whose colour was mauve. The other three Houses were Troy (light blue), Sparta (yellow), and Tuscany (red).

Curiously, Tuscany was the odd one out because it had nothing to do with ancient Greece.

I did have a prior House allegiance of which I was oblivious at the time. My cousin Donald Hudson had earlier attended the school (1945 - 1947), his time there terminating when he became ill, possibly with glandular fever, which forced him into a prolonged absence. I was later to understand that he became so far behind in his studies that his parents took advantage of an assisted place as a boarder at Felsted Public School where Donald remained until going to Birmingham University. His brother Paul also attended Southend High school (1949 - 1955), and therefore overlapped with me for one year. I cannot say I was aware of any of this because, although cousins, we were not close. If I had claimed allegiance to the House of my cousins I would have been in Troy rather than Athens. My other cousin, Douglas, the youngest of the three Hudson brothers, opted for Westcliff High School for reasons unknown.

The House you belonged to was reflected in the colour of your football shirt and the stripes on your tie. I was to discover that your house had an important bearing on school life, particularly when it came to sporting activities. It helped to give pupils a sense of belonging and also encouraged competitiveness, which was regarded unquestioningly as a good thing.

Mr Wyatt then issued us with Homework Diaries. Homework was a new experience for me. We were given a schedule of the due days for each homework covering all the subjects to be studied. The schedule covered one week and remained the same throughout the school year. For every weekday there were two or three homeworks due. Important subjects like Maths and English had more than one homework per week. This was really serious.

The next big eye-opener was morning assembly, a rather formal event lasting some 20 minutes. It took place every morning at 09:00 immediately after the class register and collection of dinner money. The whole school assembled in the Great Hall. We sat on seats instead of on the floor as at

Junior School. We had our backs to the stage and sat in rows facing the organ and Headmaster. Sixth formers were allowed to sit in the balconies. The prefects sat in prominence at the front in a single row facing the boys. Either side of the organ, about 10 feet above floor level, were two wooden pulpits set into the wall. Access to each pulpit was through a private door at the rear. The deputy head would stand motionless in his pulpit on the right of the organ surveying the boys as they trooped in to take their seats.

Only when everyone was seated and totally quiet would the door at the back of the left-hand pulpit open and the headmaster appear. He was a tall, imposing man with great presence. His name was Mr Price, a classics scholar from Oxbridge. On his appearance the whole school stood and waited for him to say "Be seated" in a deep booming voice.

We sang a hymn and a prefect read a Bible lesson. We all said the Lord's Prayer. The Headmaster then made the daily announcements. These ranged from lists of the names of boys selected for the school sports teams playing the next weekend, to mundane notices such as the school playing fields being out of bounds during break time due to heavy rain. Occasionally, he finished on a more serious note concerning some misdemeanour of sufficient importance to pronounce upon, such as smoking behind the cycle sheds or misbehaving in the streets on the way to school. This would usually start: "It has come to my notice that ....", followed by: "I wish to make it abundantly clear that ....", all spoken with gravitas. At the end of the announcements the Headmaster would turn on his heel and depart through his private door. This was all most theatrical and impressive.

The Headmaster's departure was a signal for the organist to start playing the voluntary and for the prefects sitting at the front to leave the hall. The boys then filed out in an orderly fashion, row by row, under the watchful eye of the sub-prefects. The whole of this time the Deputy Headmaster remained standing, motionless and silent in his pulpit.

Attendance at morning assembly was compulsory. However, Jews and Catholics (there were no Muslims) had their separate meetings in some other part of the school. They numbered in total no more than about 30. At the time I did not know where they met or what they did, and I never gave it much thought. I later learned from Danny Linehan, with some surprise and disappointment, that the Jews and Catholics just sat doing nothing in the Dining Hall. Towards the end of our assembly they filed in to the Great Hall, standing in the wings to listen to the notices.

Once a week on Wednesdays we split up for individual House assemblies. Mine took place in the geography room. It was a bit of a squash and some boys always had to stand. The House that was "Cock House" (of which more later) had the privilege of using the Great Hall for their assemblies. My House, Athens, never achieved this distinction during all my seven years at the school. As someone once joked, the Athens motto "Nihil Secundi", should have been translated "Always Fourth".

Early in that first term of 1954 at my new school my maternal grandfather died of a stroke, aged 78. I had been fond of him. I think he must have been ill for some time because I remember visiting him in his bed at Highfield Grove before he died. My parents clearly thought I should be shielded from the death of family members. It was not discussed and I was not taken to the funeral. I felt no real sense of loss or grief because I did not appreciate what had happened. I had no understanding of death. My grandfather was just not around any more. But his passing must have had an effect on me. The day after he died I forgot to report after school for a minor misdemeanour. I was punished with a weekly detention, as noted in my first end of term report.

In my first year we actually had exams at the end of the first term. This enabled the school to make its own assessment of the abilities of its new entrants. Remarkably, I came top of the class in Manual Work (Metalwork & Woodwork) and $4^{th}$ in Art. Overall I came $9^{th}$ in the class. After these exams there was a small reshuffle of boys between the A, B, C and D classes.

During the summer term of my first year all the boys in my form went on a weekly coach trip to Westcliff Swimming Pool (then called Baths) for swimming instruction. As I could already swim I easily achieved my 25-yard certificate, the main objective of the course.

Westcliff Swimming Pool was of Olympic proportions with a full range of facilities. I often swam there in the summer months with my friends. At weekends we were entertained by beauty contests, high diving exhibitions, and demonstrations of synchronised swimming from the Water Follies. Sadly, circa 1970, the pool ceased to be viable and was bought by former boxer George Walker who converted it into a casino and restaurant complex that my father occasionally visited.

During my time at Southend High School I was always in the "A" class, but this was not automatic. At the end of each school year in July we had a week of exams. The marks for each subject were added together and the boys in the class ranked in order of performance. I was never top - this was always Brian Wiggins. But I did once come second and I was usually in the top ten. The system operated like the football league. The bottom two or three boys dropped down to the lower class and the same number at the top of the lower class were promoted to the higher class. There was always a main caucus of boys who never changed class.

My best friends in class during my time at Southend High School were Peter Lee, Michael Pratt, Andrew Sargent, Doug May, Danny Linehan, Chris Harding, Dave Patrick, Alan Bates, George Farrow, Brian Wiggins and Dave Beadle. Other boys in my class were Dennis Tilbrook, Michael Beckerman, Frank Smith, Chris Brown, Barry Nichols, Philip Mercer, Brian Bence, Roger Herbert, Stuart Butler, Jeremy Blandford, Roger Holbrook. Norman West, Colin Green, Colin Moore, David Goldsmith, Colin Sadler and Phil Robinson. I cannot recall the other two or three.

I remember quite a few of the teachers. I have already mentioned the Headmaster, Mr Price. He rarely did any teaching and was seen only occasionally in the school corridors. He struck terror in me as a new boy to the school.

The Deputy Head was a rotund history teacher called Mr Smith. However, Mr Smith was elderly and retired at the end of my second year. The position of Deputy Head was taken by Mr Coakes who taught Latin and Greek, in which subjects he had gained a double first at Lincoln College, Oxford. Mr Coakes was a thin, wiry man with a sharp barking voice. He walked extremely fast as if he was always in a hurry.

Mr Coakes commanded instant attention and had no problem keeping discipline. He had only to show his face and a whole class fell silent, even the notorious yobs in 5D. (A contemporary of mine in 5D was the notorious and talented Viv Stanshall, later to become best known for his work with the Bonzo Dog Doo-Dah Band. He died in a fire at his London home in March 1995). Although Mr Coakes was strict he was respected for his fairness. Various myths surrounded him, like his vaulting over the 6ft school gates when they were shut one day, and his parachuting exploits in France during WW2. With regard to the latter, the story was that, unable to speak French, Mr Coakes communicated with the local priest in Latin thus demonstrating to those in any doubt that Latin is a useful language.

Mr Coakes was also the father of one of my best friends, Richard Coakes. This meant that I frequently met Mr Coakes socially at Richard's house. I found him to be a much warmer person in his home environment but he always maintained a distance until I was much older and had left school. He was totally professional, and even though I had a close friendship with his son he never showed me the least favouritism at school. Even more coincidental, the Coakes' lived in Hobleythick Lane right next door to my maternal grandparents. My grandmother got on well with Mr and Mrs Coakes who were always good neighbours. They had similar strict attitudes and maintained a mutual respect for each other.

I particularly remember our maths teacher in my first year. He was Mr White, an elderly and kindly man whose nickname was of course Chalky. He was a fanatic about numerology (the study of mystical or esoteric relationships between numbers and physical objects or living things) and could be deflected easily from the topic he was supposed to be teaching to talk at great length about his most recent discoveries. For example, he told us that if you divided the length of all the tunnels inside the Great Pyramid by the cube root of its volume you obtained the value of π to 6 decimal places, or something like that.

In my second year our form master was Mr Tregunna, a stocky Cornishman as wide as he was tall. He was our history teacher, but he also took us for the extra-curricula topic of current affairs. This included the development of our skills in public speaking, something that I detested. I remember having to prepare two short talks at different times, one on the Seven Ancient Wonders of the World and another on Apartheid in South Africa. This latter topic was totally incomprehensible to me. I could not understand the antipathy between Whites and Blacks or what the Dutch Reformed Church had to do with it, and anyway, I had rarely seen a black person except in films. We had to give these talks without notes. The only way I could manage the ordeal was to learn my talk off-by-heart.

In my second year I dropped Art having been told by the teacher, Mr (Tom) Willcocks, that I was no good at drawing. Unfortunately, I believed him in spite of my earlier excellent exam result. I have to conclude that, for me at least, Mr Willcocks was an uninspiring teacher, because I am now a moderately good artist. Interestingly, Mr Willcocks' son was none other than David Willcocks (later to be knighted), the celebrated choral conductor, musician and composer best known for his work at King's College, Cambridge, where for several years he was Director of Music and chapel organist.

I started to learn Latin in preference to German. We were first of all taught Latin by Mr Coakes, a strict and respected teacher as already mentioned. In my fourth and fifth years our Latin teacher was Mr Cleaver, a tall awkward character with a nasal voice. We played him up quite a lot. When he was agitated he had the habit of walking up and down in front of the class, pressing his hands together through the tips of his fingers in a strange bouncing motion, his voice becoming more and more nasal.

With Mr Cleaver we played the classic prank of all time, or so we thought. One of my friends, Pete Wyss, was studying at

Southend Technical College and seemed to be able to take time off when he wanted. John Roff in our class lent Pete an old Southend High School blazer and tie. The plan was for Pete Wyss to enter our Latin class pretending to be a member of our school and to tell Mr Cleaver that the Headmaster wanted to see him immediately. We set this up for a day when we had our Latin lesson in one of the "huts". These were temporary classrooms built on a grassy area just inside one of the school gates, so giving Pete a quick escape route.

On the appointed day we sat in our Latin class not really believing that Pete would turn up. But we were not disappointed, and we sat transfixed as he entered the room and calmly delivered his message. Mr Cleaver stared at him for a moment, barked at us to continue in private study, then marched out of the door. We watched in amazement as Mr Cleaver sped across the playground on his way to the Headmaster's office, his gown flowing behind him, to disappear through a side door by the gym into the main school building. Pete Wyss made a rapid departure through the school gates. After about five minutes Mr Cleaver returned, his face bright purple. He started his nervous routine of marching up and down making the usual church spires with his fingertips. After a few moments he turned to the class and shouted in his nasal voice, "Who was that boy?" at which we could hardly contain our mirth. The lesson somehow continued and Mr Cleaver never mentioned it again.

After this episode Pete Wyss' status rose considerably and "Who was that boy?" became a catch-phrase. After I left school I heard that Mr Cleaver had fallen down a mountain in South Africa and broken both his legs. An alternative story I heard was that he had been walking by an 8ft high brick wall that collapsed on him. Whatever the truth, I was sad about it.

We had several French teachers including a Mr (Buzz) Watson. He spoke English normally enough, but when he spoke French it was very thick as if he had a bad cold. He seemed to think that this was the way to achieve the genuine French accent. Maybe it was for all I knew. So to gain his approval we tried to imitate these strange nasal sounds. Chris

Brown was the best exponent. When asked to recite something in French, Chris would hide his face behind his book and speak with one hand clamped over his nose. He was congratulated by Mr Watson on his fine accent. About once a week Mr Watson got excited about what he proclaimed to be "the silliest rule in French". This was the gender agreement required between the past participle of a verb and any direct object that preceded it. I think he was right. I was never good at French while at school. It became a living language only after I met Lucienne.

I had several English teachers, both for language and literature. I never did well at these subjects. In English literature we studied Shakespeare which I hate to this day, and we read set books like Cranford and Little Women which I think were inappropriate for teenage boys. As for poetry I could not find the enthusiasm to make the enlightened criticisms that were expected. It was totally over my head. I was equally useless at learning poetry and always made a fool of myself when asked to recite something from memory in front of the class. Possibly I was not helped by the teachers. One of them, a Mr Harding, was a source of great amusement. He shook uncontrollably and dribbled all over our homework books. I later understood that he was suffering from Parkinson's disease. He must have been a brave man to continue teaching and submit himself every day to the merciless jibes of the boys. He died soon after I left the school.

English language comprised writing essays, précis, and grammar. I always scored low in essays because I was too restrained and reserved in my writings. I now realise that this was due to a lack of confidence. I simply had to let my imagination, of which I had plenty, run free. I would then have done well. I was saved in English language by précis and grammar because they were more logical. An important part of grammar was "7-column analysis". This involved identifying the structure of complex sentences in terms of their verbal clauses, subject and object clauses, and adjectival and adverbial clauses. I recall that there were four types of adverbial clause: time, place, manner, and reason (when, where, how, and why). This was all a piece of cake for me. It was just mathematics in a different guise. If it had not been for grammar I would have failed "O" level English.

There was one good English teacher, Mr (Flossy) Webster, who sadly never taught me. He was a flamboyant character with a shock of wavy blond hair. Outside teaching his main interest was cricket, in which he tutored the whole school.

In my second year we had an English teacher, a Mr (Joe) Harper, who was particularly strict and severe. On 1 April Michael Beckerman decided to play an April Fools prank on him. This was most out of character because Michael was a quiet and retiring boy. It was also bad judgement to play a trick on this particular teacher. Michael smeared tomato ketchup over the door handle. The teacher opened the door to enter the classroom and stared down at his hand in disgust. He was not amused. Michael was sent straight to the Headmaster where he received "six of the best". As it happened, Michael actually excelled at English. He won first prize in the class for the best short story that had to end with the words "It was all because the clock stopped". Michael's story was about a man who went to heaven and was admonished by St Peter for being late. This was particularly poignant because Michael was a Jew.

I had several physics teachers while at school. The most notorious was Mr (Boris) Freeman, referred to by some boys as the caveman. Mr Freeman looked rather like an ape and had bad breath, no doubt due to heavy smoking. He had a ferocious and uncontrollable temper that we boys would goad him into displaying. When he lost his temper he gave a magnificent performance, throwing things about the room and shouting at the top of his voice. He would scream out some wonderful lines such as: "By the time I've finished banging your heads together you won't have enough wit left to make a half-wit". These were not meant to be jokes but we would collapse with laughter. The event that had us in fits was when a boy on the school field sent a discus crashing through the window just as Boris was demonstrating a delicate experiment.

True to form he ranted, raved and shouted, eventually tearing out of the room to try and catch the culprit. Sadly, Boris later died of lung cancer.

Chemistry was not my favourite subject. The principal teacher I had was Mr (Basher) Griffiths, the Head of the Chemistry Department, a dour character with a dry sense of humour. He once informed us, much to my astonishment, that a basic law of science was "however careful you are, the last drip always goes down your trouser leg". I do not think that my lack of ability in chemistry reflected too much on Mr Griffiths who was probably a good teacher. There just did not seem to be any logic in the subject. Everything had to be learned off by heart and I was not good at that. Maths on the other hand was totally different. Once a concept was grasped everything followed according to an inevitable logic that did not need to be learned. This was particularly true of mathematical proof. I have to conclude that chemistry is a boring subject. It still has the aura of Bunsen burners and bad smells buried in the Victorian era. This is probably why so many university chemistry departments are having to close. Yet when it comes to the study of living things, chemistry is more fundamental than biology.

There was a father-and-son pair of teachers, Mr (Wally) Wyatt who taught physics and his son Mr (Cliff) Wyatt who taught biology. I was taught by both at different times. The biology department had a good reputation because it was headed by a Mr (Di) Rees who had written several university texts and school course books that at the time were still in print. However, one of his supporting staff, a Dr Bismati or something, was not quite up to it. He was supposed to have a doctorate from the University of Calcutta in India. But we suspected something was wrong when he mis-pronounced many of the technical words. For example, "oesophagus" is a 4-syllable word with the accent on the second syllable. This teacher pronounced it o-e-soph-*arr*-gus, i.e. 5 syllables with the accent on the fourth. Our suspicions were well-founded because it turned out he was a fraud with no qualifications. He quickly disappeared. I quite liked biology until we started dissecting worms and frogs in the third year. I later gave it up for some other preferred subject.

I did all the regular subjects including Woodwork and Metalwork that were both compulsory in the early years at school. I was good at these subjects. We learned planing, sawing, drilling, filing, scribing, bending sheet metal, and annealing. Woodwork was taught by Mr Pickstone and metalwork by Mr Hamer. In woodwork I made several things including a letter rack and a tie rack. While planning out the letter rack I asked Mr Pickstone where the waste bin was because I wanted to sharpen my pencil. This was met with hoots of laughter from the other boys (the floor was always covered in wood shavings) and I wanted to be swallowed up by the floor. In metalwork I made several objects, including a poker, a sugar scoop and a pencil box. Metalwork was more scary than woodwork because you sometimes had to use the furnace, for example, to case-harden my poker. I obtained top marks for my pencil box, which Mr Hamer held up to the class as an example of fine workmanship. I actually did better than Brian Wiggins who I think was a bit peeved. Some of these objects, like the tie rack, sugar scoop and pencil box, are extant.

We had an inspirational Music teacher called Mr (Reginald) Foxwell. He would squeeze the whole class into his tiny music room at the back of the Great Hall and play us stirring music on his gramophone, like Walton's Balthazar's Feast. He was a fine organist and often played on the BBC's Light Programme. He also used to tour giving popular concerts on theatre organs round the country. I knew nothing about these activities at the time. He once said that he would give free lessons to anyone who wanted to learn to play the organ. Unfortunately I did not take up the offer. I used to enjoy the voluntaries he played at the end of morning assembly.

Mr Foxwell ran the school choir in which I sang until the age of 14. A great event in which I enjoyed performing was the Christmas Carol Concert of the Nine Lessons. We did two performances, one at school and one at a local church in Southend. It was a wonderful prelude to Christmas. Shortly after this Mr Foxwell quietly disappeared from the school scene. I learned much later that he had left to die of cancer, still a relatively young man.

My friend George Farrow was in the choir at St Mary's Church, Prittlewell. George was being taught to play the organ by Mr Penny. This elderly gentleman was still organist and choirmaster at St Mary's just as when my father was head chorister there 35 years previously. So George was familiar with the voluntaries that Mr Foxwell played after assembly, which introduced me to the organ music of César Franck and Louis Vierne. But it was not until the sixth form that I had the freedom to stay behind and listen each morning to the complete voluntary, then played by the new music teacher. It was here at school that I first heard Franck's "Pièce Héroïque" and Vierne's "Carillon de Westminster".

In my third year our form master was Mr Harvey who taught German. At the beginning of the year we were presented with the strange choice of having to drop either geography or biology. I chose to drop geography.

At the beginning of my fourth year I dropped History and Biology. Our form master in my fourth year was a Mr Saunders. He was head of the Maths Department and was also our maths teacher at that time. Mr Saunders must have been in his 50's. He was a tall, courteous man with spectacles and greyish hair. He had a physical handicap with one leg shorter than the other. He wore a special built-up shoe but still walked awkwardly with a pronounced limp.

In my fifth year our form master was again Mr Saunders who continued to teach us Maths. Mr Saunders was respected by all the boys. He was a kind and patient teacher and we enjoyed his maths lessons immensely. He encouraged all of us in Set 1 to take "O" Level Pure Maths early, entering us for the January exams. We were all successful. He went on to prepare us for the Additional Maths exam in June. I particularly remember his lessons on calculus and the duality between differentiation and integration, all beautifully explained.

Then one day soon after the start of the summer term Mr Saunders failed to appear for the morning register. The headmaster stood in for him, which was most unusual. I can hardly ever recall a teacher missing school, and it was extremely rare for the headmaster to step into a classroom. A week went by and there was still no sign of Mr Saunders. The school made no announcement about where he was or when he was coming back. We speculated and asked questions but received no response. Then it all came out about two weeks later in a newspaper report in the Southend Standard. Mr Saunders had been arrested for indecent exposure. Apparently, he had been kerb-crawling in his car distributing explicit photographs of himself to young children on the pavement. The photographs did not show his head, but police had identified him by his physical deformity.

We boys were of course shocked by this news but were too innocent to understand what it really meant. The school made no comment whatsoever about the arrest. There was a blanket of silence on the subject. All we knew from a circulated leak was that Mr Saunders would never be returning to the school. This seemed to us a great injustice. Whilst Mr Saunders' behaviour was a bit strange it did not seem particularly serious to us. But we had no notion of paedophiles and had never even heard the word. I have no idea what happened to Mr Saunders.

A supply teacher was rapidly enrolled to take over as our form master and maths teacher. His name was Mr Briggs, a tall man with a military bearing in the same mould as our headmaster.

In slightly slower time a new head of maths was recruited. He was Mr (Alan) Drummond, a rotund man who was always beaming with pleasure. Mr Drummond began taking us for maths and I never looked back. He was an excellent teacher and an inspiration to me. Soon after, another new maths teacher was recruited, a Mr (Ray) Fretton. Mr Drummond became our teacher for pure maths and Mr Fretton for applied maths. Their lessons were a great enjoyment for me and I learned a lot.

Much later after I left the school I discovered that my father knew Mr Drummond because he was a mason. More surprisingly, I learned that Mr Drummond was an organist and played at the Masonic meetings. Yet more surprisingly I discovered that my father knew Mr Fretton, and also the headmaster and many of the other staff members because most of them were masons. When I was about 16 my father asked me if I might one day be interested in becoming a Mason. He said what a great honour it was for a father and son to be in the same Lodge. I am sure he was saddened by my lack of interest and he never mentioned it again.

Once a week we had a 45-minute lesson in the gym and 80 minutes of games on the school playing fields. This was during school time and was compulsory, as was the communal shower afterwards, unless you could plead having a bad cold. In addition, boys would represent the school playing games against other schools out of hours, usually on Saturdays.

The sports master who took us for gymnastics was Mr Webb. He enjoyed all sporting activities and was captain of Southend rugby team. We often saw him playing in Priory Park on Saturday afternoons. My enthusiasm for gym was dampened after being relegated in my second year to the dumb-dumb group. We were each given two simple tests, head-over-heels and vaulting over the "horse". I failed both and so was branded as a "group 4", almost equivalent to being classed as physically handicapped. Again, I think it was all due to my lack of confidence, which was not helped by my fear of breaking my glasses. This marred my progress in sport but did not spoil my enjoyment. I have since understood that I was physically strong and capable, possessing good control of my body with a sense of balance and timing. I think I could have been a much more successful sportsman.

The euphemisms "physically handicapped" and "disabled" did not exist when I was at school. We would say that someone was deaf, blind, crippled, dumb, mute, retarded, according to the condition. Whilst this was accurate, I think that handicapped people were generally treated as inferior when I was young.

Sport was an important part of school life because it promoted physical well-being, competitiveness and team spirit. These were all considered to be good things. A further incentive was that the house with the most points for sporting activities had the honour of being "Cock House" for the following year. In addition to inter-house football, hockey, cricket and tennis matches, all of which earned points for your House, there were "standards" in which everybody had to take part. Standards were about field and track events such as running, jumping, discus, shot-putting and javelin. For each age-group a certain "standard" was defined which set the target to be reached. So you were competing against the standard rather than against another boy. The standard was set quite high to give a success rate of no more than about 20%. After school in the summer months every boy had to turn out on the school field to try and meet the standard in the various field events. You had only one attempt at each standard. If successful in an event you earned one point for your House. I think I only ever succeeded in the hop-step-and-jump.

One further important sporting activity in school life was the annual cross-country race. This event took place in Belfairs Woods just before Easter. There was some pressure on boys to enter this race but there was never actually a shortage of volunteers. Any boy could enter provided he took part in the training sessions. This involved running on the roads and pavements each week after school making several circuits round the school perimeter. A complete circuit was about two miles.

I entered this race at least twice. I was encouraged by my father who told me about his own accomplishments and about the prowess of his own father as a runner. On the eve of the first race my father gave me a pep talk. He told me that mental determination to win was as important as physical fitness. I was not convinced. The course, based at Belfairs Woods, was devised by the teachers with sixth formers acting as stewards. It took us along lanes, on paths through densely wooded areas, through bogs and across streams. The sixth form stewards erected barriers at certain places of difficulty in order to channel the runners and force them to attempt jumping across muddy swamps. The demise of the runners maximised the pleasure of the spectators. The race was usually won by Dave Floodgate or Alan Bates. I think the best I did was 21$^{st}$ in a field of about 60.

At the end of my second year at Southend High I left St Peter's Church choir, having drifted apart from my older friend Trevor Williams who had left the choir when his voice broke. I no longer had any close friends in the choir. But my main reason for leaving was that the vicar had opened a Youth Club that met Friday evenings in St Peter's Church Hall, and this clashed with choir practice. The choirmaster, Mr Herbert, tried to persuade me to stay but the attractions of the Youth Club were irresistible. All my friends went to the Youth Club, even those who did not go to St Peter's Church. There was table tennis, music, and dancing with girls, although I rarely managed the latter. The Youth Club was a great success.

An implicit condition of joining the Youth Club was to attend at least one church service each Sunday. A group of us would sit in the back row at Evensong and generally mess about talking and laughing. The vicar, Mr Reading, was often supported by an eccentric lay reader who always managed to include in his sermon an admonition that every member of the flock was "cheap and nasty", which had us in fits. We were not popular with the rest of the congregation. A second implicit condition of joining the Youth Club was that we attend weekly classes with the Vicar in preparation for our confirmation as full members of the Church of England. I was confirmed by the Bishop of Colchester at St Peter's Church on Sunday, 10 Mar 1957.

At about this time the Williams over the road bought a television. They would invite us over as a family on Saturday nights to watch. We all sat in darkness watching the flickering screen and enthralled by such programmes as "Dragnet", Hughie Green's "Double Your Money" and Michael Miles' "Take Your Pick". These two game-shows were the first on UK television to offer cash prizes.

By this time I had developed two separate sets of friends. The first set comprised my friends in my class at school. I got on well with them but they all played school sports on Saturday (football, hockey or cricket), which restricted the opportunities for me to see them out of school. So I became close to other boys in the neighbourhood. My main friend in this second set was Richard Coakes, the son of the Deputy Head and Classics teacher at my school. Richard had failed the 11-plus much to the sorrow of his father, so Richard was sent to a private school called Eton House in Wakering Road, Thorpe Bay. (It now goes under the name of Thorpe Hall School). Richard cycled the several miles there and back every day, including Saturday mornings, in both summer and winter.

Richard became my closest friend. He had a similar off-the-wall sense of humour to me and he was not a sporty person like my school friends, which for me was an attraction. I felt at ease with Richard. We would go into town together, walk down the pier, or drink beer in pubs in the rural villages around Southend. Mostly we used our bicycles. We went off on long cycle rides together, far out into the country to places like Stambridge, Paglesham, Canewdon, Great Wakering and Fambridge. These were often places where there was water and boats to mess about in. Once we cycled up the old London Road to Canvey Island, shortly after the 1953 floods[10].

---

[10] High tides and heavy storms caused severe floods on the night of Saturday, 31 January, 1953. The devastation on Canvey Island cost the lives of 59 people and led to the temporary evacuation of the 13,000 residents. Altogether 307 people died in England, 19 in Scotland, 1,836 in Holland, 28 in Belgium, and 230 at sea. This disaster followed on the heels of the great 1947 Thames flood that affected much of the Thames Valley and elsewhere in England. Never believe that the recent heavy rains and floods, dreadful though they are, are anything but the normal vagaries of the weather. There have always been extreme weather conditions since time began, some of them much more severe than we have recently experienced. The main difference is that we are now so densely populated that many more people are affected.

Canvey Island is a curious sea-walled piece of flat land jutting out into the Thames estuary, about 15 miles from Southend, with a gaunt skyline of oil refineries and storage depots. We found this to be a desolate, windswept place and I have never returned.

A favourite venue was Hockley Woods, an area of dense woodland sufficiently large that you could actually get lost. It had many attractions including streams and colonies of ants, a large Spanish variety up to a centimetre long that had been imported many years before to feed game birds. They had since multiplied and firmly established themselves in the woods in large ant-hills often a metre high. The sport was to goad them in various ways such as urinating over them or, even better, by dropping burning twigs onto their nest and watching their mass suicide as they desperately tried to eliminate the threat. When stressed in this way the ants emitted a bitter smell of formic acid. The woods contained masses of bluebells, not the now common Spanish version, but the more delicate indigenous variety. We transported huge bunches of them home for our mothers.

We also cycled at night, for which my bike was equipped with front and rear lights. One dark winter's evening Richard and I were cycling to St Peter's Church Youth Club when we were chased and stopped by another cyclist. He announced pompously that he was "a member of Southend Constabulary". I failed to understand that he was telling us he was a policeman. He warned me for having no rear light. I checked, and the battery was flat. Policemen today do not bother about bicycles or cars with faulty lights. They have more important tasks.

One Sunday Richard and I cycled all the way to London to visit Plaistow engine shed in the East End, a round trip of about 70 miles, made easier by the fact that the main arterial road to London had a wide cycle track most of the way, completely separate from the road itself. This amenity was way ahead of its time. Richard had a puncture on the way back and we had to walk the last few miles. Surprisingly, most of the cycle track still exists and remains in good condition. In spite of my closeness to Richard Coakes I always felt I needed him more than he needed me. I often knocked at the door of his house. I do not remember him ever knocking at my door.

Another friend called Dave Hall also failed his 11-plus and went to Fairfax High School. Dave lived at the top of the road in Henley Crescent. He was an amusing friend with a great love of piano boogie. We gave him the nickname Kidshot. He taught himself to play various classics like "Honky Tonk Train Blues" and spent hours perfecting his technique. I was inspired by his enthusiasm. Dave was to become a successful newspaper reporter on the Southend Standard.

A particular incident springs to mind. It was dusk in winter and there was thick snow on the ground. Richard, Dave and I were at the top of Henley Crescent rolling massive snowballs when in the failing light we espied in the distance Terry Absalom, nicknamed Podge Kid, trudging up the road under the weight of two large shopping bags. I was now less close to Terry, partly because of his annoying pranks and also because although he passed the 11-plus and went to Southend High School like me, he had finished up in the C class with different friends. We hid from Terry behind two huge snowballs that we had rolled together to form a wall, and rapidly made a large arsenal of snowballs. As soon as he was within range we opened fire with a continuous volley that knocked poor Podge Kid to the ground scattering the contents of his shopping bags. Still under heavy fire he scrabbled around in the snow and managed to recover his shopping before scurrying up his front path to safety. We hooted with our victory. Podge Kid's mother then came out very angry and scolded us saying that the tins in the shopping were now too bent for her to open.

Dave's friend Pete Wyss, noted earlier for his prank on our Latin teacher Mr Cleaver, was another 11-plus failure who went to Fairfax High School. Both Dave and Pete went on to study at Southend Technical College. Pete was to become an electronics expert. On one occasion he showed us his latest electronic device, a radio interferer that fitted inside a matchbox. By tweaking the screw at the end you could produce a loud whistling interference on any radio in the immediate vicinity. We spent a whole Sunday afternoon playing with this device on Southend seafront, and watching the annoyed portable radio listeners on the beach trying in vain to retune their sets. I have since learnt that our jamming activity was illegal.

At about this time the Daily Mail (or it might have been the Daily Herald, the newspaper my parents took) began an interesting publicity campaign during the summer months. They started a promotion with a stunt involving a character named Lobby Ludd. His silhouette would appear in the paper the day before his visit to a named seaside town. The idea was that you had to try and spot him. If you thought you saw him you had to approach him and issue a challenge using the exact words: "You are Lobby Ludd and I claim my five pound prize." We would walk along the seafront and dare each other to challenge the most unlikely people to see their reaction. This was great fun. This sort of promotion would not work today. You would likely receive a punch in the face.

A fourth friend was Paul Archard, a rather strange boy who went to Westcliff High School, the other grammar school in the town. Paul lived in Hobleythick Lane a few doors up from my grandparents. He was the eldest of four children. His parents, both teachers, were rather dissolute characters who spent most evenings in the Bell Hotel public house until closing time. Paul held occasional parties at his house that he called "social outlets". Richard Coakes and I sometimes attended. Paul provided food that we hesitated to eat, having seen the state of the family kitchen and the communal frying pan. We had many laughs with Paul Archard, mostly at his expense.

Paul was a devout Catholic with a great musical talent. He played the organ at his local church. My shared appreciation of music with Paul brought us closer together. He introduced me to well-known orchestral pieces like "The Planet Suite" and "The Four Seasons" that we listened to for hours in his front room on his record player.

The five of us, Richard Coakes, Dave Hall, Pete Wyss and me, with Paul Archard on the periphery, were a rather motley bunch. We generally had a good time together. I think my other friends at Southend High School were mystified by my choice of associates.

Most of my friends went to the St Peter's Youth Club, even those who never went to church. I played in the Club table tennis team and was quite good. The other members of the team were Michael Pratt, Alan Bates, and occasionally Peter Lee. We cycled to other Youth Clubs in the area for evening matches as part of a local league. I cannot recall ever winning a match.

In order to improve our game we obtained the vicar's permission to use the church hall to practice on Saturday mornings. All went well for a few weeks until other friends started attending our practice sessions just to generally mess about in the hall. One Saturday morning Dave Hall brought along a pack of cards and we started a poker school. This was a new and exciting game for us and made a change from table tennis. Even though we played for small stakes it was possible to win (or lose) several shillings in a session. Then one week the vicar put his head round the door to check on us and was appalled at what he saw. He swept into the hall and overturned the table scattering all the cards and coins over the floor. It was most impressive, just like Jesus in the temple with the money-changers. This put an end to our Saturday morning table tennis.

When I was fourteen my parents gave me a Diana air pistol for Christmas. This seems a rather dangerous present to give a boy, but maybe my father was keen to play with it as well. It lacked power but it certainly would have hurt someone at close range. It fired small lead pellets or feathered darts. For some weeks my father and I practised shooting at paper bulls-eye targets in the garden. I then graduated to more interesting targets. If I fired it from my bedroom window I could just reach the houses opposite about 80 yards away, to be rewarded with a satisfying ping if I hit glass. I also aimed it at the next-door cat with good results. My pistol was less powerful than Trevor Williams' who had a Webley air pistol. Even more powerful was Terry Absalom's air rifle, which he used for killing birds in his back garden. It is strange that we were allowed to be armed with these weapons, for which no licence was required.

I have mentioned the bicycle with the funny-shaped handlebars that my parents had given me for Christmas when I was about nine. This bicycle served me well for a few years. In fact, it was my father who helped teach me to ride, running along the road beside me and holding the back of the saddle to help me balance. I kept my bicycle in the sideway alongside my father's, under an old sheet of tarpaulin to shield it from the weather. Later my father built a wooden shed to house our bicycles and other garden items.

Whereas my father had a bicycle, my mother did not. I never saw her ride a bike. My father sometimes took me out cycling on Sunday mornings to pick blackberries on the cliffs near Leigh, and sometimes in late summer to the misty fields behind Eastwood Church to gather mushrooms. The mushroom expeditions started at six in the morning to catch them at their best, returning home in time to cook them for Sunday breakfast. Their strong, gamey taste did not really appeal to me, although I love them now.

When I was sixteen I resolved to get a sportier bike more like those of my school friends. One or two older boys had a Claud Butler, the absolute top of the range. There was no way I could afford such a bike and I accepted the fact without rancour. I saved up some money and sold my flute and piccolo for a knock-down price at a second-hand shop in Leigh. After visiting all the local cycle shops I finally bought a Dayton Consort for £19.50, reduced from £21, from a shop at the back of Prittlewell. I was over the moon with this bike. It had thin light-weight tyres, modern "derailleur" gears (10 in total) instead of the old-fashioned but reliable Sturmey-Archer gears, drop handle-bars, calliper breaks, and a narrow seat that cut into my back-side. It looked very professional. It went no faster than my old bike but felt 100 times better.

Early on at secondary school I developed an interest in train spotting, or rather steam engine spotting. I may have picked it up from Michael Pratt, or possibly Richard Coakes. The purpose of this hobby was to spot the number on a steam engine that uniquely identified it. Each steam engine belonged to a class and there was some satisfaction in spotting, or "copping", every engine in a particular class. Books published by Ian Allan listed every engine in service in the UK. He had calculated the business to a fine art. The basic book was called the Locoshed Book costing 2s/6d (12½p) comprising almost 100 pages of nothing but columns of engine numbers. There was a code next to each number indicating the home shed for that engine. For example, the Shoeburyness shed was 33C, and Weymouth shed was 71G.

For more detailed information with photographic examples you had to buy individual books for each railway region of which there were four – London Midland, Eastern (including North Eastern and Scottish), Southern, and Western (including Wales). Each book cost another 2s/6d, or you could buy a combined hard-back edition for 10s/6d. They were reissued regularly as engines were scrapped or changed their home base. When you copped an engine you underlined its number in the book. For us train spotters these books had the status of bibles. The hobby was in its prime for no more than 20 years, from the end of WW2 to the end of the age of steam, circa 1965. It was a hobby of great romanticism with roots in the technology of the Victorian era. Apparently, there are now enthusiasts who "cop" diesel engines, but I find this incomprehensible.

Of course, you could go to railway stations or stand on railway bridges to see steam engines and many boys did this, but it was a slow process and relatively unrewarding. You did much better to visit the sheds where the engines were housed, maintained and fired up each day before work. That way you saw hundreds.

I often cycled with Richard Coakes to the local Shoeburyness shed and to the sub-shed at Southend Victoria, but they were both fairly tame. More worthwhile sheds were further afield.

The classic tour was the "Seven Sheds" in London - intensely exciting. The correct procedure was to write to the manager of the region in which the sheds you wanted to visit were located in order to apply for a visitors permit. This process was tiresome because you had to write letters weeks in advance, and even then you could not be sure that the permit would be granted on the day you requested. This made it difficult to organise a trip if you wanted to visit several sheds. So we usually ignored permits and simply trespassed on railway property, known as "bunking" the shed. We exchanged information with fellow train spotters on the best way to bunk a particular shed without being caught.

I did the Seven Sheds tour several times, first with Michael Pratt, then with Richard Coakes, once with my cousin John, and twice by myself. We caught the 5 a.m. "Milk Train" or the 6 a.m. "Workmen's Train" from Southend Victoria to Liverpool Street, then took the Underground to Kings Cross. This was where the adventure started.

Kings Cross shed, known as Top Shed, was in the Eastern Region of British Railways (only later called British Rail) and was notoriously difficult to bunk. It was out of the question to walk through the main gate because it was guarded by lines of offices. You would be caught and thrown out immediately. So entry was by a circuitous route involving a climb over a 6-foot brick wall behind the main line station followed by a scramble through a network of martialling yards, a walk parallel to the main line exposed to detection from workmen in the signal box, and finally into the sheds proper where there was some cover between the lines of engines. The smell of steam and coal was intoxicating as was the fear of being caught.

Kings Cross shed was home to many of the great A4 class of locomotives, like Mallard that held the world speed record of 126 mph. They were massive, powerful engines with graceful lines. It was quite an experience to see these giants at close quarters and walk between them as they rested, still warm. The engines towered above us, our heads barely level with the wheel axles. If we were daring we would climb up into a cab, an operation known as "cabbing", but usually there was no time. We were desperate to go up and down the lines and record all the numbers in our notebooks as quickly as possible in order to minimise the risk of being caught. Of course we often saw men working on the engines but they usually ignored us. The man to avoid was the foreman who normally sat in his office. So we kept clear of offices. The way out of Kings Cross was the same as the way in. A successful visit to Kings Cross shed gave a great sense of achievement and relief. It was a good omen for the rest of the day.

We then took the train to Willesden Junction to visit Old Oak Common and Willesden sheds, both big depots. Old Oak Common served the Western Region and supplied engines for all the trains out of Paddington. Willesden was an important Midland Region shed supplying Euston and St Pancras. Access to Old Oak Common was along the footpath by the Grand Union Canal and through a hole under a wire fence. In 2010 my daughters gave me the excellent birthday present of a day out on the Cholsey & Wallingford Railway with the opportunity to drive the steam engine. I talked to the instructor about experiences with engine sheds 50 years previously and mentioned Old Oak Common. He immediately recalled the hole in the wire fence! Access to Willesden was straight through the main gate. There was never any problem bunking either of these sheds, but it took a long time to walk round them taking us up to lunchtime.

The next two sheds were Neasden and Cricklewood. These were smaller Eastern and Midland Region sheds reached from Willesden by short bus rides, giving us time to eat our sandwiches. Neasden shed was in the middle of a field with weeds growing through the tracks. It had a sleepy feel about it. Cricklewood was more bustling close to a main road. Neither of them posed any real problem.

The last two sheds on this tour were Camden Town and Kentish Town, both Midland sheds of moderate size reached by another bus ride. Camden Town was quite difficult to bunk. It was in a heavily built-up shopping area and adjacent to the main line. The only access was through a small door in a 10 ft high brick wall on a busy pavement that opened directly into the shed. The foreman's office was immediately on the other side of the door. So you just had to open the door and take your chance, which was about 50/50. If the foreman caught you he usually threatened prosecution for trespass then threw you out. If this happened you did not risk going back in again. It was just too bad. On the other hand, Kentish Town was relatively easy to bunk, after which it was usually time to start for home. These Midland Region sheds were filthy and the engines were not kept clean like on the Western and Southern Regions, but they still smelt good.

It is interesting that there were two designs of engine shed, straight sheds and round houses. In a straight shed the engines are in parallel lines that run into the shed at one end and terminate at the other end. This is the simplest design, but it means that an engine at the end of the line cannot be taken out without first moving all the engines in front of it. The problem is solved in a roundhouse by locating the engines on radial tracks like the spokes of a wheel. A single track runs into the shed with a central turntable. This system allows an engine to enter or leave the shed without disturbing any of the others. Round houses were less common than straight sheds probably because they were more expensive to build and took up more space. Kentish Town was a good example of a round house; in fact there were two adjoining each other. When Kentish Town was eventually closed the round houses were converted into a modern theatre.

If we finished the Seven Sheds early, we would maybe make the journey to Clapham Junction main line station. This still claims to be the busiest railway station in the UK with trains from both London Victoria and Waterloo. With so many platforms, 17 altogether, there was never a dull moment with trains constantly passing through. If we were lucky we would cop a big Battle of Britain Class locomotive or a Merchant Navy Class as it thundered through the station. The point about Clapham Junction was that it enabled you to see Southern Region engines, whereas the Seven Sheds were on the Western, Midland and Eastern Regions. Of course, there were Southern Region sheds but the main ones in London, like Stewarts Lane and Nine Elms, were like fortresses and impossible to bunk.

I should say that there was a certain protocol in copping steam engines. Ideally you had to see the number for yourself. This was located in large numerals on each side of the cab and in smaller numerals on the front of the engine. If someone else saw the number as a train went through a station you could still cop it provided you saw at least something of the engine. If you happened to be looking the other way then it was cheating to cop it. On the other hand, if you were chased out of a shed it often happened that different people in the group had copped different numbers. So notes would be compared outside the shed. The general rule was that an engine number noted by one person in the group was deemed to have been spotted by everyone else. This was because an engine in a shed is generally stationary so everyone must have seen it, unlike a train going fast through a station. This may all seem rather pedantic, but to us it was deadly serious.

The geography and geology teacher at Southend High School was Mr (Wally) Allen. I gave up geography after my second year, but I maintained contact with Mr Allen through the holidays he organised for the boys. Every summer he and Mr Harvey, a teacher of German, took 30 to 40 boys away for a week's holiday to some interesting part of the country. The week always involved a lot of walking with some instruction on the local geography and geology, and invariably included a boat trip of some sort.

I first became aware of these annual holidays in my second year when the location was the West Country (Dartmoor and Exmoor), but I found out about it too late to go.

The first school holiday I went on was to North Wales in the summer of 1957. I was 14 years old. We stayed at Lledr Hall in Pont-y-Pant, just a couple of miles from Betws-y-Coed. It was a comfortable youth hostel with lots of amenities. The instructions issued by the school in preparation for the holiday were to bring stout walking shoes, a pullover to keep warm, a rucksack, and in case of bad weather a plastic mac !

It rained most of the time so I took few mountain photos. But the bad weather did not stop us climbing Cader Idris and

Mount Tryfon in thick mist. The plastic macs did not fair too well in the rough conditions and were torn to shreds. The day we were due to climb Snowdon the weather was so bad that the plan was abandoned. Instead we visited Conway to see the sights, including the castle and a house that was claimed to be the smallest in Great Britain.

On another day we did a low level walk that took in a series of stunning waterfalls, including Swallow Falls, located between Betws-y-Coed and Capel Curig. On our rest day we travelled to Llandudno to take the ferry for a day trip to Douglas, Isle of Man. We climbed Snaefell, the highest peak at just over 2000ft.

I discovered much later that Mr Allen and Mr Harvey had both spent time over the Easter holidays to reconnoitre all the walking routes and finalise the plan for the whole holiday, including the drop-off and pick-up points for the coach (we tended to do linear, not circular walks). They did this every year, a measure of their dedication.

One of the boys in our party was George Farrow, in the same Form as me. George was a sort of junior Les Patterson although he was not Australian. He would boast and demonstrate how he could produce the longest fart. Once on a school holiday in our dormitory in Betws-y-Coed he bent over and actually ignited it with a match to produce a blow-torch effect. George was also a bully towards a certain boy called Michael Pratt who seemed to attract ragging, or perhaps George was just over-exuberant. One evening George made Michael an "apple-pie bed". This is usually understood to mean a bed made with the sheets folded so as to prevent the person from entering it. But in this case it was literal, using the remains of George's evening pudding. Michael was upset but to his credit did not tell Mr Allen.

It is difficult to understand why Michael was teased so much. He was an excellent sportsman, fairly bright, good-looking and popular with the girls. By way of example, when Michael was 15 he started an affair with a rather attractive married woman who lived opposite the school gates. He visited her regularly before coming to school in the morning.

This affair lasted about two years, and today would have earned the lady in question a prison sentence if caught. So Michael had healthy interests and was generally popular with both girls and boys. So why was he the butt of so many pranks?

This triggers a memory of a dramatic incident with Michael. One Saturday afternoon he and I went to play football in Priory Park. Michael was in goal and I was kicking the ball trying to score. At a certain point I kicked the ball hard from long range. The ball sped through the air in a swerving arc. Michael misjudged it and the ball hit him hard in the groin. He doubled up in pain. When he got up his white shorts were soaked in blood. Michael jumped on his bike and raced home screaming. I followed in hot pursuit. Once home Michael's father took control and rushed him to the doctors' surgery. It turned out that Michael had an erection while we were playing and the impact of the football had split the end of his penis. Pretty scary stuff. Michael recovered and never bore me a grudge. The incident did not appear to affect his success with girls.

I enjoyed the Wales holiday so much that in the following year, 1958, I went on the school holiday to Scotland. We stayed in a suburb of Glasgow in a rambling Victorian tenement hostel. We made daily excursions to places in the Trossachs National Park. This holiday was enjoyable but not exceptional, which is probably why I recall little about the walks.

The highlight of this holiday was a ferry trip down the Clyde to the Isle of Arran where we climbed the highest peak, Goat Fell (2870 feet), in hot weather. I know from my photographic records that this was on Saturday 19 July, which implies that we must have travelled up and down to Glasgow mid-week.

One evening a few of us train spotters, including the two Pratt brothers, Dave Patrick and George Farrow, took a tram ride to visit a local engine shed in Eastfield, North Glasgow. We were quite successful in copping engines not normally spotted down south.

Next day, Sunday, was a free day. A small group of us decided to take the train from Glasgow to Edinburgh. The purpose was to see the sights but not the tourist ones, our principal interest being the engine sheds. That day we managed to bunk both the Dalry Road and the St Margarets sheds. Unfortunately, we failed to do the big Haymarket shed, home to some of the magnificent A4 beasts.

A particular memory of this holiday is seeing our respected teachers naked in the communal shower. This was a quite a shock, and confirmed that nakedness reduces us all to the same level.

My maternal grandmother died in the spring of 1957, aged 76, when I was 13. She had been in a nursing home in Crowstone Avenue, Westcliff, for some weeks. As with the death of my grandfather it was all kept low key and I was not made aware of the funeral arrangements. The subject was taboo. My parents obviously thought I was still too young to handle the emotion. I think they were embarrassed to give me any details.

Some months after my grandmother's death I became aware that my mother had inherited some money from the sale of her mother's house. My mother was frugal and no doubt wanted to save the money, but my father obviously persuaded her to release some of it. The first thing he bought was a television set. After that we hardly ever listened to the radio again.

My parents also started going on holidays. They went several times to a small, upmarket Holiday Camp just outside Torquay called Barton Hall, later taken over by Pontins. I went on the first of these holidays and thoroughly enjoyed myself. It was at the end of July 1958, immediately after my holiday in Scotland with the school.

All of us, including my brother Nick, stayed in a hillside chalet overlooking the site. My father was sociable and made friends easily. As usual, he identified a mason and they became buddies for the rest of the week. He often kept in touch with these holiday friends and would meet up with them again in future years. This holiday gave me my first experience of horse riding. It was a bad experience because the saddle had not been properly secured, and when I tried to mount the horse I finished upside down beneath it. This put me off horse riding thereafter.

One day we went on an outing to Newton Abbot. I took time out to visit the local engine shed and works, which used to be quite an important repair depot for Western Region locomotives.

We went on this holiday by train. On the return journey I had been playing with my father's old but reliable camera. Regrettably I left it on the train seat at Paddington Station. My father was upset, but I suspect he was secretly pleased for the excuse to buy himself a new reflex camera.

That same summer of 1958, in August, my paternal grandparents celebrated their Golden Wedding anniversary. A party was held in the back garden of their bungalow in Hobleythick Lane. Most of the family was present for this happy occasion. The cake was prepared by my Aunt Cissie with her natural artistic flair.

The following year, immediately after "O" Levels, I went on the third and last of my annual school holidays. It was July 1959. The destination this time was Cumbria. We stayed in a hostel in Keswick on the Penrith Road opposite the River Greta. On return visits to Keswick many years later I have never been able to find this hostel. I think it must have been demolished to make way for the new Fire Station.

The weather for this holiday was excellent with five days of walking. The coach we had at our disposal to ferry and collect us enabled us to do linear walks. My recollection of the precise routes is vague, but I know roughly which fells we climbed because I still have the photographic evidence in the form of annotated colour slides.

We did Scafell Pike, England's highest mountain, via the Corridor Route. I believe the return was via Great End, Angle Tarn, Thunacar Knott, Stickle Tarn by Pavey Ark. and down into Langdale, an exceptionally long trek of some 18km.

We did Helvellyn from Glenridding by way of Striding Edge. I remember vividly reading the plaque just below the summit commemorating Charles Gough and his dog, who both perished on Striding Edge in 1805. There have since been many more fatalities on this ridge. One of our number suffered from vertigo and fainted, and had to be helped along the most precipitous section of the ridge. I am not sure how we came off Helvellyn. I think we finished up at Thirlspot.

We also did Skiddaw from South to North, which I recall was an arduous slog over a lot of slate. On another day we walked from Buttermere over Whiteless Pike, Grasmoor, Coledale Hause, Cauldale Beck, finishing at Braithwaite. On yet another day we took the launch from Keswick to Ashness Gate and did Ashness Bridge, High Seat and Bleaberry Fell, returning on foot to Keswick.

With today's health and safety rules it would be impossible for two teachers to take a group of 30-40 boys on adventurous walks such as these. We were very fortunate to have such dedicated teachers and freedom from today's restrictive legislature.

On our rest day in Cumbria we visited the nuclear power plant at Calder Hall on the West Coast. It had been built only a few years before, officially opened by the Queen on 17 October 1956. It was heralded as a landmark in scientific achievement and energy generation. We spent the best part of a day touring the plant, viewing the nuclear piles and being instructed on how it all worked. It was an educational visit. I understood that this was how all our domestic and industrial energy would be supplied in the future. We were not told that the main purpose of the plant was to make plutonium for nuclear bombs. Neither were we told that just two years before our visit, on 10 October 1957, a serious fire had broken out in the radioactive core of Pile No. 1 in the adjacent nuclear site of Windscale.

This is now recognised as the world's first nuclear accident, probably as serious as Chernobyl. In the absence of any visible smoke or flames most people were oblivious to the disaster. The Prime Minister, Harold Macmillan, imposed a blanket of secrecy. Many years later the truth came out that a radioactive cloud had been released that spread contamination over England and Wales, and as far as Europe. The contamination in hotspots on the West Coast of Cumbria and the decommissioning of the burnt-out pile remain issues to this day. Windscale became such a symbol of hate for opponents of nuclear energy that in 1981 the public relations people changed the plant's name to Sellafield in an effort to banish the bad memories. Today, Sellafield is one of the biggest nuclear dustbins in the world.

In the evenings of this Cumbrian holiday we were free to roam round Keswick and wander down to the Lake. This was a wonderful time with vivid memories. It sowed the seeds of my love for the Lake District that were to rest dormant for the next 30 years.

There was a great spirit of adventure on these walking holidays. Some boys like Alan Bates and Michael Pratt would race ahead with Mr Allen not far behind. I was usually in the vanguard. Mr Harvey took up the rear to encourage stragglers. We returned exhausted each day to the waiting coach in which we sang the latest pop songs on the drive back to the hostel. At that time Buddy Holly was all the rage. I am greatly indebted to two these teachers for introducing me to the joys of walking and climbing.

We always travelled to our holiday destination by train, usually by express from a London main line station. The principal pastime on the journey was train spotting. In those days it was possible to open a small sliding window above the main window in the seating compartment, or alternatively slide down the complete window in the exit door at the end of the carriage. This enabled us to stick our heads out of the window in readiness to cop the number on the front of trains speeding past on the adjacent track. The exercise had to be finely judged. You had to wait long enough to discern the number, but not long enough to risk having your head knocked off. Trains have since been redesigned to stop these dangerous antics. If we happened to pass an engine shed we all frantically wrote down numbers and compared notes afterwards. George Farrow would smother Michael Pratt with his jacket so that he could not claim to have copped these engines.

In early 1959 my father bought our first car, a blue Ford Anglia, no doubt dipping into my mother's inheritance. My father had a driving licence but had never taken a test. He had been given his licence towards the end of the war while working in the Post Office and had hardly driven since. Consequently, he was a poor driver and did not improve with practice. The gear stick on the Anglia was quite stiff which made gear changing hard work. The clutch was also fierce. So initial sorties in the new car involved a lot of kangarooing. My father developed a habit of looking out of the side window when changing gear as if to dissociate himself from the car's antics. He had several accidents over the years, fortunately none of them serious.

Our house in Henley Crescent had no garage or front drive, and my father did not like leaving the car outside our house on the road. So he parked it most nights in his mother's garage in Hobleythick Lane involving regular late-night journeys. The car was liberating for my parents because it made travel so much easier. My mother appreciated it because it reduced the burden of shopping.

At the weekend we often went away on day trips to visit relatives. On one occasion we drove to Farnborough to visit Auntie Florrie (grandfather Parsons' sister) and her husband Uncle Billy (my godfather). I recall that Auntie Florrie was quiet and kind whereas Uncle Billy was stern and remote. It was rumoured that Uncle Billy was rich although their bungalow did not give that impression. According to my mother, he was a miser and never installed a proper bath in their house, forcing his wife to bathe in a tub in their kitchen right up to the day she died.

On another occasion, we drove to Hemel Hempstead to visit my mother's cousin Molly and her Aunt Milly who lived at picturesque Water End Cottage, a property that figured in Shell advertisements of that era.

We often drove up to London, usually on a Sunday, to visit either Auntie Haidée and family or Uncle Harold and family.

Our route took us up the Southend Arterial Road as far as Gants Hill then usually round the North Circular. Sometimes my father detoured through Central London to attend Sunday morning service at the church of St Clement Danes at the end of the Strand. This grand building, built by Sir Christopher Wren, is situated on an island in the Strand close to the Courts of Justice. I think my father liked it because he was working in London at the Air Ministry and this particular church had been adopted as the central church of the Royal Air Force. There may also have been a Masonic connection.

On one such visit to Haidée's when I was much older we met my Uncle Billy (my godfather, not Haidée's Billy) after his wife had died, while he was staying with Haidée for a few days. He mistook me for my cousin Paul Hudson, but this was because he was going blind. It was always rumoured that Uncle Billy had a great deal of money, so my father had often encouraged me to write to him just to keep in touch because he had promised to leave £100 to each of his godchildren. I think I complied. But when he died, on 8 October 1985, his relatives saw none of his wealth, not even Auntie Haidée who, with her usual generosity had taken great care of Uncle Billy in his old age when he was losing his sight. He left all his money to the local church.

That summer of 1959, shortly after my Cumbria holiday with the school, I went away with my parents and Nick in the new car to a holiday camp near St Austell, Cornwall. I believe it was the Duporth Holiday Village. We left home at four in the morning for the long journey to Cornwall, there being no motorways in those days. This was another enjoyable holiday. We went mackerel fishing out of Charlestown and drove sightseeing to many places including Mevagissey, Looe & Polperro, Newquay, Fowey, St Mawes, Falmouth, and St Ives.

I made friends with a boy of similar age. One morning we went exploring in the country lanes and came upon a small slaughter house in a farmer's back yard. Such places were still legal. The farmer was about to slaughter some cows and sheep. He used a pistol with a retractable bolt. It was gruesome and shocking to see the poor animals executed.

The steaming innards were cut out of the dead twitching creatures, the guts falling to the floor in a stinking heap. At dinner that evening there was roast lamb, but I could not face eating meat for the rest of the week. This was the last time I went on holiday with my parents.

In January of my fifth year at Southend High School I took "O" Level Pure Maths. At the end of my fifth year in June I took "O" Level English, Latin, French, Physics, Chemistry, Applied Maths, Additional Pure Maths, and "A" Level Geometrical and Mechanical Drawing (the latter for some reason counting only as an "O" Level). I passed in all these subjects. In those days there were only two grades at "O" Level: Pass and Fail. There was no grade for Distinction except at "A" Level. However, the school was given the marks and some teachers gave us an inkling of how we had done. I learned that I had done particularly well in the science subjects.

I was awarded a school prize at the end of the fifth form for my achievements at "O" Level. Prize winners had to choose a suitable book up to a certain value. Lost for inspiration I did not know what to choose. We had just made a rather boring school visit to May & Baker's chemical factory in Dagenham. Maybe this is what led me to select a dreadfully boring book on chemical engineering. It was presented to me by the guest speaker at the Annual Speech Day and Prize Giving ceremony held the following March. I have never opened it.

Relatives often asked me what I wanted to do as a career. I had no idea, so I would simply reply that I wanted to be a Civil Engineer, or occasionally, a Chemical Engineer. I did not know what these careers involved but my answer seemed to be satisfactory.

# Secondary School to "A" Levels (1959-61)

In September 1959 I moved into the sixth form to specialise in the sciences. My teachers were Mr Drummond for Pure Maths, Mr Fretton for Applied Maths, Mr Barton for Physics, and Mr Griffiths for Chemistry. I particularly enjoyed my maths lessons with Mr Drummond and Mr Fretton. I looked forward to them and relished solving the problems we were set. I began to realise that I had a real talent for mathematics. Physics was also good fun, especially the laboratory work. Chemistry I detested but I was stuck with it.

We also had to do one period each week studying a new foreign language. This was the headmaster's idea to stop us scientists getting too narrow in our studies. I chose to do German. I hardly learned anything. Curiously, the headmaster did not make the classicists study a science subject.

Life in the sixth form was different from the lower school with much more freedom. We had a heavy work load with free periods for private study in the library. Sometimes the teachers actually treated us with respect.

I began to spend a lot of my spare time with my friends Richard Coakes, Dave Hall, Pete Wyss and Paul Archard. We frequented the local pubs, even though we were under age, and became regulars at the Bell Hotel. These activities were funded from money I saved from occasional holiday jobs, plus the 5 shillings (25p) a week pocket money that my father was then giving me.

I should say a bit more about pubs, or public houses. There used to be many more pubs than there are today, but they tended to be smaller establishments relying for most of their trade on "locals". Opening times were typically 12:00 to 14:30 and 18:00 to 22:30. There was no such thing as theme pubs, gastro pubs or family pubs. Children under 16 were not allowed in pubs and there was usually no food available. The main purpose of the pub was to drink, an activity confined mainly to the male population. You could buy crisps and nuts which increased the thirst of the punters, but meals, if available at all, were usually confined to lunch times when cheese ploughmans or pork pies might be served. Every pub was divided into two sections: a Public Bar and a Saloon Bar. The Public Bar was usually a bit scruffy and intended for men in dirty boots and working clothes. The Saloon Bar was posher with a carpet and the beer costing a penny a pint more. Occasionally there was also a Private Bar, usually rather small, where you had extra privacy. One or two pubs attracted visitors from greater distances because of their special charm and character. These were the forerunners of today's theme pubs. A good example was The Peterboat by the Thames in Old Leigh, surrounded by fisherman's cottages and cockle sheds. But I found The Peterboat too up-market, noisy and expensive.

That first Christmas in the Lower Sixth a large group of us gathered together on Christmas Eve for a pub crawl. About 20 of us met at 18:30 in the Royal Stores in Southend at the top of Pier Hill, resolving to walk the three or four miles back to the Bell Hotel with a drink in every pub on the way. After the Royal Stores we went to several pubs in Southend High Street including the Royal Hotel and the Middleton Hotel, then on to Prittlewell where we did the Blue Boar, Golden Lion and Spread Eagle, finally arriving at about 23:00 at the Bell Hotel. By this time we were feeling somewhat worse for wear, but some of us still managed to attend Midnight Mass. Danny Linehan went to St John Fisher Catholic Church near Earls Hall Parade, and I went to St Mary's in Prittlewell to join my parents.

One year my hangover was so bad that I could not eat my Christmas dinner and it ruined my day. Nevertheless, this first reunion was such a success that we solemnly resolved to meet up every Christmas Eve at the same time and the same place to reenact the pub crawl. We would do this for ever and as long as our health allowed. We did succeed in repeating the event for a few years and then it just faded. In any case, the Royal Stores was pulled down when Southend High Street was rebuilt.

After Easter at the beginning of my last term in the Lower Sixth the new sub-prefects were announced by Mr Coakes. All my friends were made subs but I was not included. This was a great disappointment to me. I assumed it was because I did not excel in sport. Although a sub-prefect had rather menial and boring duties, it made promotion to full prefect in the Upper Sixth a virtual certainty. So I was definitely put out at not being made a sub-prefect.

On Friday 6 May 1960, Princess Margaret married Anthony Armstrong Jones at Westminster Abbey. It was the first royal wedding ever to be televised. I can't say it particularly interested me, but a friend Faye Hockey who was a royalty fanatic wanted someone to accompany her in the evening to the end of Southend Pier to watch the Royal Yacht Britannia pass by with the newly-weds on board on their way to a honeymoon in the Caribbean. I was pleased to oblige. We walked down the pier and waited ages with a large crowd in the gathering dusk. The yacht was late. All we eventually saw were its tiny lights on the horizon. It meant nothing to me but I liked being with Faye.

At that time there still existed conscripted National Service of two years in the Armed Forces for all healthy males from age 17. I recall occasional visits to the school from former boys in their uniforms. If at age 17 you were still at school or attending university it was possible to delay conscription, but it was not something that concerned me unduly. As it happened, conscription was ended on 31 December 1960, so I just missed National Service by one year.

At the end of my year in the Lower Sixth I took "A" Level Pure Maths which I passed without any problem. It was becoming clear to me that I was quite good at maths and I certainly enjoyed it. For my examination achievements I was awarded the Lower Sixth Form prize for mathematics. The book I chose to be presented with was a Manual of Photography. This was a useful book which I read thoroughly.

At this time in my life I was on top of the world. Most things were going well and prospects were good. Five of us school friends celebrated by spending a summer holiday at Butlins holiday camp in Filey, Yorkshire. A photo (Plate 24) shows us relaxing in the bar, under age, on the Saturday night of our arrival. It will be noted that we all wore our best suits and ties. This was how you dressed for a night out.

That evening was the one and only time I ever became paralytic with alcohol. In fact we had not really drunk any more than usual. We had returned late to our chalets and turned in a little worse for wear. Soon after going to bed I needed to go to the toilet. Our chalets did not have toilets. The toilet complex was a communal building about 100 yards away in the middle of a field. There was no moon and it was pitch dark, but I found my way to the toilets without any trouble because they were illuminated like a beacon. On my way back I was suddenly hit by the cold night air which sent me into a dizzy swoon. I lay on my back in my pyjamas on the wet grass fully conscious but unable to move. I eventually managed to get to my feet but had no idea where my chalet was in the blackness. I returned to the illuminated toilets to take my bearings. I finally found my chalet and the safety of my bed.

Compared with my friends I was a bit of a geek with ears that stuck out. I wanted to be one of the lads but I wasn't quite up to it. Each of my friends excelled both academically and in at least one sport, representing the school in matches each Saturday during term time. Doug May was the goalkeeper in the school football team. I was told that he actually went to tap-dancing classes to improve his balance but kept this fairly secret. Danny Linehan was in the school cricket team and was captain of the school hockey team. Peter Lee was in the school football and cricket teams. Chris Harding was in the school hockey and cricket teams. I did not play in any of these teams. In addition, they were all rather handsome compared with me and had no trouble getting girl friends, whereas my success with girls was minimal. So maybe it is not surprising that I had developed a second circle of friends round Richard Coakes.

But one or two things happened in the sixth form to boost my confidence. Firstly, some-one suggested we should start lessons in ballroom dancing. It was not uncool to do that sort of thing in the late 50s because it was a good way of meeting girls. So a group of us from school, plus my friend Richard Coakes, went to lessons every Tuesday evening with a female teacher who had a small dance studio over a shop in Hamlet Court Road in Westcliff. We did not go with dance partners. We just danced with the spare females who happened to be at the lessons. We learned the basics of the quickstep, waltz and social foxtrot. We learned a bit of Latin as well. I picked it up easily and soon realised that I was a much better dancer than any of my sporty friends. This boosted my confidence.

After the instruction there was a period of social dancing to practise our new steps. We were encouraged to dance with as many people as possible. The females present were of all ages. There was one middle-aged woman who had a little notebook. If one of us asked her for a dance she made an entry in her notebook so that we were on her waiting list. Very 1920s! All she needed was a chaperone to complete the effect.

There were several girls of my own age. One of them was the afore-mentioned Faye Hockey. Faye was a few years older than me but she always acted like a young teenager. Faye lived at the top of Hobleythick Lane and went to our Church Youth Club. She was good fun and most of my friends had been out with her at some time or other. One evening after our dance lesson I found myself walking home with Faye, just the two of us. I found Faye attractive and I walked her into a dark alley to try and snatch a kiss. She gently pushed me away saying that "she liked me too much for that sort of thing". I could never work out what that meant.

There was another girl called Pamela at our dance lessons who I found absolutely stunning. Pamela had dark hair, a curvaceous figure and bouncy personality. I think she danced with me because I was good but otherwise she treated me with disdain. She was much more interested in Alan Bates even though he was shorter than me and a rotten dancer. But I persevered with Pamela and she was gracious enough to accept several of my invitations to take her out.

Before each meeting I would have a bath, a rather rare event, and sing "When I fall in love" by Nat King Cole in my best crooning voice. I would then cycle to her house in Herschell Road, Leigh, a distance of some 6 miles, and we cycled together from there. We once went to a dance at a ballroom in London Road where I remember dancing the quickstep with Pamela to "The Lady is a Tramp". Another time we went to see a film at the Metropole in Westcliff. At our last meeting we spent a sunny Saturday morning walking down Southend High Street and along the sea front. I can pinpoint this event precisely to June 1960 because the Everly Brothers' record Cathy's Clown had just come out and was being played everywhere. My relationship with Pamela was innocent and never developed to the point where I could refer to her as "my girlfriend". I think Pamela found me rather boring. I was heart-broken when it finished.

Soon after this my friend Dave Hall gave me a rather remarkable piece of news. He said that Susan Pillsworth had told him that she was mad about me and "had the hots" for me. This seemed most unlikely and I did not believe it until it was confirmed at a face-to-face meeting. Susan Pillsworth lived in Hobleythick Lane down the road from Richard Coakes. Her father was a consultant anaesthetist at Rochford Hospital. The adults would remark upon Susan's perfect complexion. She should have been the ideal girlfriend but unfortunately I did not find her attractive. She was a bit plump and was best friends with Jane Archard, sister of Paul Archard, which was definitely uncool. I was flattered by her feelings towards me which I found incomprehensible, but Susan Pillsworth just did not do anything for me. I may have groped her once but that is all. In later life she married a man who became conservative MP for Bournemouth East. I believe she had an unhappy life when she discovered after many years of marriage that her husband was having a gay relationship with a black footballer.

Just down the road from Susan Pillsworth lived the Steggles family. The eldest son called Irwin was very clever. He went to Southend High School but was two years younger than me. Irwin had a rather remarkable walk. Instead of his arms moving in anti-phase to his legs they moved in phase. So when he swung his left leg forward his left arm also moved forward, and when his right leg moved forward so did his right arm. This strange walking technique did not appear to impede his forward progress. It became known in my vocabulary as the "Irwin Steggles Walk". Top models on the cat-walk appear to have been taught a variant of this walk, but 90° instead of 180° out-of-phase. They would be more elegant and stylish if they learned the rhumba walk so that their hips move after taking the step, not at the same time.

My parents went on holiday with Nick that summer of 1960 to revisit Barton Hall. I stayed at home and relished the new freedom with my friends. This was the first time I had fended for myself. I cannot imagine what I ate.

I returned to school for my second year in the sixth form and my last year at school. I was delighted at the announcement that Dave Patrick and I were to be made sub-prefects. We had both excelled at "O" levels and early "A" levels, and I assume this was the reason. Soon after I was promoted to full prefect, putting me on an equal footing with my friends. Being a prefect was prestigious and meant several things. First of all you had special gold braid sewn round the edge of your school blazer and cap. You were also entitled to wear the full ornate version of the school crest on the breast pocket of your blazer instead of the smaller version that everyone else wore. I think my mother had to sew on all these accoutrements. Secondly, you had access to the Prefects Room where nobody else was allowed. Even the teachers had to knock before going in. Most important, the prefects were responsible for maintaining discipline in the school playgrounds and corridors.

A prefect's authority even extended to outside school. For example, it was a school rule that every boy had to wear his school cap at all times on the journey between home and school. Some boys flouted this, putting on their cap only at the last minute as they approached the school gates. So about once a term the Head Boy secretly arranged a "cap check" which involved prefects hiding at various key locations on the routes to school to try and catch boys not wearing their caps - a bit like police speed checks. Miscreants received a prefect's detention. This involved standing in silence after school outside the Prefects Room at the pleasure of the prefect in charge. I enjoyed being a prefect but was not good at keeping discipline, especially with the yobs in 5D.

The group photo in Plate 25 shows me and my fellow prefects. Curiously, on the back of the original photo, still in my possession, is a draft letter Andrew Sargent wrote while at Bristol University to Mr D B Bartlett, the Chief Education Officer in Southend (yes, Bossy's dad was still in post), accepting an invitation to discuss university life.

We prefects did not always preserve our dignity. There was a tradition that once a year in winter, if it snowed, the prefects would run *en masse* round the perimeter of the school playing field and allow themselves to be snow-balled by the rest of the school chasing after them. After this act of humility our stature rose in the eyes of the boys who thought we were not too bad after all.

It was of course strictly against school rules for boys to smoke, at least on the school premises, but there was a strange inconsistency here. If ever we had to knock on the door of the staff room on some errand, a teacher would half-open the door giving us a brief glimpse of the interior which was always through a dense smoky fug. It was also known that the headmaster, Mr Price, was a smoker but we never actually saw him at it. Certainly, he died much later of lung cancer.

The presentation of school prizes was made on Speech Day. This was an important event in the school year. It was always held at the end of the spring term following the academic year of the awards. It was an evening affair attended by boys and their parents. The Headmaster gave a "state of the union" address. Boys who had excelled themselves in sport or exams were awarded prizes, presented by a minor celebrity or bigwig.

One particular year the celebrity at Speech Day was Sir Stanley Rouse, Secretary of the Football Association (FA) and President of FIFA. The event required the Head Boy, in this case John Ling, to give a "Vote of Thanks" to the guest speaker. John's theme was something to do with the way experiences at school fashioned a boy for life. It started well enough until for some inexplicable reason he began saying that the boys were treated too much like children and should be given more responsibility. He became even more outspoken suggesting better ways of running the school and criticising the teaching staff. We boys sitting in the front rows could not believe our ears at this audacity. What were all the parents thinking? The headmaster, sitting on the stage next to his honoured guest, began fidgeting and turning shades of purple. He twice muttered "Sit down boy", in a low booming voice but John carried on to the bitter end. After this embarrassing episode the headmaster insisted on vetting all speeches that Head Boys had prepared.

As I have said before, Brian Wiggins excelled at everything, both academic and sporting. The only subject in which he may have lacked ability was music. Possibly he was tone deaf. He was a keen golfer and introduced me and a few other school friends to the game. One Saturday Brian suggested we all go to Belfairs Golf Course for a game, my first experience of golf. It was most satisfying to hammer the ball down the fairway but rather less interesting once you got to the green. I was not very good. We arrived at one of the last holes and sized it up. Although the green was a relatively short distance away, it was hidden from view being high up on a hill. All you could see to indicate the location of the green was the top of the flag that marked the hole. I struck the ball and it flew high up in the air disappearing from sight over the top of the hill. The others did likewise. But whereas they found their balls with ease, mine had disappeared. We eventually found it in the hole - a "hole-in-one". This became a feat much talked about for a short time. I never played the game again, except once with Richard Coakes. I found it to be a great time-waster.

At 17 years old I was now legally allowed to drive a car. I started lessons in the summer of 1960 with the Royston School of Motoring in a Ford Anglia just like my father's. I had ten lessons paid for by my parents. They went well and in November 1960 I took my driving test passing first time. This was a better achievement than most of my friends. After this success my father sometimes generously allowed me to take his Ford Anglia out by myself.

Brian Wiggins had passed his test some months before me because he was that much older. His father was a builder and must have had money because he bought Brian an old Morris 10 Cambridge. One weekend in winter Brian drove Danny Linehan, Chris Harding, Doug May, and me up to Tottenham to watch an evening Spurs football match at White Hart Lane. We were excited to watch Danny Blanchflower, the Spurs captain. After the match we went for drinks before the long drive home.

Another weekend Brian arranged a Saturday night party at his house for his prefect friends, generously paid for by his father. Brian's house turned out to be a modest bungalow. His younger brother Colin was there. He was the complete antithesis of Brian. During the party Colin suddenly collapsed on the floor and started writhing frantically and foaming at the mouth. His brother Brian, who understood what was happening, jumped on top to restrain Colin and prevent him doing himself injury. Brian explained that his brother was an epileptic and was having a fit. I had never witnessed anything like that before and was quite shocked.

In that first term of my second year in the sixth form, Mr Drummond my maths teacher entered me for a London University Intercollegiate Scholarship in mathematics. This involved taking special exams set by the University. For three consecutive mornings I was shut away for three hours by myself in a small store room where I sat the examination papers under the supervision of the school secretary. I did not excel myself in these papers. I had done enough preparation but I was not mentally ready. However, I did do well enough to be called later for interviews in London at University College and Kings College.

The first interview at University College was a disaster. I was nervous and messed up all my answers. They asked me to sketch the graph of "xy = 0". My mind went blank and I started drawing hyperbolae. At the end of the interview they suggested I should think carefully about whether reading mathematics was the right choice for me. I was rather deflated. I was not surprised that I did not win a scholarship and was not offered a place at University College.

My second interview at Kings College went much better. The questions were less technical with questions about books I had recently read. I was offered a place. But I had not enjoyed my experiences with London University, so I turned the offer down in favour of a place at Bristol University which also had a strong Maths Department. They offered me a place without interview, dependent only on my achieving three passes at "A" Level. My maths teachers said that the more they wanted you the less they asked for.

It had occurred to me to apply to Oxbridge but I was not keen after my bad experience with London University. In any case, the Entrance Exams for Oxford and Cambridge required you in those days to spend a third year in the sixth form, because their high standard and breadth entailed a lot of additional preparation. I did not want to delay university for another year. I think that if my parents had pushed me I would have gone for it, but there was no history of university education in my family and so no real appreciation of what it meant to go to Oxbridge.

I was awarded the Upper Sixth Form prize for mathematics. For some ridiculous reason the book I chose for my prize was "The Magic of Alistair Crowley" by John Symonds which I previewed in a shop in Southend Arcade at Victoria Circus. This was a rather dull book about a perverse practitioner of witchcraft and was wholly inappropriate as a school prize. I never read it but still have it.

On the very last afternoon of the autumn term before breaking up for the Christmas holidays, a long-running tradition was the Prefects' Concert. This was attended by the whole school and was popular with its pantomime atmosphere. The concert always comprised a series of puerile sketches aimed at making fun of individual teachers. They were never referred to by name but were recognisable by the way they were mimicked. The sketches were sometimes cruel parodies. It was the one time that the boys could make fun of the teachers with impunity. It is surprising that our strict headmaster allowed this tradition to continue every year.

Our concert that year was most successful, with several of the teachers who had dared attend walking out in disgust. Others took it in good spirit. The main sketch writers were the head boy Brian Wiggins and Andrew Sargent. I had several small parts, the most successful being my impression of the school's deputy caretaker who had a strange walk, was always pushing a broom, and muttered "All right 'en" to the boys he passed.

During the Easter holidays of my last year at school I travelled in our blue Ford Anglia with my parents and Nick up to Harrogate in Yorkshire for a week. My father was attending a Civil Service conference in the town. We stayed in a comfortable hotel. It rained incessantly and I spent most of my time in the warm hotel lounge revising my physics work for the imminent 'A' Level exams.

My eventual success in school sports came rather late. In my early teens I had developed a keen interest in tennis and became reasonably good at the game. I played socially on the tennis courts at St Peter's Church and with friends on other public courts. The school tennis team was particularly strong. The captain was Bob Hammond, an Essex county player and the same age as me. He was partnered by a young 11 year old, David Lloyd [11], at that time the junior county champion.

---

[11] David Lloyd (born 3 Jan 1948) became a professional tennis player. He and his younger brother John Lloyd (born 27 Aug 1954) were two of the most successful British tennis players in the 1970s and 80s. David captained the British Davis Cup team and founded the David Lloyd Tennis Clubs and David Lloyd Leisure. He also coached Tim Henman. John

(David Lloyd's younger brother John joined the school only after I had left). This tennis pair was unbeatable. The second pair comprised my form mate and friend, Andrew Sargent, and another player whose name I have forgotten. This second player was better than me, but he was unreliable and often failed to turn up for matches against other schools. These matches were lost by default. So Bob Hammond sacked him and asked me if I would take his place. I was stunned to be asked and accepted gratefully. This was half-way through the summer term of my last year at school in the second year sixth. I did not excel myself in my tennis performances but maybe we won a few matches.

My efforts were enough to earn me tennis colours, a great honour. They were presented to me at morning assembly by the Headmaster at the end of the summer term. Unfortunately, I left the school almost immediately after and so never had the opportunity to wear the colours on my school blazer.

The end of my period at Southend High School in the summer of 1961 was a time of great euphoria. I did well in my exams, obtaining four distinctions with marks over 90% in "A" Level and "S" Level Pure Maths, and in "A" Level and "S" Level Applied Maths. In order to obtain a distinction, what is called an A* today, you had to achieve a mark of 75% or more. In my year, a particularly strong one, distinctions across all subjects were achieved at the rate of about 5%. It was not easy to achieve a distinction in those days. I also obtained a good pass in Physics which pleased me, but only a pass at "O" Level in "A" Level Chemistry. This failure of which I was slightly ashamed did not really surprise me since I loathed the subject.

But my failure in Chemistry was over-shadowed by my success in maths for which I was awarded a State Scholarship. In those days, anyone who obtained a place at

---

Lloyd was to reach one Grand Slam singles final and won three Grand Slam mixed doubles titles. He was the first husband of the former top woman player Chris Evert. John went on to become a BBC tennis commentator.

University was entitled to a Local Authority Scholarship that paid for all tuition fees and included a grant to cover basic subsistence. State Scholarships were awarded to those who had done well at "S" Level. It was more prestigious and the subsistence grant was slightly bigger. The award of a State Scholarship meant that my name would appear in gold letters on the Honours Board in the Great Hall. Most of my friends had similar successes. We were a very strong year. At the end of term we left school for the last time.

So what became of all my old school friends? I occasionally saw some of them during university vacations, but after that I never kept in touch. From snippets of information from recent school reunions I have gleaned the following.

Doug May did not go to university. He married and became a computer consultant now living in Exeter. Danny Linehan went on to study Botany at Birmingham University and stayed on to do an M.Sc. in brewery. He then worked for a brewery before emigrating to Canada where he played hockey for Canada's national team eventually becoming their captain and then the team coach, after which he became a professional golfer. Andrew Sargent was sent down from Bristol University at the end of his second year having failed his intermediate Law exams. He married a long-standing girl friend who lived by the spinney in Hobleythick Lane. He later divorced and remarried. He went on to work in Public Relations with his own company and now lives in South Wales. Chris Harding completed his degree in Civil Engineering at Bristol University, and after a successful career is now living somewhere in Suffolk. Peter Lee took a year out working with poor children in India before going to university. He became a dentist specialising in orthodontics and for a time had practices in Oxford and Chichester before settling in Bognor Regis. Brian Wiggins studied medicine at Corpus Christi College, Oxford, and after qualifying as a surgeon went on to become a GP in Moulton, Suffolk. After a fulfilling career he sadly died of cancer in January 2016. Dave Beadle qualified at Bristol University in Zoology and became a lecturer at the University of the South Bank. Dave Patrick gained a BSc and PhD from Bristol University in geology and went on to work in industry. George Farrow studied geology at Durham and became a university lecturer in Glasgow. Michael Pratt qualified as an engineer at Queen Mary College, London. For many years he worked for Rolls Royce Engines in Germany. He returned to the UK where he now lives in High Wycombe.

My other group of friends I saw more regularly, at least until I left university.

Dave Hall worked as a successful newspaper reporter and spent the whole of his career with the Southend & Basildon Standard. He married and lived with his wife in a former fisherman's cottage in Old Leigh where Lucienne and I once visited him a few years after we were married. I had the impression that he always remained "a bit of a lad". He divorced and as far as I am aware had no children. Many years later I learned that he was remarried to a French woman and living happily in Châteauroux in Indre, central France. I had a brief, rather unsatisfactory telephone conversation with him in which he said he had been inspired to marry a French woman after my own example.

Paul Archard went on to Teachers Training College specialising in music and French. In spite of his idiosyncrasies he was a faithful friend to me. Paul has had a difficult life. He actually married, which was a great surprise to his friends, yet he had always yearned to be a normal, happily married person. Richard Coakes once met Paul's wife whom he described as attractive and charming. But Paul's marriage had the misfortune to end in divorce. His wife initiated proceedings against him on the grounds of mental cruelty. Paul refused to accept divorce on account of his strong Catholic convictions. The case went to the High Court and made the national press. The divorce was granted and no doubt Paul was devastated, particularly as they had a small son. I lost contact early on with Paul and never actually saw him after his marriage. The next time I heard about Paul he was a supply teacher in Southend. But Paul did eventually remarry. He currently lives near the town centre in Southend with his second wife Sylvia. He contributes a lot to local musical life. In particular he still plays the organ at two churches, one of which is St Peter's in Eastbourne Grove, the Anglican Church that I used to attend.

Richard Coakes was my best friend in my teens. He left school at 16 to start work at the Phoenix insurance company in London, commuting every day. He stuck this for some years before making a career move to start a management training course with Trust House Forte. He learned the catering business from scratch by working in Little Chef restaurants around the country. He did just about every job in the business and made a great success of it, eventually rising to become Area Manager for the North East. During this time he married and had two children. He took early retirement from Trust House Forte and started helping his wife in her bakers shop and tea rooms, The Lunesdale Bakery, in Kirkby Lonsdale, a small market town in Cumbria.

Some years ago in the mid-90s I called in to see Richard on my way back from a walking holiday in the Lakes. He was busy in the shop but found time to have a quick beer with me in the pub opposite. He told me he was getting divorced and had joined the Masons, both of which shocked me. He also said that if I gave him notice he could arrange to take a day off from the bakery and go fell-walking with me. So in May 1998 prior to another of my trips to Cumbria I contacted Richard and arranged a day to go walking with him up Bow Fell. When I arrived at my hotel in Ambleside Richard rang to say that something had come up and he had to cancel, but he would meet me at my hotel that evening and we could go out together for a meal. I did the long Bow Fell walk by myself in sweltering heat and met Richard that evening. We ate in the Glass House Restaurant (later to feature in one of Gordon Ramsay's "Kitchen Nightmare" programmes). We had a good meal but a stilted conversation. The only things we had in common to talk about were the events of our youth. He was not quite the same person I had known so well 40 years earlier. We were no longer comfortable together. After his divorce, Richard remarried and retired to live with his new wife in the Yorkshire Dales.

This meeting with Richard convinced me of the folly of trying to rekindle the friendships of our youth after a long period of lost contact. It only tarnishes happy memories and leads to disappointment. On the other hand, I have found that contact by email is satisfactory.

And what of my erstwhile friend Terry Absalom who used to live at the top of Henley Crescent? I have painted him as a rather spiteful character, but I recently discovered a side of Terry I never foresaw. After a short spell of teaching in Southend, Terry became an airline pilot, a rather astonishing turn of events. He worked for various airlines and latterly flew for British Midland out of Heathrow and Luton. Terry was one of the founders of the Nancy Blackett Trust in which he served in various roles to become its first Vice-President in 2002. For those not familiar, *Nancy Blackett* was Arthur Ransome's favourite cruising yacht. He named her after his popular character, the adventurous leader of the Amazon Pirates in his "Swallows and Amazons" books. The aim of the Trust is to help young people up to the age of 24 who wish to develop an interest in sailing. In about the year 2000 Terry contracted motor neuron disease from which he eventually died on 19 August 2006.

It will be noted that a high proportion of my friends had failed marriages ending in divorce, certainly higher than the national average.

# Holiday Jobs

I had a series of holiday jobs in my final years at school and while at university. When I was fifteen I started work in the run-up to Christmas at the Royal Mail Sorting Office in Southend. I did this for several years. I could have applied in the normal way, but my father fixed it through a friend called Mr Spirling who was a manager in the Post Office, and probably a mason. Mr Spirling lived in Carlton Avenue and had an allotment next to his house. My father also had an allotment there for a short time, before moving to a bigger plot just off Rochford Road.

I enjoyed working in the Post Office at Christmas because there was a great festive atmosphere. For two years running I worked at the Southend sorting office in the parcels section. I stood in front of an enormous matrix of sacks supported by a metal structure into which we had to toss the parcels according to the delivery area. This was quite rough on the parcels and I can't say that anyone bothered at the noise of breaking glass. In fact this would generate loud cheers and applause, all in the festive spirit. Another year I worked on parcels delivery accompanying the driver in his van in the Leigh area. This was more interesting than sorting and had the prospect of tips.

One Christmas I noticed that one of the part-time workers in the sorting office was severely mentally handicapped. It turned out he was the brother of Mr Drummond, my maths teacher. Mr Drummond, who never married, shared a house with his brother whom he looked after. They lived close to Jones Memorial Park.

One year, during the Easter holidays, I found a job at the old Gaumont cinema in Southchurch Road. It had just closed down and was stripped out to be made into a warehouse. My job was to build and install Dexion storage shelving which was to fill the whole of the interior. I did this job with George Farrow, a close friend from school. There was a mini heatwave and it was sweltering work in the confined cinema building. We emerged at intervals to buy bottles of fizzy

Corona drinks from a nearby shop to quench our thirsts. Over a short period I sampled all the flavours, my favourites being Cream Soda and Dandelion & Burdock. The job lasted only a few days, finishing as soon as we had installed all the Dexion.

In the summer holidays I worked on the deckchairs. The headquarters was an office behind the bandstand on Southend Cliffs. Because the sea front was so long, about five miles, it was divided into several pitches called Southchurch, Southend Central, Southend West, Westcliff, Chalkwell, Cliffs, and Bandstand. Each pitch had an Inspector in charge supported by two or three students. At different times I worked on all these pitches. I became adept at folding and unfolding deckchairs, two at a time and at great speed. It cost one shilling (5p) to rent a deckchair for an hour, and half-a-crown (2 shillings and 6 pence) for all day. This was quite expensive especially for a large family. I got used to hearing all the old excuses like, "We're only sitting for a few minutes", or "I'm an old-age pensioner", but it made no difference. I never had any serious confrontations.

We were paid reasonable wages at an hourly rate with time-and-a-half on Sundays and double time on Bank Holidays. So we hoped it would not rain on these days or we would be sent home without pay. On a good week I could earn about £14, including tips.

The noisiest pitch was Southend Central because it included the "Golden Mile" that ran from the Pier to the Kursaal. The inspector in charge was an interesting character called Denny who had a large hole in his forehead as a result of an accident. He had been a double bass player with Cy Laurie's Jazz Band [12] in the 50s. He talked about only two

---

[12] Cy Laurie (1926-2002) was a clarinettist, bandleader and club owner. He was a central figure in the revivalist movement in British jazz during the 1950s leading to the so-called Trad Boom. Cy Laurie's interest in mysticism and philosophy led him to disappear suddenly from the scene in 1960, in pursuit of enlightenment in India with the Maharishi Mahesh Yogi, rather like the Beatles some years later. He did not return to playing until the late 1960s when his career had a revival.

things: his jazz experiences with fellow musicians like Humphrey Lyttelton, Chris Barber and Ken Colyer, and his female conquests many of whom still seemed to be current. I once went to see him playing with a local band in a pub in Leigh. He was very good.

A particularly entertaining pitch was Southchurch because the inspector-in-charge, one Maurice Alder, a drop-out from Southend High School, ran his pitch like a Gestapo commandant both in speech and style. He treated difficult deckchair customers like prisoners-of-war and was a source of great amusement to us workers, including one "Cuts" Morris who was another drop-out. Cuts travelled to work every day by bus from his home in Hawkwell, some 10 miles away.

On the deckchairs each of us had a one hour lunch break, staggered so there was always someone left to cover the pitch. I usually rode home on my bike for lunch, cycling the 5 or 6 miles at top speed on my trusted Dayton racing bike. One day I returned from lunch to my pitch at Southend West, and was just leaving my bike parked against the railings in the seafront car park when the driver of a big Jaguar that was parked close-by started his car while it was still in gear. The car leapt forward and crashed into my bike bending the front wheel. I was horrified. The driver was apologetic and gave me money for the repair.

Another holiday job was car park attendant at Southend Airport. I found this job by a circuitous route. My friend Peter Lee was friendly with Faye Hockey who in turn rather fancied a boy called Terry Gregson. In fact, at that time Faye was besotted with my friend Richard Coakes, but he was not interested. Faye eventually married Terry Gregson. His father was a manager at Southend Airport and it was he who found the holiday jobs for Peter and me. Southend Airport was expanding rapidly with regular flights to northern France and the Channel Islands. Peter and I worked shifts in the car park to cover the full week during the daylight hours when there were flights.

Plane spotting at Southend Airport was becoming popular, so the management used their initiative to place deck chairs in a special cordoned-off viewing area on top of a flat roof over the café and charged people to sit there. It became an additional responsibility for us to issue the tickets for the deckchairs. The job was really easy and soon became tedious. This was in July or August 1960 when I was just 17. I can pin-point it because the new record "Apache" by the Shadows had just been released.

That summer my admirer, the aforementioned Susan Pillsworth, had a German pen-friend staying with her called Freda. Although she was rather plain I found myself smitten with her. Possibly it was something to do with foreigners. I only ever met her in the company of her host Susan, usually by chance in Hobleythick Lane, or at the house of Jane Archard who was a friend of Susan and who was also the sister of my friend Paul Archard. One day Susan announced at short notice that her friend Freda would be leaving to return home to Germany. I was shocked and hastily arranged for Peter Lee to take my afternoon shift at the airport so that I would be free. Susan, Freda and I took a bus to Hockley Woods where we spent the afternoon walking in the sunshine. After that I pined for Freda but never saw her again.

I graduated from airport car-park attendant to become an airport cleaner and baggage handler. This was a more interesting job. The cleaning part involved swabbing down the labyrinth of airport corridors, first with a soapy mop then with a clean water mop. This was done first thing in the morning from 7 a.m. before the first flights. The rest of the job entailed meeting the planes as they came in, the biggest being the Elizabethans. We took the plane's bar box into safe keeping, unloaded the luggage onto trolleys that we wheeled to the customs area, and most important of all, waited around the exit of the customs area hoping that some elderly traveller would ask for their case to be carried to their car with the prospect of a tip. It always seemed to be those who looked like they could least afford it who gave the biggest tips.

I enjoyed having the free run of the airport, but this did not last long. For reasons I never understood the foreman took a dislike to me and accused me of shirking. This was really not true. I think he considered me too educated. But it meant I was finished at the airport. I went back to deckchairs on the beach.

By the time I finished at Southend High School in the summer of 1961, I and all my friends had secured university places. University term did not start until October so we had several weeks to earn money on a holiday job. But that year jobs seemed to be scarce.

For a short period my friend Peter Lee and I became ice-cream salesmen. We reported to the depot in Southend behind the Ritz cinema and proudly presented our new driving licenses. We were each given a van with a tank full of petrol. We had to buy the ice-cream that was supplied in large stainless steel containers. The deal was that we would make our profit from selling the ice-cream.

I was assigned a pitch some 15 miles away in the new town of Basildon. I was most unsuccessful. I did not attract many customers and I dropped half the ice-cream on the floor in trying to ladle it onto the cones. Peter had a similar experience. I met up with Peter at the depot that evening of the first day and we both decided to throw it in. The owner of the business was angry and would not refund us on the unsold ice-cream. My friend Pete Wyss, on the other hand, landed himself an ice-cream pitch at the top of pier hill, a prime site for trippers and holidaymakers. Pete earned a lot of holiday money and became a dab hand at ladling out the ice-cream into cones making it hollow in the middle.

In some desperation Peter Lee, Andrew Sargent, Danny Linehan and I presented ourselves at an employment agency in Southend. The man wanted to know what qualifications we had. We were able to reel off an impressive list of 'O', 'A', 'S' levels and scholarships. The man said he had nothing suitable in Southend for young men of our calibre. He rang up a colleague in London who said that maybe there were some "temping" jobs in the City. Our train fares were paid for us to visit the man in London at his bureau close to Liverpool Street station. We made the journey and were each successful in landing temporary jobs in different offices.

It was exciting travelling up to London. One lunch hour I met up with my father who was proud to take me to his office in the Air Ministry and introduce me to his colleagues. Once or twice I met up with my friend Richard Coakes after work for drinks. Having left school after "O" Levels, Richard had already been working for two years. He had a steady job in the Phoenix insurance company in Fleet Street. We would meet in a pub called The Clachan in Mitre Court opposite his place of work. After these sessions we got home quite late making it hard for me to get up for work the next morning in time for the 7:30 train.

I felt quite important travelling in the train to London with all the other commuters. As for the actual work I did, I have just a dim recollection of shuffling papers and making tea for people in a busy insurance office. The four of us found our jobs boring. It was also tiring getting up early and arriving home late. News came through that there were now job vacancies on Southend deckchairs. We handed in our notice and ceased to be city gents after only three weeks in post.

I obtained my job on the deck chairs and was assigned to the Cliffs pitch. This turned out to be my downfall. The Cliffs pitch was hard work because it covered half a mile of cliff gardens with lots of walking up and down steps. Also, it was close to Headquarters so the Inspector for this pitch was the Chief Inspector himself, an elderly and bad-tempered tyrant. Once, I wandered off my pitch probably chatting to a friend I had met. I saw the Chief Inspector in the distance and quickly returned to my duties. But he had seen me and when I returned from my shift he gave me a serious telling off. He said that when he spotted me I ran away and "he couldn't see my arse for dust". This was a wonderful expression that I committed to memory.

I made good friends with a fellow student worker on my pitch called Dennis O'Leary, a lanky bearded character who was very laid back. For some reason the Chief Inspector liked Dennis whereas he did not like me. He had given Dennis the coveted job of operating the Cliffs Lift. This was a single unit that could carry a maximum of 30 passengers up the cliffs from the foreshore to the bandstand and gardens above. The length of the track was 130 feet. All Dennis had to do was collect the money from the passengers, close the sliding doors, push the lever to make the lift start, wait until it engaged with the buffers at the other end, then open the doors to let the passengers out. What you might call a cushy number.

Of course, to ease my aching legs Dennis would let me ride on board to save me climbing the steps to the top of my pitch. This was strictly against the rules. One day I asked Dennis if I could have a go at driving the lift and he agreed wholeheartedly. We took on a full load of passengers and set off. Dennis said that whatever happened I must not let go of the drive lever. Of course I did, and the lift stopped mid-way. Alarm bells started ringing and the safety circuits cut all power to the electric motor. The lift was stuck half-way up the track. The people in the lift started screaming and shouting about cable-car disasters in Austria. Dennis and I tried to calm them down.

After what seemed an eternity the lift started moving slowly upwards. It was being cranked up by hand. We arrived at the top and the doors were opened. There stood the Chief Inspector and the Bandstand Manager holding fire axes. The spineless passengers bleated about never coming to Southend again saying they would be making official complaints. That was the end for me and I was sacked on the spot. I never worked as a deckchair attendant again. Dennis O'Leary was let off with a reprimand and resumed his duties forthwith.

With the end of my career as a deckchair attendant my main line in holiday jobs became "on the dust". My mother had seen an advert in the local paper for temporary dustmen. I applied and got the job. It was really hard work but was enlivened by a great team spirit in the gang. This comprised a lead dustman called the ganger, four ordinary dustmen, and the driver of the dust cart. We worked five days a week, starting at 7 a.m. and finishing at 4:30 p.m. with breaks for breakfast and lunch. The best break of the day was always breakfast at about 9 a.m. We would sit in a steamy workmen's café where I usually had a sausage sandwich, a large mug of tea, and a roll-up. What bliss. Over the course of two summers and Easters while at University I worked on just about every round from Chalkwell to Thorpe Bay and Southend to Eastwood. I even emptied the bin at my own house.

Every house had a dustbin which was collected weekly from wherever the owner of the house kept it, usually in their sideway outside the back door. There was rarely any problem of access. There was no recycling so everything went into the one dustbin. Each dustman had a lightweight plastic skip. You emptied the contents of the dustbin into the skip, and carried the skip to the dust cart for emptying. Sometimes a dustbin would be so full that it was too much for one skip. In that case you had to carry the dustbin out to empty it into the dust cart, then carry it back and collect your skip. This entailed an extra journey.

I soon learned the knack of lifting a heavy bin onto my shoulder and carrying it. If it was really heavy you would shout to a mate to "gimme a lift". The heaviest bins were made of cast iron. In the standard carrying technique the bin was balanced on the shoulder using only one hand, leaving the other free to open and close gates and accept tips.

The ganger knew how to distinguish the "good" houses from the "bad" houses. Good houses had nice clean bins and often left a small tip on the dustbin lid. Some house owners, usually elderly ladies, wrapped up every item of rubbish individually in neat newspaper parcels before putting it in the bin. In other houses the refuse was just thrown straight in. This was alright if the bin was cleaned regularly, but in bad houses the bins stank in hot weather and were full of maggots which we made a point of spilling down the garden path.

Quite often a house owner would ask if we could take away additional rubbish. The procedure in these circumstances was to fetch the ganger. He would survey the rubbish and scratch his head, saying how we were not really supposed to do this but maybe he could help out as a special favour. This technique usually elicited a generous tip. All tips were pooled and shared out at the end of the week, amounting in a good week to about £2 each. This was in addition to my basic weekly wage of about £10.

On one occasion the ganger was called in by a householder who asked him if he could get rid of a grand piano. The ganger looked at it and said, "Do you think my name's Tarzan?" This caused great mirth. A satisfactory arrangement was reached by which the instrument was disposed of after normal working hours.

In those days central heating was rare and most houses had a coal fire of some sort. Coal fires generated cinders that usually finished up in the dustbin. We had instructions not to collect any bin that we suspected had hot cinders in it. It once happened that hot cinders slipped through the net and finished up in the back of the dust cart, generating a pocket of fire deep inside the compressed rubbish. The first we knew about it was when smoke began pouring out of the back of the lorry and the metal walls started to feel hot. This was clearly a dangerous situation. The first reaction of the ganger was to have the driver discharge the whole contents of the lorry onto the road. This would have been a very entertaining spectacle, but it would have fallen to us to clear up all the mess. So instead the ganger instructed the driver to take the lorry to the tip. It departed down the road at high speed leaving behind a trail of acrid smoke. We waited half-an-hour for its return before we could resume our work.

On another occasion a new man joined our gang. The ganger became suspicious that he was pocketing the tips instead of pooling them, so he laid a trap. He made sure that it fell to this man to empty the bin of an old lady who was known to leave a tip every week, without exception. This caught him out and he was "sent to Coventry". This may seem rather mild but it was in fact humiliating for him and a just punishment.

I got on well with the men in all the gangs I worked with. I think they respected the fact that I never complained about the toughness of the job and always did my bit. I was very fit by the end of it.

In one gang, a short brazen Scotsman known as Jock took a special interest in my well-being and acted as a sort of minder. In my last week he invited me round to his house one evening after work. He lived with his wife on the ground floor of a small terraced house, since demolished, somewhere at the back of Tylers Avenue near Southend town centre. They had no children. To my embarrassment he and his wife had gone to the trouble of preparing a meal for me. This came as a surprise as I had already eaten and I could not do it justice. We later went out for a drink at The Railway. Jock initiated me into the strong Whitbread brew called Final Selection that came in small nip-sized bottles. Jock was asking me about my education of which he had none, and what I planned to do. I somehow felt ashamed of my privileged life.

# University Years 1 and 2 (1961-1963)

These days people go to university to improve their job prospects. In the 1960s they went to become better people, in the sense of being better educated. Jobs were not a big issue because there were plenty of them. The unemployment rate was at an all-time low of about 2%.

In the autumn of 1961 I began my studies at Bristol University reading Mathematics. My State Scholarship provided me with an annual grant of £240 per year, slightly more than the standard Local Authority grant of about £210. This grant was for my maintenance to cover food, accommodation and books, all the tuition fees being paid for separately by the authorities. My grant, together with holiday earnings, was enough for me to live on. When I eventually left university I had no debts.

Three of my school friends had also accepted places at Bristol: Andrew Sargent to read Law, Dave Patrick to read Geology, and Chris Harding to read Mechanical Engineering.

It was Sunday, 1 October 1961. The four of us travelled together by train from Southend to Bristol via Paddington. I carried a single, heavy suitcase. Curiously, in that era, parents did not accompany their children to settle in at University. There was considerable excitement and some trepidation at starting our first term. We were going to be away from home in a completely new environment for the next 10 weeks. I had never even visited Bristol and so had no idea what was in store for me. I did not appreciate that 11 Henley Crescent would never be my "home" again.

We arrived at Bristol Temple Meads station late in the afternoon. The university accommodation department had placed all four of us in the same digs. We took a taxi to St Johns Road, Clifton, to meet our landlady. Our digs were on the top floor of a large Victorian house in a residential area some 15 minutes walk from the University. We soon nick-named our landlady Mrs Q***.

It was about 5 o'clock and Mrs Q had not prepared a meal for us, because she did not do evening meals on Sundays.

Elated by our new freedom we decided to eat out and took the opportunity to explore the city. We walked in the evening sunshine down Whiteladies Road, past the Students Union then housed in the Victoria Rooms, round The Triangle, past the main University buildings with its imposing Wills Tower, then down the steep hill of Park Street towards the Cathedral, city centre and docks. Everything was exciting and on a much grander scale than Southend.

We found a curry house called the Koh-i-Noor in a narrow side street called Denmark Street close to the city centre. I ate a Meat Vindaloo that was extremely hot and satisfying. Remarkably, this was to be the venue for many future festivities. The restaurant comprised three floors, with the food transported to the higher floors by a miniature, hand-operated lift contraption. The Koh-i-Noor was closed several years later by the Public Health Authority as a result of unsanitary conditions in the kitchens including rat and cockroach infestations, but their curries were always the best in town.

The following week was Freshers Week when we were introduced to the Students Union, all the student societies, and the layout of the University. I was rather overwhelmed.

We then began our lecture courses in earnest. I soon realised that university mathematics was on a different plane to school mathematics. In pure maths we started to address such fundamental questions as continuity, limit processes, and the meaning of number. Nothing could be assumed. Everything had to be derived from first principles or axioms. The logical arguments of proof employed were new to me and rather subtle, and the degree of rigour came as a great shock. I was astonished at the power of pure thought. Applied maths was similarly rigorous but at least I could understand the problems it was trying to address. What I had studied at school seemed childish in comparison. I began to realise that not all maths problems had solutions. I also began to appreciate the power of generalisation.

There seemed to be no compulsion to attend lectures and no requirement to take notes. The lectures were generally delivered rapidly and with little opportunity for questions, even if we dared ask any. It was all down to us to sink or swim. I am not even sure that we had any tutorials.

The mathematics department at Bristol had the distinction of having not one, but two, FRS's, which was unusual outside Oxbridge. The first was Hans Heilbronn, a former student of the great Landau at Göttingen and an expert on number theory. He had fled Germany for Britain in 1933 and eventually became Head of Department at Bristol and Professor of Pure Mathematics. The second FRS was Leslie Howarth who had studied under Goldstein and von Karman. His field was fluid dynamics and turbulence. He was the Henry Overton Wills Professor of Applied Mathematics and later Head of Department when Heilbronn left for Canada in 1964. These two men attracted exceedingly high quality research staff. Sadly, as an ignorant student I never appreciated their stature in the field of mathematics.

In the first year we had a multitude of courses including Mathematical Methods, Functions of a Complex Variable, Analysis, Number Theory, Projective Geometry, Electricity and Magnetism, Newtonian Mechanics, Potential Flow and Viscous Flow. Some of the pure mathematics I found particularly obscure. Having since studied the history of mathematics I would now feel much more at home because I can see the origin of the ideas. A History of Mathematics course would have been useful in the first year to tie everything together.

The applied maths was less obscure because it was concerned with modelling physical phenomena about which I had some prior notion. In retrospect it could have been made far more interesting and relevant if we had been encouraged to view some of the phenomena that the mathematics was trying to model. For example, we could have gone to watch the Severn Bore when later we studied solitary waves, or simple experiments could have been set up in the adjacent engineering labs. As it was, the applied maths courses were too academic and seemed divorced from the physical phenomena they were trying to model and explain.

Throughout my undergraduate career I never once had an afternoon lecture. They were always in the morning. We had 8 lectures a week each lasting one hour. Typically, we had lectures at 10 and 12 o'clock on Mondays, Wednesdays and Fridays, and at 11 o'clock on Tuesdays and Thursdays. I sometimes had difficulty getting up in time for the 10 o'clock lectures. The afternoons were supposed to be for private study but I spent most of them playing snooker.

The standard of lecturing at Bristol was generally good but fell far short of what is expected today. A large rotating blackboard was used but no notes were provided to support the lectures. We had to scribble down our own notes and try to make sense of it all afterwards. I still have some of my notebooks. Several are surprisingly neat, written in my acquired Gothic script, whilst others are unintelligible.

In addition to the two FRS professors, the academic staff included the usual collection of eccentrics and misfits. There was Dr Bill Chester (later Professor), a bad-tempered Scot, who was said to have a chip on his shoulder because he had been passed over for an FRS.

Then there was the pure mathematician Dr (later Professor) Marstrand who worked on "very difficult problems". Dr Marstrand broke a leg while surfing on the Severn Bore. He walked around with a crutch while his leg was healing. Then one day he arrived for his lecture balancing precariously on two crutches with neither foot touching the ground. He explained that he wanted to be prepared in case he was unlucky enough to break the other leg as well. Forty years later I was staying at the Scafell Hotel in Borrowdale, Cumbria, on a walking holiday when I heard a loud booming voice that I recognised in the restaurant. It was an elderly Professor Marstrand. I introduced myself to him and of course he had no recollection of me. He told me he had been a fell runner in his youth and done his knees in. He had knee and hip replacements but could no longer climb the fells.

There was also Dr (later Professor) Phillip Drazin, a generous and approachable man who was an expert on the stability of fluid flows and their transition to turbulence. He wrote several seminal papers and text books. He served for many years as the mathematical consultant for the Oxford English Dictionary.

Another lecturer of note was Dr Derek Moore, a fluid dynamicist who wrote rapidly on the blackboard in a meticulous and perfectly aligned script. His completed blackboard looked rather like a page out of one of Wainwright's "Pictorial Guides" except of course it was all mathematical. Dr Moore was a "trad jazz" fanatic and played the clarinet semi-professionally. He was later to become a professor at Imperial College where he gained his FRS.

I recall two other applied maths lecturers, Dr Patterson and Dr Powell, who were both good lecturers and more involved with teaching than research.

Lastly I should mention Dr Mike Rogers, another fluid dynamicist who had the vision to anticipate the overwhelming importance of computers for solving mathematical problems. Numerical calculations in those days were still done by hand, often taking days or even weeks, and numerical weather-forecasting was unheard of. Dr Rogers founded one of the first departments of Computer Science that started life as a sub-division of the Maths Department. He later became Professor of Computer Science and eventually took over as Head of the Maths Department. Dr Rogers was later to become my PhD supervisor. Coincidentally, many years later he was a panel member on two of my Individual Merit promotion boards.

Professor Rogers died in 2003. I believe that all the other mathematicians named above have also died, with the exception of Professor Marstrand who in 2018 was still listed as an Emeritus Professor at Bristol.

I and my three school friends remained close because we shared the same digs, but we made non-overlapping sets of friends through the different courses we were following. In those first weeks of that first term at University we fell into our individual routines that were to mark the pattern for each of us over the next three years.

Andrew Sargent joined the University Footlights and became deeply involved with the arty people and the development of his latent impresario talents. This became so consuming for Andrew that he neglected his Law studies, had to re-sit his exams at the end of the first year, and at the end of his second year was sent down. He then worked for a short period putting on shows at Bristol's Coulston Hall.

Dave Patrick, studying Geology, was a swot and never deviated from his objective of obtaining a First, which he achieved, followed by a PhD.

Chris Harding made close friends in the engineering crowd and became heavily involved in university hockey. He was to obtain an upper second in Engineering. The engineering courses were intensive with lectures and practical work that filled the whole week, the complete opposite of my course. Consequently, nobody saw very much of Chris Harding because he was always working.

As for me, I hovered from one interest to another without commitment to anything. I joined the speleology society for a short period and scared myself silly clambering through narrow subterranean tunnels and "chimneys" with Dave Patrick in the Mendip Hills, sometimes up to my waist in freezing water. I joined the Footlights with Andrew Sargent and confirmed to myself that I had no talent for acting, but I nevertheless appeared in walk-on parts in a Student Revue, building on my earlier experience as an impersonator of the school caretaker. I played tennis but was not good enough to play for the University. I spent a lot of time with some dissolute mathematicians who, rather than study, were more interested in snooker, gambling, and Art Blakey's Jazz Messengers.

Meanwhile, things were not going too well at our digs. Half-way through the term Andrew Sargent received news that his parents were in the process of divorcing. Andrew took it badly and became severely depressed. One evening he fell on the floor in Mrs Q's kitchen and prayed that his parents would get back together. It seems that they did and Andrew's belief in the Almighty was renewed, but only briefly, because after a few months I believe his parents separated permanently.

There was also trouble on the Mrs Q front. We found her to be rather critical and uncompromising. We fell into a routine with evening meals, gathering in the small dining room to be served our two courses. Mrs Q's son worked for Brains, the West Country pie manufacturers. One of their principal selling lines was "Mr Brain's Faggots" which were served up regularly each week. They were quite tasty and are still going strong to this day.

We had regular servings of Mrs Q's home-made shepherds pie. I wrote to a friend in Southend, one "Cuts" Morris, a former compatriot on the deckchairs, mentioning in passing the poor standard of some of Mrs Q's cooking. Cuts replied with a long letter, sympathising with us for having to endure pie "containing shepherds of dubious origin". I put the letter in my bedside drawer.

Then Mrs Q started acting coldly towards us and, in particular, would not talk to me. She would serve our dinner banging the plates down on the table. At the end of our first term, the week before we were due to return home for Christmas, one of us asked Mrs Q about the arrangements for next term. She replied that we need not bother to come back. Then it all came out tearfully that she had read my letter while prying through our drawers. I was not sorry to leave Mrs Q. She should not have been so nosey.

The university accommodation office fixed us up with new digs for the start of the next term after Christmas. They were below Kingsdown Parade in Somerset Street on the steep escarpment overlooking Bristol City Centre. It was one of a number of narrow, cobbled streets lined on both sides with Georgian and Victorian terraced housing that looked as though they were due to be condemned, half of them already derelict. This area of Kingsdown is now a conservation area described as follows. "The streets follow the contours of the slopes, crossed by narrow paved lanes protected by traditional bollards. The buildings are now Grade II listed and of historic and architectural interest. Somerset Street is quiet with a charmingly diverse arrangement of terraced houses of different periods." I hardly recognise this quaint description.

After Christmas, Andrew Sargent, Dave Patrick and I arrived at our new digs on a cold and grey winter's afternoon. Chris Harding had made alternative arrangements for accommodation with an engineering friend. The feeling of the area was Dickensian dilapidation with its oily flagstone pavements and old-fashioned yellow street lighting. It was a Sunday so the landlady had not prepared an evening meal for us. We left our cases in the hallway and went straight out for a curry in nearby Stokes Croft.

We returned to our digs later that evening to find that our cases had disappeared. They were no longer in the hallway and not in our bedrooms. A student called Carl appeared swaggering from the sitting room. Andrew Sargent knew Carl vaguely because he was a fellow law student. Carl had been educated at Marlborough School. We were to discover that he was a pompous, offensive character, a bully who loved playing cruel practical jokes. A lot of the time he was drunk. It was clear that Carl ruled the roost in these digs.

Carl said he had no idea where our cases were but seemed most amused at our plight. After playing us along for about an hour he said that maybe we would find our cases next door. We went outside to find that "next door" was a derelict house in a poor state of repair. We went in through the open door but could see nothing in the pitch darkness. We returned and managed to borrow a torch from the landlord. Carl followed us back into the derelict house and watched with glee as we stumbled through the empty rooms over rubbish and broken floorboards. We finally found our cases in a back room. This was the first of many annoying incidents with Carl.

Our landlady was a fat Austrian woman with an English husband who was quiet and gentle. By coincidence their name was Brain, but no relation to the erstwhile pie manufacturer. She came from the old school of cooking. She had an enormous soup pot continuously simmering into which all the leftovers went. So the first course at dinner was always soup, which changed slightly from day to day as new items were added to the pot. This was a good example of the smoothing process achieved by temporal averaging.

The meals were wholesome and filling but this did not stop Carl complaining. One morning at breakfast he declared that the eggs were too greasy. He took his plate and emptied the contents out of the window onto the bonnet of Mr Brain's old car. Pleased with the effect he took our plates from under our noses and emptied those onto the car as well. Mr Brain was uncomplaining. I think that both he and his wife were a bit scared of Carl.

One of the other students in the house was Richard who was studying history. Richard was quiet, well-mannered and a Christian. He was always working hard at his studies. He was Mrs Brain's favourite and she called him her model student. Carl despised Richard and thought he was a wimp, so he did his best to get Richard into trouble.

One meal time we were all sitting down to our helping of meat and two veg. The meat was a bit tough and the vegetables a bit soggy so none of us finished our plates, except of course Richard who always ate up everything. So in a mad flourish Carl took each of our plates and emptied all the slops onto Richard's. Mrs Brain came in to clear away and Carl made a big speech about how badly brought up Richard was for leaving his food. Richard left the room in tears. It was all very cruel.

Carl really was a disgusting character, the nearest I have ever met to Flashman in real life. He would return home drunk most evenings and was usually sick. We would hear him throwing up in the toilet and moaning, "Ooh, never again, never again".

One evening shortly after his return after midnight we heard a lot of scuffling and banging overhead. A few days later we began to detect the smell of paraffin in the house and someone noticed stains on the ceilings. It turned out that Carl had collected a quantity of oil lamps used by workmen at night to indicate road works. He had deposited all these lamps in the loft and they were leaking. Mrs Brain reported Carl to the university authorities, but subsequent events were to overtake any response from them.

Carl went out the following Saturday to the Rummer, a popular pub in Bristol town centre frequented by students and medics. It specialised in serving "schooners" of Harvey's sherry. He got totally drunk and announced to everyone he met that there was a party that evening at his address and all were invited. At around 11 p.m. when the pubs closed a crowd of about 30 drunken students turned up at our digs, headed by Carl who let them in. They were disappointed not to find a party. There was a great commotion and Mr Brain summoned up the courage to throw them all out, which he eventually succeeded in doing. However, the next morning Mrs Brain discovered that a silver plate of great sentimental value was missing from the hall table. The only explanation was that the plate had been stolen the night before by one of the revellers. She was distraught and called the police. Carl denied all knowledge and there was not much the police could do.

This marked the end of our short stay in these digs. Mr and Mrs Brain gave us a week's notice to leave, all of us except Richard. We had been there only three weeks, but they were eventful three weeks. We urgently needed to find alternative accommodation.

The university accommodation service again came to our aid. The three of us were split up. Andrew Sargent moved to digs in a run-down slum area of Bristol known as Montpelier. Dave Patrick moved to digs in Bishopston, a moderately attractive residential area, where he was to stay for the rest of his time as an undergraduate. I also moved to Bishopston to digs in Morley Square.

I saw less of my old school friend Dave Patrick but still met up occasionally with Chris Harding. I saw a lot of Andrew Sargent in his new digs and became friendly with his fellow inmate, Robert Perry. Robert, whom we renamed Robbers, was a cheeky character and a student vet in proud ownership of an old Austin. We all got on well together.

My new landlady was quite different from the previous two. I shall call her Mrs T. Her husband had a clerical job at Avon docks. They had one daughter who lived at home and worked for an insurance company in the centre of town. The house itself was a small Victorian semi-detached that Mrs T kept clean and tidy. She considered herself to be middle class and going up in the world, but I think she felt held back by her husband's dockland job. She took in students only on sufferance because she temporarily needed the extra money. She had just two students, myself and a medic called Robert. He was Welsh and I never understood a single word he said. We shared a downstairs room where we ate and slept. I did not see much of Robert because his course was quite intensive.

Of all my various accommodations whilst at Bristol, these digs in Bishopston were the furthest from the University, a distance of over two miles. Every day I walked to and from Queens Building in all weathers to attend my lectures. I never used public transport and I always wore the same summer clothes.

Two incidents will serve to sum up Mrs T. When I first joined the household she made a long speech about how she wanted me to be part of the family and how I could use the house and garden as if they were my own. But she did not mean this. She actually wanted us to stay in our room. This became clear when one sunny afternoon I wandered into the garden planning to read a text book. It happened that Mrs T was sitting in a deckchair sunning herself. When she saw me she flew into a tantrum and accused me of being bad-mannered and abusing her hospitality.

On another occasion I was coming down the stairs one evening, having been to the toilet, when for some inexplicable reason I became disoriented. Instead of opening the door on the left to the student room I opened the door on the right to their drawing room to behold Mr and Mrs T and their daughter sitting on the sofa watching television. I suppose it could have been worse. I was shocked at my mistake and mumbled an apology. Mrs T was furious and said that my behaviour was outrageous and I was imposing on their privacy.

At the end of spring term I returned home for the Easter break. I attended Speech Day at my old school, at 7:30 pm on

Thursday 22 Mar 1962, to collect my prize and tell my maths teachers how I was getting on. The visit to the school was uneventful except I noticed that my name on the Honours Board in the Great Hall was incorrectly recorded as A. J. Parsons. I mentioned this to a teacher who promised to look into it. I noticed a second error. The official school programme incorrectly recorded that I had passed Chemistry at "A" Level, but I said nothing about this.

On returning to the school some months later I noted that the error on the Honours Board was still not corrected. I mentioned this to my father who complained to the Headmaster, Mr Price, at the next Masonic meeting. The correction was made immediately.

Meanwhile I was greatly enjoying student life. I did not work too hard and had plenty of spare time. In the summer term I entered into the spirit of rag week by trying to break the world handshaking record. I went to the town centre dressed in some silly garb comprising a white sheet. My friends, headed by Andrew Sargent, herded crowds of shoppers into lines that I could quickly walk down, shaking hands as I went. I failed to beat the world record by a large margin.

I then took part in the student procession on one of the floats. I was looking forward to the big event in the evening. Acker Bilk was playing at the student dance in the Victoria Rooms, the Students Union [13]. On returning to my digs in

---

[13] The Victoria Rooms is a complex of buildings that first opened in May 1842 as an Assembly Rooms. For many years it was the most important cultural centre in the West of England. In 1848 the renowned 19th century soprano, Jenny Lind, appeared on its platform. In 1852 Charles Dickens delighted a large audience with a selection of readings. From 1873 the venue housed a large organ originally built for the Royal Panopticon of Arts and Science in London's Leicester Square, from where it was removed to St Paul's Cathedral and then to the Victoria Rooms. In 1899 it was replaced with an electric organ built by Messrs Norman & Beard, the inaugural concert given by the famous organist E H Lemare on 31 October 1900. The organ was destroyed by fire in 1934 along with the original hall. The restored building became the University of Bristol Students Union until 1964 when a new union building was built close by. The Victoria Rooms

Bishopston to change clothes I began to feel unwell. I was sick and shivering and went straight to bed with flu. I was devastated to miss Acker Bilk.

Towards the end of term we had weekends away for a couple of reunions with other school friends who were spread around the country at different universities. For both reunions Robbers transported us in his old car. The first was in Oxford where Brian Wiggins had won an Open Exhibition to Corpus Christi to study medicine. The event comprised large amounts of alcohol and curry followed by an unsatisfactory sleep in the University Parks. The proctors chased us out of the Parks early in the morning. We returned to Brian's College Rooms and tried to sleep on the floor. Unable to sleep we got up early and went for a drive in Robbers' car. We stopped on a big roundabout on the Oxford ring road. We observed one of those new-fangled traffic strips stretched across the road that counted cars. We spent the next hour laboriously stamping on the strip and clocking up about 1000 cars between 7 and 8 a.m. on a deserted Sunday morning, hence influencing Oxford's future traffic policy. We found this hilarious. Later that morning we went punting on the Isis before driving lazily back to Bristol.

The second reunion was in Birmingham where Danny Linehan was studying Botany. I was struck by the fact that the bar in the Student's Union at Birmingham did not serve alcohol. This was the result of an edict from Cadburys, the principal benefactor of the University, who were Quakers and teetotal. However, the University had found ways to circumvent this rule and our visit was not disappointing. I was not to know of my future close association with Birmingham University.

On return to Bristol, Andrew Sargent and I spent a day watching the tennis tournament at Bristol Lawn Tennis Club in

---

then became a Conference and Exhibition Centre until 1996, when it became the home of the University Department of Music. I devote this amount of space to the Victoria Rooms because it was where I met Lucienne in February 1964.

nearby Redlands. This was one of the warm-ups for the top players prior to Wimbledon. In an intimate atmosphere we stood in warm sunshine round the edge of the courts watching such stars as Margaret Smith and Chuck McKinley, who both went on to win their respective Wimbledon singles titles the following year. We also watched Ramanathan Krishnan, the legendary "touch player", who never smashed the ball and won all his points by gentle and accurate placement.

End-of-year exams were fast approaching. I did not excel myself, but fortunately they did not count towards my degree. Mrs T announced that she would be happy to see the back of me. The feeling was mutual. At this point I reflected that some students seem to get on well with their landladies, some learn how to get on, and others never acquire the art. I thought ruefully that I fell into the last category. Nevertheless, I had stayed in these digs for a total of one and a half terms - a record for me.

Andrew Sargent and Robbers had also had enough of their digs in Montpelier. We had all had our fill of landladies and resolved that next term, the start of our second year, we would find a self-contained flat for the three of us.

At the end of that first year Andrew Sargent who, like his parents, was a political animal, got himself voted in the student elections to the position of Director of Entertainments. In this elevated position he was responsible for organising all the dances and events for the forthcoming year at the Students Union based in the Victoria Rooms. This suited his impresario leanings admirably. He asked me if I would like to be his Treasurer of Entertainments. I agreed and prepared myself to be interviewed by the Students Union Committee. Andrew primed me with questions I would be asked, like whether student events should make a profit. I gave the right answers and got the job.

That summer I returned to Southend and worked hard at holiday jobs to earn some money. Towards the end of that summer I decided to sell my precious Dayton Consort bicycle to fund a more interesting hobby. I had long been envious of the tape recorder owned by my friend Michael Pratt. I was not interested in record players, but with a tape recorder I would be able to record and erase music at will. So with the money from the sale of my bicycle and some of the funds earned on the bins I bought a Grundig tape recorder for £35, a lot of money. By today's standards this was old technology. It comprised a heavy tape deck about the size of a small suitcase on which you could record either from a microphone or directly via a lead from a radio. This machine satisfied my musical interests for the next two years.

I began to think about what I wanted to do as a career when I left University. I really did not have much idea. My father suggested that I might want to be an actuary and suggested I go and talk to one of his Masonic friends, a Mr Bernard Bransden, an actuary working for a London firm of stockbrokers. He set up the meeting and I went to visit Mr Bransden one evening at his posh house in Chalkwell. He was a tall, thin man with a bullet-shaped head and an enormous Adam's apple. I learned that an actuary was someone who dealt with the financial impact of risk and uncertainty in fields such as investment, life assurance, pensions, and expert assessment of complex financial systems. Actuaries evaluate mathematically the likelihood of events and quantify the contingent outcomes in order to minimize losses. I also learned that it took several years to qualify as an actuary, that not many people made it, but those who did made a lot of money. I decided that in the absence of any better ideas I would go for this.

In the autumn term of 1962 I started my second year at Bristol. Andrew Sargent, Bob Perry and I moved into a first floor flat at 1 Codrington Place. The best thing about the flat was the address. The flat itself had only the barest of essentials and furniture was non-existent. The owner of the building lived on the ground floor. She filtered all the mail that was delivered. In order to impress her with the importance of her tenants we posted letters to ourselves addressed to such personages as "Lord Robbers of Hastings". I am sure this had the desired effect.

In our hovel of a flat there was a free-standing gas cooker in the corner of the living room and that was about it. We used the gas cooker for heating baked beans and also for heating the room. The poverty of our living conditions did not detract from the freedom we now had. We brightened the walls with cheap prints of Renoir paintings and pretended we were sophisticated.

I settled into my student job as Treasurer of Entertainments. This took up a few hours of my time each week. At the beginning of the week I would issue the tickets for the next Saturday dance to the Porters Office in the Students Union. The porters sold the tickets to the students. The Monday after the dance I collected the proceeds of the ticket sales from the porters and the profits from the Bar Manager and entered it all in the ledger. I had the use of a small office at the back of the Students Union. I then took the money to the bank. I also wrote out cheques for all the expenses. At the end of each term I took the ledgers to an accountant in the Economics Department for auditing.

None of this was difficult. I just had to be careful to keep a strict record of all transactions. The major benefit was that I had free access for myself and any friends of my choice to all the Union dances. The bands were sometimes well-known, like the Dave Clark Five and Ken Collier's Jazz Band. Once a term there was a Soirée which was a black tie and gown affair where we drank wine and did ballroom dancing. Soirées were for couples so each boy had to invite a girl.

These Saturday night dances were always popular and often sold out before the event. They were also a source of friction between "town and gown" because many of the local youths thought they should be allowed in. One Saturday we had advance warning that a gang of yobs planned to gate-crash our dance. So we enlisted the help of Dave, an engineering student who played rugby for the University and the England under-21s. He was an enormous hulk with a neck as thick as his thigh. He stood on the door and we told the porters to stand clear in the event of trouble. The gang arrived and barged in through the entrance doorways. The ringleader was threatening. Dave simply stepped forward, picked him up in a bear-hug that stopped him breathing, and deposited him outside. We had no more problems.

An important event near the beginning of that autumn term was my acquisition of a car, possibly part of the master plan in selling my bicycle. Robbers already had his old Austin which he regularly took to pieces, and Andrew had a Lambretta scooter. I did not want to be outdone. A car was a great status symbol because very few students owned one.

I studied the classified ads in the Bristol Evening Post and homed in on a 1936 Morris 10 for sale in Ashley Down, East Bristol. On a bleak, rainy night in November I went to test-drive the car with Robbers. It seemed to work so I bought it for £36. I was proud of my car but soon noticed it had problems getting up steep hills, particularly Park Street. Sometimes a passenger had to get out and push to help the car up the incline. But you could not expect everything from an old banger. I had no inkling of the troubles ahead.

The swinging sixties were a bit late to arrive in Bristol which retained its traditional values for some time. This brings me to my first, rather unsatisfactory romance at university. She was called Sue Ison, a fresher studying English and Spanish. Sue was from (East) Cheam, Surrey. She was well-spoken, demur and rather sophisticated compared with me. I first met her in a coffee bar called the Berkeley opposite the main university buildings. For some reason I was greatly attracted to her, but as usual my feelings were not reciprocated. I think she found me rather dull and scruffy. In fact I think she found everything about me a bit disgusting, including the self-inflicted cigarette burns on my wrist and my nick-name Snoz.

But the real problem was Dave Beadle. Dave had stayed on for a third year in the sixth form at Southend High School in order to sit the Oxbridge Entrance Exams in the Autumn Term. He failed the exams and settled for a place at Bristol to study Zoology. This meant that he was a fresher like Sue. Dave also liked Sue, and for some inexplicable reason she preferred Dave to me.

At about this time I started to have serious problems with my car. It just seemed to have lost power and had increasing difficulty climbing the gentlest of slopes. My engineering friends advised that the O-rings on the pistons had worn out. They reached this conclusion because the engine offered no resistance when turned on the starting handle.

I drove my car with Robbers to a scrap yard that he knew in Stokes Croft. We found an old Morris 10 engine buried in the mud. I bought it for £5 thinking it could never be worse than my current engine. With the help of a small crane we managed to manoeuvre the muddy engine onto the back seat of my car. We then drove slowly to the car park at the students union where I left my car until the weekend.

Some friends who claimed to know about cars turned up at the car park on Saturday afternoon to switch the engines round. This turned out to be rather difficult. Not only were we unable to undo the rusty bolts that held the old engine in place, we could not even extract the new engine from the back seat. In the end we pushed my car round to the garage of a car mechanic just behind our flat in Codrington Place. He did eventually replace the engine, charging me several pounds for his efforts, but it did not really work any better than before. On reflection it did occur to me that possibly he had never actually changed the engine - how was I to know?

In view of the state of my car I did not trust it to get me to Southend for Christmas so I left it parked on the road outside 1 Codrington Place. As a result of the severe weather my car was to remain there for the next three months in a state of hibernation. I think I had a lift home with Robbers who stayed with Andrew Sargent in Southend for a few days before driving to his parents' house in Guestling, near Hastings.

That Christmas of 1962 was the start of the winter now known as The Big Freeze. The snow and freezing weather started on Boxing Day to herald one of the coldest winters on record in Great Britain. Temperatures plummeted and lakes and rivers froze over. In blizzard conditions road and rail transport were severely disrupted and airports closed. The Navy managed to keep Chatham dockyard open by using an icebreaker but the other London docks remained closed with ice floes and mini-icebergs on the Thames. Some schools had to close but for the most part they managed to remain open. I returned to Bristol in the New Year to find my car completely buried under a mountain of snow. We were bitterly cold in our flat and wore extra clothes to go to bed, although during the day I always wore the same clothes as in summer.

I had a visit one weekend from my Southend friends Richard Coakes, Pete Wyss and Paul Archard. We had a good time, but I don't think they were too impressed with the living conditions in our bitterly cold flat where they slept on the floor.

I persevered with Sue Ison but to no avail. I took her to the Rag Ball in March but she was not impressed and was rather distant. I even wrote her a poem inspired by the hanging of Billy Budd in the recent film I had seen, but she totally failed to appreciate this. The sickening Dave Beadle was always hovering in the background.

One week a girl friend of Sue's from London came to stay with her. I met this friend and it was immediately obvious, and rather astonishing, that she was smitten with me. She had a lively personality but looked like a stick insect. She stayed overnight in our freezing flat at Codrington Place but I was attracted to her about as much as to a bean pole. So this followed the now familiar pattern of all my relationships with the opposite sex.

About the second week of March the Big Thaw set in. The roof of my car began to emerge from the snow drift. I shovelled away the snow to force an entry and found the battery was flat. I got it recharged at the garage and managed to start my car. What a relief.

It was approaching the end of Spring term and the harshness of the winter in our miserable flat had taken its toll on our spirits. Andrew Sargent had already decided to move out to live with the arty crowd he was now close to. Robbers and I could not afford to stay on by ourselves, so we set about finding some comfortable digs in time for the summer term.

At the end of Spring term I resolved to drive home for Easter in my car. Bob Perry was going to drive his separate way to Hastings. Andrew Sargent stayed on in Bristol over Easter. He had fallen madly in love with a girl studying classics who was staying on as well. Possibly, Andrew was discouraged to return home because his parents were going through a divorce. So I had space in my car. I asked Sue Ison if I could give her a lift back to Cheam. I was delighted when she accepted. Unfortunately Dave Beadle then asked me to give him a lift back to Southend. I really had no wish to give Dave Beadle a lift while I had Sue in my car, but I could hardly refuse.

We set off early on a fine day. There was still snow on the verges and it was bitterly cold, but generally the roads were clear. The M4 was not yet built, so the route to London took us along the old A4 through Chippenham, Marlborough, Hungerford, Newbury, Reading, Maidenhead and Slough. We stopped at a pub in Marlborough and ordered Minestrone Soup to warm us up. The landlady pronounced it with two syllables, as Mine-Strone. Sue thought this was hilarious.

Progress on the road was slow. I found the steering to be increasingly stiff. It began to require a great effort to turn the wheel to get round corners. On occasions I even had to do a 3-point turn just to negotiate a roundabout. What had begun as a jolly caper was becoming an ordeal. To top it all Sue began to question whether we would ever reach her house - the selfish bitch. We eventually got there and I dropped her off around 7 p.m. close to the busy South Circular. Dave and I carried on resolutely aiming for the Old Southend Road.

Darkness had set in and my headlights were rather dim. Somewhere near Grays a bend in the road took me by surprise. I wrenched the steering wheel but the car did not respond and we hit the kerb. Both nearside tyres burst and we keeled over to one side. I managed to park the car on the grass verge. We were about 20 miles from Southend and it was getting late. I found a telephone box and called my father. He came to pick us up in his Ford Anglia.

The next morning there were still remnants of snow around. I had to work out how to recover my car. By good fortune my friend Richard Coakes was free and his father generously agreed to Richard borrowing his car for the morning. It was a two-tone blue Hillman Minx and rather smart. We went first to a car scrap dealer somewhere behind Southend United football ground where I managed to buy two suitable wheels complete with fitted and inflated tyres. We made the journey to my abandoned car where I fitted the replacement wheels. This enabled me to get my car home.

A helpful neighbour advised that the problem I had with the steering was probably due to the king pins seizing up. He suggested that an application of heat might ease the problem. So I borrowed my father's blow torch from part of his house painting kit and gently applied heat to the area of the king pins by the front wheels. This seemed to do the trick and my car became driveable again.

However, my enthusiasm for car ownership was waning. It was becoming a drain on my finances, so I resolved to sell my car. I placed an advert in the following Thursday's Southend Standard. The day after the advert appeared a well-dressed man arrived at our house with his son to inspect the car. They seemed to like it and we agreed a price of £25. As they were leaving with the car I asked in passing what they planned to do with it. They said they were funeral directors.

On 23 March 1963 my cousin Jean married David McCarthy at Union Church, Totteridge. My cousin John and I were both ushers. There was still the remains of snow on the ground. Jean had met David while they were working for British Rail. A year or two after their marriage David joined the Metropolitan Police and subsequently spent much of his career in the Special Branch. They had two children: Sarah born in 1968 and Matthew in 1970.

In spite of my clash with Dave Beadle over Sue Ison we remained good friends. His mother had started an evening job as catering manager at the newly-opened casino at Southend Kursaal. Such a gambling enterprise was verging on impropriety in 1963. Dave said that his mother could get us in without having to pay the membership fee. I had never before been to a casino and had no idea what to expect. Dave and I turned up one evening in our best clothes. It turned out to be an extremely exclusive establishment with luxurious fittings and deep pile carpets, like something out of Casino Royale. Most astonishing of all, the food and drink was all free and could be ordered at the playing tables in unlimited quantities. The drinks were all shorts and the food comprised expensive canapés and delicately cut sandwiches, presumably prepared by Dave's mother. Dave and I were astonished at the opulence and took full advantage.

I did not enter into this adventure without some preparation. I had worked out what I believed was a foolproof method for winning at roulette, and I had managed to convince Dave as well. The method employed betting on red and black only, and was a variant of the old system of doubling your bet until you won. However, with the old system of betting on just one colour your net gain was one unit only when your colour finally came up. My system was to bet on both black and red at once, doubling up on the colour that had just lost. This ensured that on average you won one unit every time. We played this system and it worked well. Other gamblers round the table could not understand how we could be winning by backing both colours at the same time. After an hour or so we had won over £100. This was a small fortune and we decided to quit.

I was proud of my system and on returning to Bristol found that my fame had spread. But of course we had been lucky. The method is not foolproof unless you have an infinite bank, in which case you have no need to gamble. This came home to me when we later visited the casino a second time to make a killing and finished up losing everything previously won. Our bank ran out after a long run of consecutive reds.

At the start of the summer term I returned car-less to my new digs at 5 Frederick Place, a Georgian terrace quite near the University. There were four of us staying in these digs: Robbers and myself, a mature student funded by his employees at Somerset County Council to read a degree in

English, and a loud-mouthed young man who worked for an insurance company in the centre of town and had to listen to The Archers every night. We each had separate bedrooms, which was an improvement on earlier digs. The food was good and we were comfortable. That summer term of 1963 was hot, idyllic and carefree, in total contrast to the appalling winter. It was the time of the exotic Profumo scandal. University work was going well, my social life was going well (except with girls), and Robbers' old Austin was behaving itself.

We developed a ritual that was enacted most evenings. In the city centre near the docks there was a pub called "The Old Duke" located in a cobbled lane called King Street. The pub was quite the shabbiest in Bristol serving stale beer in dirty glasses. It was the sheer grottiness of this pub that attracted us. After dinner at our digs Robbers and I would jump into his car and drive at breakneck speed through the busy streets of Bristol to "The Old Duke". After downing a pint we would race back to our digs taking the corners round The Triangle on two wheels. This ritual was re-enacted most evenings in an effort to create a new record. Our fastest time there and back was about 10 minutes. Amazingly we were never once stopped by the police.

There was a man next door to our digs who annoyed us intensely. He would put notes on Robbers' car asking him not to park his wreck in front of his house. He particularly did not like Robbers taking his vehicle apart in the road. So we played a mean trick. One evening we put a penny coin inside the metal cover of the front nearside wheel of his car. This made a low level rattling noise that was worrying for the driver. The source of the noise was almost impossible to locate, particularly because it ceased as soon as the driver stopped to investigate. After we had extracted sufficient pleasure from this trick we removed the coin.

As Treasurer of Entertainments I took part in organising the Students Summer Ball held in May. At the last minute I realised I had no partner to invite. So on an impulse I rang up a girl called Leslie who lived in Leigh near my home town of

Southend. Leslie had a job in London and often travelled to work on the train with Richard Coakes. I had met her briefly at a party in Southend over Easter. Amazingly Leslie accepted my invitation. I collected her in Robbers' car from Bristol Temple Meads station on the Saturday of the Ball. I had booked her into a room in a hotel near the university, the Hawthorns Hotel which was quite extravagant. (The Hawthorns Hotel was later bought by the University and converted into student accommodation). Leslie was good fun and vivacious. We had a good time at the Ball. Both of us could do basic ballroom dancing because Leslie and I had attended the same dance classes back home in Westcliff. We danced all evening to "Clem Gardiner and his Orchestra" and "Her Majesty's Royal Marines Dance Band". I thoroughly enjoyed it.

The next morning I overslept with a hangover. The landlady knocked on my bedroom door to announce disapprovingly that a girl was waiting at the front door. It was 11 o'clock and I quickly dressed. Leslie had an hour or so to spare before her train home. So we went in Robbers' car to show Leslie the view over the Avon Gorge from the Downs before driving to Temple Meads station. We had a cup of tea in the station café. The train arrived and there was a poignant moment when we embraced on the platform. Leslie said something like, "You will probably never ask me out again". I said that was nonsense and I would contact her over the summer vacation. I drove back to Frederick Place feeling rather low. I had missed Sunday lunch and the landlady was not pleased. I never saw Leslie again.

Robbers and I were frequent visitors to Bristol Downs where we often walked along the cliff-top path with spectacular views overlooking the Avon Gorge. One Sunday afternoon, for no obvious reason, Robbers took with him a coil of rope that he always kept in his car in case he ever needed a tow. We hit upon the idea of a mountaineering spoof. We climbed over the safety railings and hid behind a bush just above the vertical drop of the cliffs below. We tied ourselves together with the rope and waited for some Sunday walkers. Just as a large group approached we emerged from our hiding place, poked our heads above the parapet feigning fatigue and breathlessness. Robbers muttered how it had been a tough climb and something about the Matterhorn. Our audience was gob-smacked.

It was approaching the end-of-year exams and I began revising in earnest. These exams were important because they were Part 1 of Finals and so counted towards my degree. They all took place over the space of one week mostly in the mornings, which was difficult for me. After a late night of revision I was awoken the first morning by Robbers shouting that it was five minutes to nine. I had an exam at 9 o'clock. I leapt out of bed and threw on some clothes. Robbers drove me at a dangerous speed, jumping red lights and mounting the kerb to drive along the wide pavement outside Mabbs department store to bypass the early morning traffic jams. We arrived outside the University Examinations Hall at about 9:05. Fortunately they let me in and I settled down quickly.

I must have done alright in these exams because I was allowed back for the final year - unlike my landlady who said she did not want to see me again. This quite staggered me because I thought I had learned how to be a model paying guest. Robbers and I resolved to return to renting a self-contained flat in my final year.

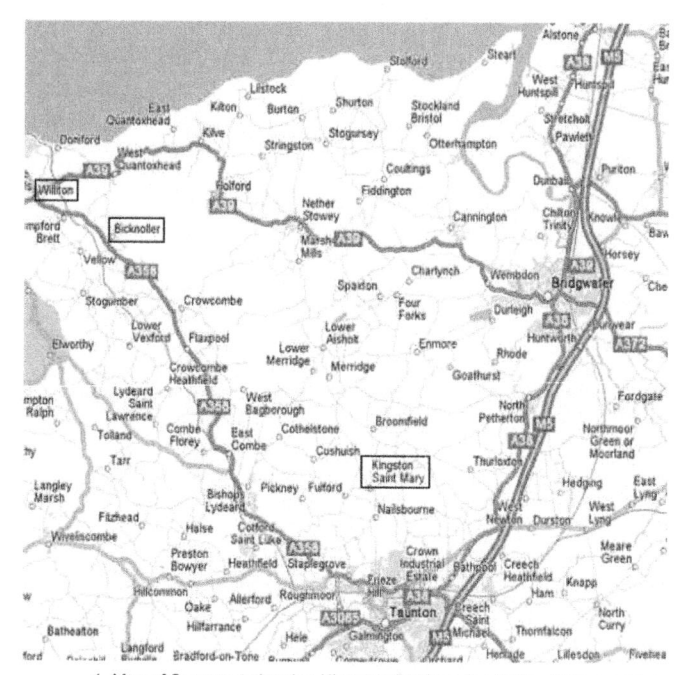

*1. Map of Somerset showing Kingston St Mary, the home of my ancestors.*

*2. Kingston St Mary's Parish Church where many Parsons ceremonies took place.*

219

3. Extract from Regimental Payroll Muster 1827/28 showing entry for Thomas Parsons.

4. No.81 South Street, Taunton, where my g2-grandfather lived with his wife Mary circa 1860.

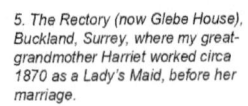

5. The Rectory (now Glebe House), Buckland, Surrey, where my great-grandmother Harriet worked circa 1870 as a Lady's Maid, before her marriage.

6. My great-grandparents, James Parsons and his wife Harriett circa early 1880s.

7. My great-uncle, Henry James Parsons (stage name Phil Parsons) circa early 1900s.

8. My grandfather, Thomas Besent (centre) outside shop in Westcliff, circa 2016.

9. My mother and her brother Harold, circa 1915.

10. Thomas Besent with wife Gertrude and children Dorothy (my mother) and Harold, circa 1919.

11. My mother aged about 20, probably in a park in Westcliff, circa 1931.

222

12. Grandma Violet Parsons circa 1930s.

13. My grandparents Thomas and Violet Parsons with daughter Margaret, circa 1938.

14. My parents before their marriage, circa 1939.

15. My parents' wedding at St Mary's Church, Prittlewell, Essex, on Wednesday 29 January 1941.

16. Baa-Lamb and me, circa 1946.

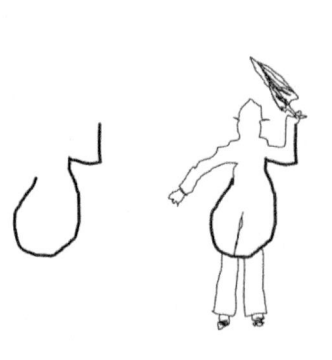

17. A recreation of my winning Cadbury's competition entry, to make a novel drawing incorporating the shape on the left, circa 1951.

18. In Auntie Haidée's back garden, circa 1951.

19. The infants class at Earls Hall Primary School, Christmas 1949 (I am in 2nd row, far left).

20. Front cover and opening page of my Junior School "London Book", Spring 1954.

21. Form 1A (1954-55) – my first year at Southend High School.

22. Christmas dinner 1957 at my paternal grandparents' house.
From L to R: my mother, my father, Jane Peters, Billy Peters, Uncle Frank, Grandpa and Grandma Parsons.

23. My paternal grandparents' Golden Wedding celebration in back garden of their bungalow in Hobleythick Lane, August 1958. From left to right:
   kneeling:   ?, Nick, ?
   seated:     Margaret, Grandma Parsons, Grandpa Parsons, Jane Peters, Uncle Bert.
   standing:   Uncle Billy, ?, ?, ?, ?, Father, Cissie, Haidée, Frank, Mother, ?, Auntie Daisy, ?, Myself, Douglas.

24. Summer 1960 at Butlins Holiday Camp in Filey, Yorkshire.
From left to right: Doug May, Danny Linehan, Peter Lee, myself (just 17) and Chris Harding.

227

25. Southend High School Prefects 1960-61 including the Headmaster, to his right the Head Boy Brian Wiggins, to his left the Deputy Head Boy John Roff, and myself behind (no significance in this position).

26. Celebrating our engagement (and my 21$^{st}$) at 11 Henley Crescent, July 1964. From left to right: Jane Peters, Auntie Jennie, Uncle Harold, my paternal grandmother Violet Parsons, John Besent, Sheila, Haidée Peters, my paternal grandfather Tom Parsons, unknown girl staying with Haidée, myself, Lucienne, Margaret Parsons, Billy Peters, my mother (photo taken by my father).

27. L'épicerie in Rue de Matel: left, Monsieur Page; right, Lucienne and me (August 1964).

28. 16 July 1965, the eve of our wedding outside house in Rue de Matel. From left: my mother, Dick Coakes (my best man), Dave Hall, myself, Lucienne, my brother Nick, M Page, Mme Page, Haidée, Frank, Margaret. (Photo taken by my father).

29. Our wedding at L'église St Etienne, Roanne, 17 July 1965.

30. Baby Isabelle with Lucienne and M & Mme Page in Bristol, June 1966.

31. Bristol University Degree Congregation for my PhD, January 1969.

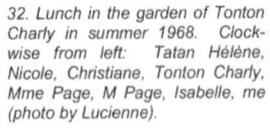

32. Lunch in the garden of Tonton Charly in summer 1968. Clockwise from left: Tatan Hélène, Nicole, Christiane, Tonton Charly, Mme Page, M Page, Isabelle, me (photo by Lucienne).

*33. Admiralty Research Laboratory, "O" Group, December 1970*
Back Row: Harvey Meyer, Brian Cross, Ron Morris, Harry Applin, Alan Parsons, Roger Panter, Mike McCann, Robert Smith, George Bond.
Front Row: Gavin Dyer, Tony White, Joan Wright, Yvonne Sale, Phil Lindop, John Gill, Jim Cook, Stan Osborne, H Warwick, J Cutter.

*34. Isabelle with baby Caroline. April 1976.*

*35. Caroline's christening at St Edmund's Church, Whitton. From left: Malcolm, Lucienne, Margaret, Isabelle holding Caroline, my mother, Joanna, my father, Grandma Parsons, Frank.*

*36. Isabelle and Caroline by the canal in Roanne, circa 1977.*

231

37. Caroline in Carlton Road North, circa 1981.    38. Dorset cliffs at Warbarrow Bay near Tyneham.

39. Schematic of towed arrays.    40. Research vessel RMAS St Margarets.

41. Lucienne, Isabelle, Antony, Emily, Natalie, Madeleine on Weymouth promenade, circa 2003.

42. Presentation of International Achievement Award by the First Sea Lord, Admiral Sir William Staveley, November 1986.

43. UK-FR trials team on RMAS Newton in Funchal, Madeira, November 1987. From leftt: Steve Pointer, Alan Morrell, Jean-Claude Rodin, Roy Baker, Gordon Murdoch, Patrick Grignon, Alan Parsons, Alan Ridyard, Nick Goddard, Philippe Ostrowski.

44. My mother, Antony, myself, Lucienne, my father at my parents' house in Lympstone Close, Southend, circa 1994.

45. Lucienne and Caroline at my retirement dinner at The Prince Regent, Weymouth Esplanade, 5 Jun 2003.

# Hitch-Hiking Holiday

That summer of 1963, after working a few weeks on the bins, I set off at the end of July on a memorable holiday. I accompanied my old friend Paul Archard on a hitch-hiking trip to France. I had never been to France before, or for that matter to any foreign country. Paul on the other hand was well-travelled having hitch-hiked by himself several times to destinations round northern Europe. As school holidays approached Paul would announce that he felt the "wander lust" coming on, and just disappeared. He usually travelled alone, as was his nature, but on this occasion he was happy to have a companion. I was elated at the prospect of the adventure ahead. The aim was to head in the general direction of the South of France. Paul gave me due warning that it would be rough going, and we would be travelling light on a low budget.

We set off on Tuesday, 23 July 1963. In the spirit of the adventure we walked from home all the way to Southend and down the pier to catch the morning ferry to Boulogne, then a regular service. Our ship was the Royal Daffodil.

We each carried just a small rucksack containing the barest essentials. These included a change of shirt, a plastic mac, a toothbrush, a small quantity of French Francs, and of course a passport. We arrived in Boulogne mid-afternoon.

Our very first objective was to make a brief visit to a pen-friend of Paul's named Brigitte who was living in the area. By the time we had found her house and spent some time with Brigitte it was close to 5 o'clock.

Paul explained that the best chance of hitching a lift was to go to the edge of town since this avoided being swamped by local traffic. The sky was now overcast and threatening rain. We walked through the depressing suburban streets of Boulogne towards the main road that led to St Omer and Béthune, there being no motorway at that time. On the way we stopped at a bar and ate a plate of chips, the first of many. This was the dish with the highest satisfaction-to-cost ratio.

We carried on walking for what seemed like hours. Eventually the houses started to thin and we were now trying to hitch a lift in earnest, but without success. It was getting dark and beginning to rain. Few cars passed by and those that did ignored us. I was beginning to wonder what I had let myself in for. We had walked a long time for several miles through dismal streets without a single lift. It was not much fun.

At about 11 o'clock at night we found ourselves cold, wet and hungry. In pitch darkness we huddled in a doorway trying to shelter from the rain that had developed into a steady downpour. We had stationed ourselves on a bend in the road close to a railway crossing. This gave us a good view of any approaching vehicle, which would have to slow down to negotiate the bend. Whenever we saw headlights one of us would rush out in the rain to the edge of the pavement and give the hitch-hikers thumbing sign. After some time with no success we became resigned to finding somewhere more sheltered from the rain where we might snatch a few hours sleep.

Then out of the darkness a big coach appeared, its headlights illuminating the slanting rain. Paul jumped out and the coach stopped. The driver beckoned us in. The coach was empty. Although Paul's French was much better than mine we neither of us understood what the driver said. We had no idea where the coach was going and did not really care. It was warm and dry inside and we stretched out exhausted on the comfy seats. We both slept soundly.

We were awoken by vibration over bumpy cobble stones. Paul told me it was Paris. It was about 5 a.m. and the coach driver dropped us off in a large square. We went to a café and had bowls of refreshing coffee with my first taste of French bread, beautifully fresh.

We were lucky to quickly obtain a lift that took us south of Paris to Orléans, arriving about 8:30 a.m. We were dropped off on a perimeter road and duly started to thumb at the exit of a busy roundabout. A car soon stopped containing two women who turned out to be a mother and her teenage daughter. The daughter explained in English that she had recently been an au pair in Bristol, where I was currently a student, and they had recognised us as English travellers. They lived locally and invited us to their house for breakfast. This was unexpected hospitality. They lived on the outskirts of Orléans in a typical French terraced house with tiled floors, high echoing ceilings and shuttered windows. We were treated to coffee and croissants with various other delicacies that were lost on me. Paul seemed to communicate with relative ease but I was pretty hopeless and did not understand what was being said.

This happy experience lifted our spirits. We continued hitch-hiking on the long stretch that took us towards Bordeaux. The going was quite difficult and all our hitches tended to be short. The next night we slept first in a garden allotment shed and then, because it was so uncomfortable, transferred in the early hours to a concrete bus shelter where we could stretch out on two hard benches.

As we progressed south the countryside changed considerably. We were leaving the cold dampness of the north behind us. We arrived in Bordeaux on a hot afternoon. Paul thought this would be a good place to settle for a few days and suggested we look for work. We used our initiative and searched out various possibilities. We were offered jobs in a Coca Cola warehouse but the conditions were not right. I think they wanted us to sign up for a month, so we turned down the job offer. After all, we were supposed to be on holiday.

As we continued further south it was becoming much warmer. We travelled through vast pine forests and lush valleys to enter the area of sultry heat and heavy aromas typified by southern France. We encountered the usual joker car drivers. These are the ones who slow down and stop for you, allow you to run up to them, then drive off just as you reach them. This could be discouraging for hot and tired hitch-hikers.

We were eventually dropped off on the outskirts of Bayonne just as darkness fell. At this point we were exhausted and needed a good night's sleep. We set off on foot through deserted streets to find the local Youth Hostel which turned out to be several miles away. We arrived late at night and were relieved to be given beds. We woke up early the next morning feeling quite refreshed.

The next leg was a short one to the Atlantic resort of Biarritz. This was clearly a rich city. We headed for the main promenade and beach, bordered by smart restaurants and casinos. We were hot and tired and slept that night on the beach just in front of the main casino. We were awakened about 6 a.m. with mouthfuls of sand. The waiters from the casino were sweeping the promenade. Biarritz was expensive. Its cafés were smart and not the sort of places to serve free glasses of water and chips to the likes of us. So we quickly left to continue southwards.

It may be asked why we were so reliant on the hospitality of cafés to satisfy our thirst. This may sound strange, but the modern habit of walking around carrying a bottle of water for continuous imbibing did not exist at that time. For a start, there was no such thing as bottled water in England, only tap water. We knew that the French had bottled water but found this rather foolish and incomprehensible. This was not for the English. People were not in the habit of carrying supplies of water with them. If we were thirsty we would drink from a public fountain or ask for a glass of water at a café. The idea of buying a bottle of water in France to carry around never occurred to us. In any case, it would have been impractical because water came only in heavy 2 litre bottles made of glass.

In that era, it was actually considered the height of bad manners akin to animal behaviour to eat or drink anything as you were walking down the street. (There were exceptions such as eating fish and chips out of newspaper, and jellied eels on the seafront). The acceptance of drinking out of water bottles as you walked along was the start of the irreversible slide that led to the widespread drinking of alcohol in the streets. This is another example of the principle of entropy at work in the decline of social behaviour.

Since we were so close to the border, our objective became to head into Spain. We obtained a hitch that took us through

Saint-Jean-de-Luz and over the border. From here we made for the city of San Sebastian and, as usual, towards the beach. It was busy with holidaymakers and extremely hot. We swam in the water and lay down on the sand feeling refreshed and exhausted. We both went to sleep and woke up in the late afternoon to discover we were badly sunburnt.

Later that evening we shared a cheap bottle of Spanish wine that we drank in a seedy basement bar. I did not feel too well after this cocktail of heat, sun and *vino tinto*. We slept on the promenade that night and it was cold. We were woken in the middle of the night by the light of flashing torches held by Spanish policemen who moved us on. We escaped to the beach and slept soundly until the early morning sun began to warm our bodies.

It was Sunday and Paul, a devout Catholic, attended Mass. I went along with him to San Sebastian Cathedral. Paul was a keen organist and I shared his enthusiasm for organ music. After the service Paul overcame the language barrier and managed to persuade the organist to take us up to the organ loft. This was quite a feat given that we were scruffy and unshaven. The organ loft was cramped and high up in one of the transepts with a panoramic view of the nave. The organist demonstrated the impressive instrument but did not offer Paul the chance to play.

After the service we saw posters in the street advertising a bull fight in the local bullring that afternoon. We debated whether to watch this spectacle because we had the idea it was rather bestial. I wanted to see what happened at first hand but Paul was adamant that he would not go. So I left Paul to wander the streets while I attended the bull fight.

I bought a ticket for the cheapest seat. The arena was shaped rather like a Roman amphitheatre and was packed with people. The fights were exciting and rather bloody. I felt sickened and sorry for the poor bulls and resolved never to go to a bullfight again. Nevertheless, I stayed to the end witnessing the death of several bulls. I began to appreciate something of the art of bull-fighting.

At the end of the spectacle I somehow met up with Paul amongst the crowds. It was hot and dusty in the narrow streets surrounding the bullring. The cafés and bars were doing good business after the bullfights. We were hungry but chose to walk towards the poorer area of San Sebastian where the tariffs were lower. We finished up in a rather seedy bar where we treated ourselves to a Pepsi and a steak sandwich. This comprised a stale baguette containing a slab of raw meat that was rather tough.

That night we sought out the local youth hostel. This entailed a long walk to the edge of town. We slept soundly on straw mattresses in a large dormitory. It cost only a few pesetas. In the morning we had to help with the washing-up.

The next day we decided to follow the coast and explore a bit more of Spain. Our lift took us to the seaside village of Zarauth that we immediately fell in love with. Zarauth was old and unspoilt, set in a bay surrounded by mountains. There were not many tourists and all the locals were hospitable. We immediately made our way to the beach and found a friendly café selling omelettes, pasta, and local fish dishes at prices we could afford. It had a piano on which Paul played a rendition of Bach's well-known Toccata in D minor. This was to be our base for the next four days. During our stay in Zarauth we explored the area and climbed the headland on the edge of the bay to take in the beautiful view.

Paul began complaining that his left ear was blocked. It had started after swimming in the sea at San Sebastian. He went to a local chemist and managed to explain the problem. They sold him some ear drops which had no effect. Paul remained partially deaf for the rest of our holiday making communication difficult.

It was hot in Zarauth. The first two nights we slept on the beach but found it rather cold in the early morning. On the third day, Wednesday 31 July, there was a fiesta with street entertainment and processions to celebrate the feast of St Ignatius of Loyola. There was a wonderful party atmosphere. That night at the end of the festivities we stood on the beach to witness a terrific thunderstorm with lightning that illuminated the sea and bay for miles.  To shelter from the rain we slept on stone benches in the medieval cloisters of the local church, Santa Maria la Real.  We did not sleep well.

We were awakened early by the noise of local devotees walking into church for the 6 a.m. Mass.  They did not bat an eyelid towards us.  At the end of the service Paul approached the priest and managed to converse in French.  He asked the priest if he knew any cheap places where we could obtain a bed for the night.  The priest said to return to see him that afternoon after 4 o'clock Mass.  We met the priest as arranged and he gave us the address of one of his parishioners.  The lady lived in an old but well-maintained house in a narrow street off the main market square.  She was most welcoming and offered us a double bed in an upstairs room overlooking the narrow street.  We slept deeply that night in a soft bed with cotton sheets.  The lady wanted nothing for her trouble and we were moved by her kindness.

The next day was Friday 2 August, 1963.  Feeling as good as new we set off back into France and along the Pyrenees.  We arrived in Lourdes late afternoon.  The first thing we noticed about Lourdes was the masses of people, many of them in wheelchairs or on crutches.  It was festive with an expectant atmosphere.  We went straight to visit the grotto of Our Lady of Lourdes, the sacred site where the 14 year old peasant girl Bernadette was supposed to have seen apparitions of the Virgin Mary.  We found ourselves in an enormous queue winding its way back through the gardens of the basilica.  After an hour or so we had our brief glimpse of the grotto before being moved on.  In the evening we attended an open air, candlelit service with a congregation of several thousand.  That night we ate chips and slept in the railway station waiting room, which was quite busy.

We spent two nights in Lourdes, leaving Sunday evening.  We hitched the short distance to Tarbes arriving around midnight.  We had an excellent sleep in a railway carriage at the station.  The following day we continued our journey eastwards to Toulouse.  If there are interesting places in

Toulouse we did not see them. It seemed a big and boring city. We did not stay long.

My opinion of Toulouse was transformed after an organ holiday based there in 2012. The city centre is full of interesting medieval buildings, fine inexpensive restaurants and, most important, many superb organs built by Puget and Cavaillé-Coll. We obviously missed all this in 1963.

Our car journey continued past the impressive fortress walls of Carcassonne that we viewed on the horizon on our way to the city of Narbonne. Our luck with hitch-hikes then seemed to run out and progress became slow. By the evening of Tuesday 6 August we were 60 km from Narbonne and in the absence of any lift we decided to walk. After 20 or 30 km we were exhausted and gave up. We slept somewhere by the roadside.

The next day we were still unlucky with lifts. We did not arrive in Narbonne until four o'clock in the afternoon. We had something to eat and looked for a place to sleep. We were right in the centre of town so went to the railway station where we settled down for the night in the waiting room on comfortable benches.

About 2 o'clock in the morning a station attendant burst in and threw us out along with other vagrants. There being no suitable resting place outside, we crept back into the station and onto the platform. We could see the dim outline of some carriages in a siding about two hundred yards down the line. We walked to the end of the platform, climbed down onto the track and approached the carriages. The doors were unlocked. We climbed in and made ourselves comfortable in one of the passenger compartments. It was a good sleep. In the morning we got up at dawn and were grateful to take advantage of the wash basin and other services on offer in the toilet compartment of the carriage. We made our way back onto the platform and walked out of the station without incident.

We spent the next three days in Narbonne exploring the area. One day we went in search of the beach. This turned out to be an 8-mile walk to the separate town of Narbonne

Plage. It was a blistering hot day and we chose to avoid the roads and walk across country which was largely sand dunes with exotic plants and strange lizard creatures. When we finally arrived at Narbonne Plage it was just blue sea and miles of white sand dunes. We could see nothing there to sustain us, but at least we had seen the Mediterranean. Fortunately we were able to catch a bus back to Narbonne to save us having to walk.

On Saturday we went in search of the cathedral but found ourselves outside a large parish church from which we could hear organ music. We went inside and approached the organist. He said he was practising for a wedding that afternoon at 3 o'clock. Paul asked if he could have a go on the organ. The organist was willing and amazingly invited Paul to play the exit voluntary for the wedding. Paul agreed immediately.

We returned to the church well in advance of the wedding and Paul made his way up to the organ loft. I hovered around the entrance to the church where I was mistaken for a beggar by the well-dressed guests arriving for the wedding. I watched the bridegroom arrive last, as is customary in France, and followed the proceedings through the open door of the church. At the end of the ceremony the married couple walked down the aisle to Paul's rendition of Purcell's Trumpet Voluntary. Everyone was happy.

That evening we wandered into the down-town area of Narbonne in search of a cheap place to eat. We were passing by a smart restaurant from which we could hear loud music and the noise of much merrymaking. We approached the entrance to discover to our astonishment that it was the afternoon's wedding party. There was food and drink in abundance and we were desperately hungry. Our first reaction was that we had been guided here by divine providence. On second thoughts we realised it would not be appropriate to gate-crash this party. We were dishevelled with over a week's growth of beard and probably smelly. The bride would not be pleased to learn it was this tramp who had played her out of the church, even if she believed it. So we continued with some sadness down the road and settled for the usual pommes frites washed down with a glass of water.

We began our fourth night in the railway carriage. In the middle of the night we were awakened by the slam of a carriage door followed by heavy footsteps in the corridor. Someone was progressing down the carriage opening all the compartment doors. The attempt to open our door failed because we had locked it from the inside. We kept still and held our breaths. There was silence for a moment then the footsteps moved on. We went back to sleep. However, at dawn we were awakened again, this time by the clunking of carriages and a sense of movement. We looked out of the window to realise we were in motion. The empty carriages were being shunted somewhere. After about 20 minutes the train stopped briefly in the middle of the countryside. We jumped out onto the line, over fences and onto a road.

We made slow progress eastwards, through Béziers as far as Montpellier. Whereas Zarauth had been the high point of our holiday, we were now at the low point. Paul was tired and still deaf in one ear. I was suffering from the heat and not feeling well. We had hardly any money left, certainly not enough for a meal and a comfortable bed. On top of all this we were at our furthest point from home. We must have been depressed because we began talking about contacting the British Consulate to bail us out.

However, we stoically decided to press on to try and reach home as soon as possible. That night we slept on a park bench in Montpellier. The next morning we started trying to hitch in earnest. After about two hours a car stopped for us. It was a dirty-looking white Simca. The driver said he was going to Charleville. This was fantastic good fortune. We were not quite sure where Charleville was except that it was in northern France.

The car was small and uncomfortable and the driver drove at high speed, but we did not mind as the kilometers were steadily eaten up. We continued non-stop through Nimes, Avignon, Valence, Lyons, possibly even Roanne, then to the east of Paris and onwards north to arrive in Charleville in the early evening, a single hitch of almost 450 miles. Charleville is on the border with Belgium, only 150 miles from Boulogne. For many weeks after this stroke of luck we worshipped the Mighty Simca. We shared a loaf of bread and slept that night in the railway station waiting room in Charleville.

The next morning we picked up two discarded railway tickets and, I am ashamed to say, used them illegally to take us some miles nearer to Boulogne on a busy commuter train with standing room only. We continued to make progress with several short hitches arriving in Boulogne around mid-day. There was a ferry leaving for Southend at 4 p.m. We had our return tickets but no more money. There was a street market where the vendors were packing away. We found several rotten tomatoes in a discarded box and ate them with relish.

With great relief we boarded the ferry for the final leg of our adventure. The ship was packed with British day-trippers tucking into sandwiches, plates of fish and chips, and chocolate bars. It was torture watching all this food being consumed. The crossing lasted about five hours. We disembarked exhausted at Southend Pier, and having no money had to walk the mile and a third down the pier to the mainland. I managed to contact my father with a reverse charge phone call. He collected us in his car from the top of Pier Hill.

On the drive home Paul was rummaging through his rucksack and suddenly produced a £5 note that he had forgotten about. He said it was for emergencies. I was really mad at him.

# Final Year at University (1963-1964)

On return from my hitch-hiking holiday, Robbers contacted me from Hastings to say he was selling his old Austin to buy a slightly newer Morris Minor Convertible. It had been imported from Germany and so was left-hand drive. I agreed to provide £25 towards the purchase and insurance in exchange for the opportunity to drive it sometimes.

We had already agreed to look for a suitable self-contained flat for the two of us. The week before the beginning of term we met up in Bristol to search the newspaper adverts. We soon settled for an upstairs flat at 52 Chesterfield Road in the district of St Andrews, midway between Montpellier and Bishopston. It was not really a flat, rather the first floor of a semi-detached Edwardian house in a suburban street of similar houses. By coincidence, it was just round the corner from where Dave Beadle was in digs, and where he stayed happily for the whole of his three years.

The landlady was elderly and lived downstairs with her spinster daughter who must have been in her forties. The daughter was mentally retarded and used to cut the front lawn on her knees at night with a pair of nail scissors. Our accommodation upstairs was Spartan but we were quite happy with it. We had a living room with a table and sofa, a bedroom with two single beds and a wardrobe, a kitchen with a sink and oven, and a bathroom in which I rarely ventured.

It was my final year and I had to get down to work. Robbers was in the third year of a five-year veterinary course and had a hectic schedule, much busier than mine. So we settled down to a routine. Robbers usually left the flat early. To save time getting dressed he perfected a rapid technique for putting on his trousers. Instead of putting first one leg and then the other into their respective trouser holes, he held his trousers out in front and jumped into them feet first, both legs simultaneously. This was an impressive feat but quite dangerous if it went wrong. Robbers then left in his car to attend 9 o'clock lectures. He was occupied all day, and even sometimes at weekends doing practical training at the

University Veterinary Farm at Langford some 14 miles south of the city.

I always got up much later and walked the two miles to Queen's Building hoping to arrive in time for my mid-morning lecture. In the afternoons I sometimes worked by myself in the library. We both returned to the flat at about 6 p.m. to have a bite to eat, then drove down to the pub. Usually this was the local pub by the railway bridge on Gloucester Road, another establishment with nothing to its credit. We would have a few beers and an enjoyable game of cribbage or darts before returning to our flat for more work. Robbers often worked into the early hours. In retrospect I realise he was under a lot of pressure that he did not cope with too well.

I could see that Robbers had the makings of a good vet. He had an easy manner with animals that he did not have with humans. Once, when driving back from the pub we saw the car in front hit a cat. The car drove off leaving the injured cat in the gutter. Robbers jumped out and quickly assessed the damage to the cat. He collected it in his arms and managed to put the creature at some ease. He said that the cat was severely injured internally and would not survive. We knocked at doors and found the owner who was distraught. The cat died in Robbers' arms.

Our weekday routine continued during that autumn term. I sometimes played music on my tape recorder while we were working in the flat. We developed the habit of drinking large amounts of strong coffee that we poured into pint-size glass beer mugs I had acquired. One evening I was carrying my mug of boiling-hot coffee from the kitchen to the living room when the glass bottom fell out of the mug. The coffee fell on top of my tape recorder and soaked it. I did my best to clear up the hot, sticky mess but the machine never worked properly after that. So I hired a radio from "Radio Rentals" at a cost of 7 shillings per month.

Sometimes we ventured further afield for our evening beers. On one occasion we returned to The Old Duke, a favourite haunt of ours in the city centre. We were in the middle of a game of darts when the landlord announced that

John F Kennedy had been shot dead. It was about 8 p.m. on Friday. November 22, 1963. Everyone was shocked. It was difficult to believe. We had another beer and finished our game of darts.

We relaxed a bit some weekends. One Friday afternoon in November Robbers and I set off in the car to visit my friend Richard Coakes who was at that time working in Oxford as a Trainee Manager with Little Chef. After spending Friday night with Richard in the local pubs and curry house, Robbers and I slept in the car on the grass verge of the Oxford bypass. We had a cold and uncomfortable night's sleep. The next day was spent punting on the Isis with Richard and wandering round Oxford. We visited a pub with Richard in the evening and on leaving at closing time were confronted by a pea soup of a fog. After driving Richard back to his digs, Robbers and I did not fancy another rough night in the car so we decided to drive back to Bristol.

The fog was so thick that we had difficulty finding our way out of Oxford. At one point we found ourselves in someone's front drive, the result of the strategy of following the car in front. We crawled along making slow progress. At about one o'clock in the morning we reckoned we were somewhere on the northern edge of Salisbury Plain. The fog was even thicker and we resolved to find a lay-by where we could stop for a rest.

We then perceived a dim light ahead penetrating the fog. As we approached it the source of light became stronger, resembling a searchlight. We found ourselves at the entrance to an army camp. We drove up to the barrier and asked the guard where we were. He sensed our predicament and made a phone call. He then said that in view of the appalling weather conditions we could spend the night at the camp. We were ushered in and shown to a warm dormitory hut where we had a good night's sleep in comfortable beds. We were woken up early to find that the fog had cleared. A soldier took us to the canteen where we were given a hearty breakfast. They even filled us up with petrol before we continued our journey. We were not charged for any of this.

Most weekends I went to Saturday night dances. I still benefited from free entry to the University dances. Sometimes I went to dances held in the city if there was the attraction of a good band playing. On one such occasion I went with Dave Beadle to a dance at Bristol Corn Exchange where Jerry and the Pacemakers were playing. The purpose of all this was of course to dance with girls. It was a magnificent evening of entertainment with a packed floor in the 1960s tradition of Beatles euphoria. I danced with a girl who was a student teacher at Fishponds teachers' training college in North Bristol. I arranged to meet her the following Saturday to take her to the cinema, but that is as far as it went. I did not really fancy her, and in any case Fishponds was out in the sticks. I continued to have no regular girlfriend.

I always found Robbers to be good company, but he did not take part in social activities like dances. He was not comfortable with girls but he was not homosexual as far as I knew. Robbers usually made the excuse that he had too much work to do, which was probably true. I did not foresee where all this was leading, until it was too late.

After spending an enjoyable Christmas in Southend I returned to Bristol. I soon fell back into the routine I had developed the previous term, but the atmosphere was less carefree with Finals on the horizon.

Near the beginning of term I went to a student dance at the Victoria Rooms. I was on the lookout for a girl to dance with. I saw two girls on the floor dancing together, one of them very attractive. When they stopped dancing I introduced myself. Her name was Lucienne....

# BOOK FOUR
## Marriage (1964 – 1968)

# Finals & Engagement

It was Saturday, 1 February 1964, the beginning of the penultimate term of my three-year undergraduate course. My chance meeting with Lucienne at the student dance in the Victoria Rooms was a changing point in my life. It soon became clear to me that I had met someone I was attracted to and, astonishingly, she seemed to like me too.

My first sighting of Lucienne was one of two girls dancing together on a crowded floor. The atmosphere was noisy and the light subdued. I was with my ubiquitous and dubious friend Dave Beadle. I said to Dave that I was going to ask this girl for a dance, and would he like to ask the other one (Lucienne's German friend, Antje) so that we could split them. Dave agreed and I began dancing with Lucienne.

We strutted on the dance floor for a few minutes to Beatles-type music and then I asked if she would like to sit down while I got her a drink. We had difficulty communicating partly because of the noise of the music and also because it appeared that we spoke different languages.

Desperate to please her I asked if she smoked. I understood the answer was affirmative but I had no cigarettes. I guided her to a chair at the back of the dance floor and said I would be back shortly. I was away for some time because I had to get change for the cigarette machine from the bar. Apparently, Lucienne thought I had ditched her. Eventually I returned to find with some relief that she was still sitting in the same place. We each had a cigarette and tried to converse. My French was non-existent as a speaking language and Lucienne's English was faltering. This did not seem to matter.

The rest of the evening is rather blurred. We danced together for a long time and afterwards I offered to walk Lucienne home. This turned out to be the public school, Clifton College, where Lucienne was working as an "au pair". We walked from the Victoria Rooms up Pembroke Road towards the back entrance of the school and across the playing fields. It was bitterly cold but I did not feel it. Lucienne was wearing her favourite light blue (turquoise) coat. As we walked she sang to me a rendition of "La Vie en Rose" by Edith Piaf, a beautiful song that I appreciated and had not heard before. We sat in the cold in pitch darkness on a bench in the middle of the college playing fields and embraced. I was ecstatic. We continued our walk across the fields guided by the lights of School House where I left Lucienne at the door. She gave me the telephone number of School House and I promised to ring.

The following week I was unusually busy with lectures and revision. Apparently, Lucienne was concerned that I had not phoned her. Towards the end of the week I called her from the public phone box near our flat. I arranged to take her to the cinema that Saturday night. Robbers let me have the car and I picked Lucienne up from Clifton College. We went to see "West Side Story" at the Odeon in Bristol city centre. It was busy and there was a long queue. I was disappointed that we had to sit right in the front row rather than in the darkened back row. It had not started according to plan. Nevertheless the evening turned out to be a great success.

I became increasingly obsessed with Lucienne and was deviated only by my concern about the impending final exams at the end of May and the need for me to knuckle down to work, a discipline that so far I had managed to defer. My life became one of study and revision, and meeting up with Lucienne at times that fitted my work schedule and her free time at School House. I was making fewer trips with Robbers to the local pub and I no longer saw much of Dave Beadle.

Soon after our first meeting, Lucienne happened to mention that it would shortly be her birthday, on the 15 February. I asked her how old she would be and she said 21. I knew that 21st birthdays were important because my own was coming up in June - keys of the house and all that. I felt sorry for Lucienne that she would be away from her home in France for this important occasion so I resolved to buy her a present. I drove with Robbers to Bristol city centre. We finished up at a small tobacconist in Union Street where I chose an expensive Ronson cigarette lighter. I gave this to Lucienne on her birthday. I think she was quite surprised, but pleased.

Lucienne later told me she had not been truthful about her age. It was actually her 22$^{nd}$ birthday. She had not realised the significance to me of saying it was her 21$^{st}$ birthday and had felt guilty about accepting my present. In France it is the 18$^{th}$ birthday that is important. Sadly, the lighter was soon lost during an afternoon walk we made over the fields of the Ashton Court Estate the other side of Clifton suspension bridge.

Our romance was a whirlwind, at least for me. Four weeks after we met I proposed to Lucienne. It was on the back stairs in the Victoria Rooms at another Saturday night student dance. Lucienne was taken aback and said she had to think about it. After all, this was a big step after such a short courtship. Lucienne pointed out that she was in a foreign country and there were many practical problems to take into account. Whilst my feelings for Lucienne were genuine, I did not fully appreciate the implications of my proposal. I did not care. I was living high on a cloud.

After a few days Lucienne accepted my proposal. I think we were both very happy, if a little scared. We began thinking about some of the practical issues, like telling our parents, getting officially engaged (something that people did in those days), deciding when we would actually marry, and where we would live.

We exchanged information about our lives to date. Lucienne told me she had studied mathematics at the University of Lyon but had dropped out after the first year. I never quite understood why. I imagined that our common interest in maths would promote interesting discussions on the subject, but this never materialised. I concluded that Lucienne must have been talented but had no real interest in maths.

I soon realised that Lucienne had exquisite dress sense. She always looked superb making English girls look dowdy in comparison. She may have inherited her fashion sense from her mother, or maybe it was simply her French culture. I have always been proud to be with her. She has maintained her quality of style to this day.

My parents were surprised and pleased when I told them about our engagement and they were anxious to meet Lucienne. But this did not happen until Easter at my Aunt Haidée's house in Kenton.

I had returned to Southend for the Easter vacation. My parents, my brother and I all drove up for the day to see Haidée and family where we met up with Lucienne and her friend Antje, also an "au pair" at Clifton College. They had both travelled by train from Bristol. My parents were delighted to meet Lucienne. They chatted together and my mother made polite conversation with Antje. On hearing that she came from Hamburg, my mother remarked how it had been so badly bombed during the war. The room fell silent for a moment.

Overall, the meeting with Lucienne and my parents was a great success. Uncle Billy drove Lucienne and Antje back to the station for their return journey to Bristol. I went back by car to Southend with my brother and parents.

I returned to Bristol for my last term and met up with Lucienne again at Clifton College. She said she had told her parents. They were not over the moon with the news and wanted to meet me. All this had to wait until I finished my exams, for which I was now studying in earnest and wondering if I had left it all a bit too late.

I was now working most nights into the early hours at our small table in the flat, with Robbers working hard opposite me. I spent any spare time with Lucienne when she was free. Robbers was generous with his car. I often drove it to visit Lucienne at School House, a distance of some four miles, where the au pair girls had a lounge for their own use. In a sense, my life was rather simplified by these two consuming interests, work and Lucienne. I neglected my friendship with Robbers.

In those long, languid days of early summer in 1964, Lucienne and I spent a lot of our spare time walking together on the Downs, just north of Clifton, where there were superb views of the Avon Gorge and Brunel's famous suspension bridge. Everything seemed to be working out well for me and we were both blissfully happy. Then something rather dreadful happened.

One Saturday soon after the start of that summer term I returned to the flat late at night after being out with Lucienne. I found Robbers in bed fast asleep. This surprised me since he usually burnt the midnight oil. I thought nothing more about it and went to bed myself. The next morning I woke up late and went to make coffee. Robbers was still fast asleep. Again, this was a bit surprising but then I knew he was tired from all his work.

I went out all day, probably to see Lucienne. I returned early evening to find Robbers still in bed, fast asleep and breathing heavily. There was definitely something wrong. Then I saw on the side table several empty bottles of aspirin. I could not believe it. Robbers had taken an overdose. I shouted at him and shook him but could not bring him round. He was in some sort of coma. I needed to call an ambulance. There was no phone in the house so I went and knocked next door but there was no reply. I ran over the road to a house with a light in the hall and this time found someone who could ring for an ambulance. They took Robbers away to Bristol Royal Infirmary in the city centre. They said he would have his stomach pumped and then maybe he would recover.

I was in a state of shock when a policeman knocked at the door. I had to make a statement but could say nothing useful. Robbers must have been under much more stress from his work than I had realised. I recall him mentioning how it was his turn next week to give a presentation to the other veterinary students. Perhaps that had been the last straw.

The next day I went to see Robbers in hospital. I had sort of assumed that he would quickly recover from this bad experience and everything would soon return to normal. But although he seemed to be physically fit he was completely uncommunicative. Perhaps he was under sedation. I went to see him again a day or two later and his condition was unchanged. The nurse gave me a message that Robbers' father would be driving from their house in Guestling near Hastings the next day to take Robbers home, and he would first call in at the flat to pick up his things. This all seemed rather final. So Robbers returned home with his father to recuperate. He left me the use of the car.

I recounted all this to Lucienne who was equally shocked. Because of my concern for him and the fact that we had shared a flat, I think she wondered whether there was any homosexual relationship between us. I was sad about Robbers, but I was too busy to let it get me down too much. I continued to work hard for my Finals which were rapidly approaching.

Around this time I followed up my resolve to become an actuary. I made enquiries through the University Careers Office and filled out application forms. I found myself with an offer of a job as an Actuarial Student, which I accepted, at the National Provident Institution (NPI) based in the City of London. It was not difficult to find jobs in the 1960s. I was looking forward to a long summer break and so told NPI I would start work in September.

Shortly before my exams Robbers returned briefly to Bristol to collect his car and some books he had left behind. We spent half-an-hour together. He was a shadow of his former self, totally blank and withdrawn into a shell. Robbers said nothing about his attempted suicide. He left in his car. I learned from the University authorities that he would be allowed back on his course the following year, but would be staying in a Hall of Residence where they could keep an eye on him. I would be long gone by then.

My father had sent me a £50 cheque in advance of my 21$^{st}$ birthday. On receiving the cheque Lucienne and I visited a jewellers shop on the steep hill of Park Street near the University where we chose an engagement ring. We spent £37 on an attractive ring with five diamonds that we both liked. We were idyllically happy and walked to the park on Brandon Hill by the Cabot Tower. We sat holding hands in the sunshine and contemplated our future together.

Lucienne returned to France for the summer shortly before my finals. I sat them at the end of May in relative calm. I had six 3-hour papers covering number theory, function theory and analysis, projective geometry, group and ring theory, mathematical methods, special relativity, mechanics, and several topics in aerodynamics and fluid mechanics.

I was fairly happy with all the papers except the one on projective geometry in which I failed to complete a single question. They all appeared to be so trivial that the answers seemed obvious and worth no more than two or three lines. I felt the examiners were after something a bit more fundamental than I was able to offer. Projective geometry is the study of geometric properties that are invariant under projective transformations. The more familiar Euclidean geometry emerges as a special case. Very much later, having read about the work of Descartes, Desargues and Pascal, I realised that I could have given a much better account of myself in projective geometry, whose methods are exceptionally beautiful and powerful. I really believe it was the fault of the lecturer, an eccentric Reader in Geometry who sadly died soon after in a mountain accident.

After my last exam, which was in the morning, a group of us mathematicians went to celebrate at our favourite Indian restaurant, the infamous Koh-i-Noor in Denmark Street. I had a very hot Meat Vindaloo. When I had finished, the waiter asked if we would like anything more. I ordered another Meat Vindaloo.

I now had the summer ahead of me before starting work in September as an actuarial student. I prepared to set off immediately for the journey to France to rejoin Lucienne and meet her family.

My parents had been asking me about our long-term plans. I said that we intended to marry in the summer of the following year but were not clear what we would be doing in the intervening period. My parents suggested that both Lucienne and I live with them at Henley Crescent with me commuting up to London like my father. This was a generous offer, especially since our house was a bit small for five people. I discussed this with Lucienne and she agreed to the idea.

At the beginning of June I set off for Roanne from Southend pier on the ferry to Calais where I caught the train. It was a long journey and I missed a train connection in Paris. I did not arrive in Roanne until about 4 a.m. the next morning. There was no way I could contact Lucienne (no mobile phones) but she had anticipated me and I found her waiting at the station with her father. His greeting was friendly enough but rather reserved. After all it was the middle of the night. Also, it was not as though we could talk much to each other because my French was still limited.

We arrived at Lucienne's house in Rue de Matel. It was almost daylight and we were all tired. I had little appreciation of our surroundings. I was ushered up two flights of concrete stairs to a dusty attic storeroom where I slept soundly in a soft bed.

I was awoken early by birds singing and sounds of activity in the shop downstairs. They had an épicerie. Bright sunshine was streaming in through the skylight announcing a hot day. The shop always opened at about 6 a.m. and Sunday morning was the busiest time. I met Madame Page for the first time and immediately found her to be energetic, talkative, and unintelligible.

Lucienne informed me that her parents were expecting me to ask formally for her hand in marriage. She primed me on what to say. At lunchtime I said the words in faltering French. I did not get much reaction but at least there appeared to be no objection so I registered a success.

Madame Page had prepared one of her special Sunday meals that we ate after the shop closed at 12 o'clock. I was astonished by the number of courses she produced, seemingly without effort from her small kitchen. I cannot remember all the dishes but one of them was blanquette de veau with champignons de Paris. I was overwhelmed by the strong tastes and smells of the dishes the like of which I had never experienced before. My mother would have described it as too rich; the French would say, why not? I decided quickly that I liked French food. English food was tame in comparison.

I should say here that I was never able to call my parents-in-law anything other than "Monsieur" or "Madame". This always seemed perfectly satisfactory. The English language has lost such appellations from its vocabulary, replaced by things like "Hi".

Later in the day Lucienne took me round to visit their nearby house in Rue Branly that her parents had recently had built. It was just five minutes walk away. The house was unoccupied but fully furnished like a show house complete with crockery, cutlery and utensils. I understood it had been built as an investment with the notion that Lucienne might one day live there. Like many modern French houses the main living area was on the first floor. On the ground floor there was a double garage, a storeroom and a small rear bedroom. This is where Lucienne's parents wanted me to sleep while in Roanne in order to keep me well-separated from their daughter.

Lucienne's parents had bought her a car, a Citroën Ami 6, which we were able to use for Lucienne to show me round the area. We became attached to this quaint vehicle. The engine was just 602cc with two cylinders. It had a push-pull gear stick that stuck out of the dashboard. The suspension was very soft, typical Citroën. Lucienne kept her car in the garage at Rue Branly. We also had the use of bicycles, rather old-fashioned but fully working machines that M Page had constructed from discarded parts he had picked up from the local rubbish tip.

The next day we drove around Roanne and some of the neighbouring districts. At midday we found ourselves in the village of Renaison, famous for its dam, the Barrage de la Tache. We were sitting in the shade on the verandah of a café in the main square when a large white van arrived and parked opposite. A man jumped out to make a delivery to a shop. With some surprise, Lucienne recognised him immediately as her brother André and called out to greet him. He came over and we shook hands. Pointing at Lucienne, the first thing he said to me in strongly accented English was, "Bad woman", which he repeated. I did not think this was the best of introductions. Was he trying to put me off?

Lucienne told me that her brother was a cheese merchant and lived with his family in Mably, close to Roanne. Soon after this we visited André's house and met his wife Janine and their young son Philippe then about six years old. His sister Catherine was not born until a year later. The house itself was single storey with a kitchen, dining room, two bedrooms, and washing facilities in a separate out-house. Lucienne's father had built it all single-handedly some years earlier during the war, using whatever materials he could find. He was a roof-tiler by trade. There was a large garden mostly laid out for vegetables and soft fruit. Part of the garden was later to be split off for André's new house. The whole area was at that time rural with no proper roads. I enjoyed playing with young Philippe in the garden and found it a good way to learn French.

During my time in Roanne I was to enjoy a lot of Madame Page's cooking. We started every meal with either a salad or charcuterie, or sometimes soup if it was the evening meal. This was usually followed by a vegetable dish and then a meat course. Finally we ate cheese followed by fruit. All this was eaten with slices of fresh baguette to mop up the juices, washed down with Vichy water and Kiravi red wine. On special occasions there would be additional courses finishing with a final course of cake or vacherin.

I think Madame Page appreciated that I enjoyed whatever she put on the table. Everything she cooked tasted magnificent. This was helped by the fine quality of the ingredients, many of which were new to me. It seemed strange at first to eat vegetables by themselves as a separate course. But they had such a wonderful taste that any accompaniment would have adulterated them. Even a simple plate of boiled new potatoes was a delight.

Some of her dishes that I recall are: pommes dauphines, sautéed ratte potatoes, creamy potato mousseline with merguez sausages, sautéed green beans with mushrooms and lemon, blettes in béchamel sauce, petits pois au lard, aubergine fritters with tomato coulis (one of my favourite dishes), tomates farcies au riz, saucisson cuit aux choux (even now impossible to achieve in UK), roti de veau aux champignons, steak au jus, roast chicken with onions (cooked in a pot until the onions melted and caramelised), gateau de foie (a sort of souflée made with the minced livers of chicken or rabbit) served with coulis de tomates, chocolate éclairs made with choux pastry. Madame Page never appeared to use cookery books. When running out of ideas she cooked boulettes, the English equivalent of rissoles. This was a sign it was time to go.

One lunchtime Madame Page served up a dish which she announced to me was fried fish. It looked like small goujons cooked in batter. It turned out they were frogs legs, and very nice too. I think she thought I would not eat them if I knew what they really were. I do not recall her ever serving snails, but I did eat them early on in a restaurant and have thoroughly enjoyed the ritual and the taste ever since.

The days that summer were blisteringly hot with occasional violent thunderstorms. Lucienne and I spent nearly all our time together. We were lazy and happy. In between our many excursions we developed a routine of visiting Les Places des Promenades in the town centre before lunch for an aperitif, usually Martini Rouge in a tall glass with ice and a slice of lemon. Of course we smoked in those days, so our drinks were accompanied by a delicious Gauloise cigarette. The Promenades café was always busy with additional outdoor seating opposite on the park side of the road. Entertainment was provided by the waiters who had to brave the heavy traffic many hundreds of times each day carrying their trays of drinks above their heads. There was a pedestrian crossing but the car drivers ignored it.

After our evening meal we often drove to the Taverne Alsacienne on the main square to drink coffee or a digestif. The Taverne had a popular restaurant, their speciality being choucroute. This is one of the few dishes that I never appreciated, but then it is not really French.

One day André invited me to accompany him on his weekly visit to Lyon to stock up with cheeses. We left at the crack of dawn in his big van. We managed to converse together at a low level during the two hour drive to Lyon. We visited several large warehouses where we sampled many cheeses, some of which André bought. I was astonished at the variety. By mid-afternoon the back of the van was full. We drove back to Roanne but the day was not yet finished. We visited several outlying villages to deliver cheeses to local épiceries and restaurants. I was surprised at how hard André worked. His business was clearly doing well.

Encouraged by my interest in cheese, André invited me to join him one Sunday morning to help out at a market in a nearby village. We set up our stall at about 7 a.m. assisted by Janine and her elderly parents. I actually managed to sell some cheese to the locals. André later told me that my success was due to my cutting the cheese without its fair proportion of rind, something which benefited the customers but not André.

One afternoon we visited Janine's parents. They lived rather humbly in a small, first floor tenement in a run-down area of Roanne close to the Algerian quarter. All this area has now been demolished and rebuilt.

Lucienne and I went on several excursions in her car. Once we went for the day to the spa city of Vichy, about 70 km away. It was Thursday 4 June 1964. The first half of the journey to Lapalisse contained a long section of road with steep inclines and hairpin bends. In the middle of this section in a remote wooded area there was a roadside restaurant-bar for lorry drivers. We were to pass this restaurant on many future occasions, always noting with surprise that the large parking area was full to over-flowing with lorries, day or night. We concluded that there must be exciting entertainment in this establishment, but we never stopped to investigate further.

To return to our excursion, just as we left the twisty section of road a car appeared from nowhere and overtook us at breakneck speed. We watched as it disappeared from view swerving dangerously from left to right across the road. We thought it would be a good time to stop for a break, particularly as it was hot and approaching midday. We pulled in at a roadside restaurant a few kilometres outside Lapalisse, where we ate an enjoyable lunch with a bottle of Beaujolais. In those days there was no conscious connection between alcohol and road accidents. In any case, there were very few cars about.

When we finally arrived in Vichy we tasted the sulphurous water from the source, hired a pédalo on the river Allier, and walked the tree-lined streets viewing the expensive shops. This was the first of many day trips to Vichy over the years.

On another occasion we drove to the spa town of St Galmier about 50 km south of Roanne where we visited the casino. For some reason, only spa towns in France are allowed to have casinos. I casually threw a 1-franc chip onto a number in a game of roulette and won immediately, a sum of 36 francs. This was a lot of money. We celebrated by buying two cokes at the bar by the pool and were shocked to receive the bill which just about wiped out my winnings.

As part of my introduction to French food Lucienne taught me how to cut French bread. I practised on a baguette holding it against me and drawing the serrated knife across. I miscalculated and cut deeply into the forefinger of my left hand. The blood flowed abundantly and I wore a bandage for several days. I still have the scar. This injury occurred at Rue Branly where we were preparing for a big meal to celebrate our engagement and my 21$^{st}$ birthday on 6 June.

One of André's cheese customers was the 3-star Michelin restaurant Troisgros located in Roanne in a rather unlikely area right opposite the railway station. It was run by the brothers Pierre and Jean Troisgros who, together with Bocuse and Guérard, were the originators of "La Nouvelle Cuisine" that displaced the classic cuisine of Escoffier by introducing greater simplicity and elegance in the creation of dishes. Pierre's and André's sons went to school together. One lunchtime André and Janine invited Lucienne and me for a meal at Troisgros. This was quite an experience. In my estimation the measure of the quality of a meal is how much of it you can remember many years later. After more than 50 years I can still remember every detail of this meal. We started with paté de grives (thrush paté) followed by salmon à l'oseille (the now classic dish created by Monsieur Troisgros), then pièce de charolais à la moëlle á la sauce fleurie with a side dish of gratin dauphinois, a dish unknown at that time in England. The service was excellent but not pretentious. Yet I was surprised to find only goat cheeses on the large cheese board, a new earthy taste that I had not yet acquired.

I noticed that several families were eating at Troisgros. Although expensive, it was well within the reach of ordinary people, unlike the inflated prices in UK today. André recounted how they often ate at Troisgros and how his son Philippe always asked for steak and chips.

It will be evident that many of our activities in France were centred around food. The French talked about it in the same way that the English talk about the weather. In those first few days I was totally stunned by the quality of the food in the shops. For example, in butchers' shops I had been used to seeing slabs of meat not too far removed from the slaughterhouse. But the butchers shop round the corner from Lucienne's house prepared their meat with great artistry that made it easier to cook and present at table. It looked and tasted fabulous and there was no waste.

There were many cake shops in Roanne, the most famous being "Au Fidèle Berger". I was used to seeing dried out scones, currant buns, Eccles cakes and doughnuts. In French patisseries each cake was like a miniature work of art. Yes, they were more expensive, but faced with a choice between mediocrity and a small taste of heaven there was no contest. Over the years, there has been considerable improvement in the range of ingredients available in UK, but the basic poor standard of cakes in our bakeries is unchanged.

One Sunday lunchtime after the shop closed we drove with Lucienne's parents along the picturesque road by the Loire to Chateau de la Roche, where there was a popular restaurant set by the river and overlooking the chateau. The restaurant was renowned for its speciality menu of fritures (whitebait) followed by fromage blanc. The restaurant interior and the large terrace outside were packed with families enjoying themselves. As far as I could see everyone was eating the speciality menu. It was certainly very good. It is difficult to produce fritures that are light and crisp without any hint of sogginess or bitterness. You have to use the best quality oil and change it regularly. It was an experience for me to eat the little fish whole, crunchy heads and all. Some years later this section of the Loire was dammed to create a new water supply for the area, with a resulting rise in the water level by several metres. The chateau is still there but the restaurant was forced to relocate to higher ground, no longer with quite the same rustic allure.

On another occasion all of us, including André and his family and two cousins Hélène and Raymond who had turned up from Paris, drove to Villerest, a small village on high ground overlooking the Loire. Lucienne and I were passengers in André's swishy Citroën DS with its comfortable soft suspension. We went to a small Bar-Restaurant where we had lunch. I think it was another of André's cheese customers. The interior of the bar was rather drab and did not look encouraging, and we had to wait ages. We drank several Pastis which were taking effect. When the meal eventually arrived it turned out to be excellent. I had a robust boeuf bourguignon, my first experience of this dish. It impressed me with its strong taste of tomato, garlic and thyme. There used to be so many of these small restaurants in France that served high quality family dishes with no frills at low prices. They still exist, but with the advent of fast food they are fewer.

I learned quickly about the many cultural differences when it came to food. A significant one was the protocol when you invited guests to eat with you. In England the highest compliment is to invite them to your house to eat. In France it is to invite them to a restaurant. This is because restaurant food in France cooked by professional chefs is always better than home-cooked food. In UK we often see restaurants advertising "home-cooked food". The French would see no point in going out to eat a meal that is no better than at home.

In England the correct etiquette at meal times is to rest your hands in your lap, whereas in France they should be rested on the edge of the table (to show you were not carrying a weapon). You do not bite your bread, but tear it then put it on the table beside your plate. The French rarely use side plates. You use your bread to mop up juices on your plate. When you have finished you do not place your knife and fork on your plate, but on the table beside your plate in case you need them for the next course.

I noticed other curious differences. Whereas the English like to open the windows in their houses, the French tend to keep them closed. This is because they have no particular liking for fresh air. It is also because closed windows on a hot day help to keep the interior of a house cool, particularly if there are shutters.

Madame Page had a brother Charles (Tonton Charly). He lived with his wife Hélène in a house in Mably on a large plot of land right opposite André's house. Tonton Charly had two daughters Christiane and Nicole, then aged about 13 and 8 respectively. The families of Charles and André did not appear to talk to each other, even though they were next-door neighbours. This may have been because Hélène was a rather difficult person prone to making spiteful comments, so I am told. It may also have been because André suspected Hélène of poisoning his beloved guard-dog which she claimed barked too loudly. Tonton Charly on the other hand was quietly spoken, with a gentle character not unlike my Uncle Frank's.

We often visited their house where Tatan Hélène was always most hospitable. She would insist that we have a glass of panaché (beer shandy), a drink that Lucienne detested. It did not matter how many times we said "Non Merci" even to the point of being rude, we always finished up with a glass of panaché. Tatan Hélène just could not accept that some people did not like it.

One Sunday she invited us all for lunch, the first of several at her house. We ate outside in the summer heat on a long table shaded by a tree. I enjoyed the food and the relaxed atmosphere. For me, the highlight of the meal was the green beans sautéed with fresh mushrooms. The supply of sautéed beans from her kitchen seemed to be limitless and I ate them with abandon. I learned they were grown in their enormous garden, which seemed to be devoted entirely to French beans and strawberries.

The épicerie of Lucienne's parents was always busy with a clientèle drawn largely from the families who worked at the nearby paper mill. These families lived on a large housing estate built by the mill for their employees. The shop was open from Tuesday to Saturday from 6 a.m. to 6 p.m. or later, with a three-hour lunch break from 2 to 5 o'clock. On Sundays they closed at midday and stayed closed all day Monday. They sold fresh fruit and vegetables, tinned food, general groceries, charcuterie, cheeses, milk, wine, mineral water, children's sweets, and cleaning products. Supermarkets had only just started and business in the shop still managed to thrive.

Mme Page worked all day in the shop and had lengthy dialogues with many of the customers - monologues in the case of Mme Ruffaud, the only woman who could out-talk Mme Page. M Page assisted in the shop and also went out in his van loaded with merchandise that he sold on his weekly round and at several local markets, where he was known as "Le roi de la banane".

All the non-consumables were stored in the extensive cellars and in "le petit magasin" that led off the back yard. The latter contained stores of cleaning products, packets of pasta and such like. The cellars were like an Aladdin's cave where consumables with a long shelf-life were stored, including arachide cooking oils, wine vinegars, mountains of tins of pâtés de foie, champignons de Paris, haricots verts and petits pois, both fins and extra fins, as well as the better wines like champagne. The everyday wines and mineral waters were stored in the stairwell.

The cellars were dark and creepy, especially at night. The entrance was at the bottom of the stairwell through a heavy wooden door that was kept locked. The key was hidden behind a pipe that ran along the lintel above the door. You had to grope for it in the dark. Once inside you could turn on the light. The cellar was to supply us with French food to take back to England for many years, long after the shop closed.

While I was in France I received a letter forwarded to me from my parents with my exam result. I opened it with trepidation. I was delighted to have been awarded an Upper Second. I would have been disappointed with a Lower Second and I never expected a First, which were rarely awarded. In mathematics you had to show exceptional creativity for a First. You did not get one by hard work and revision. In fact, if you had to work hard you were probably not First Class material. In some years no Firsts were awarded. In my year of about 30 maths graduates there was just one, obtained by a student called Eric Towers whom I had never really noticed.

I can say categorically that degree grades have since been dumbed down. To get an idea of how hard it was to obtain a First Class degree in the 1960s, I quote the following figures from Bristol University for my graduation year of 1964: Arts Faculty, ~200 Arts graduates, 2 Firsts awarded; Science Faculty, ~100 graduates, 6 Firsts awarded; Law Faculty, ~50 graduates, no Firsts awarded that year. Today, 30% of students across UK are awarded Firsts. Are students now cleverer or have standards dropped?

So I was happy with my result. The Degree Congregation to which I was invited took place on Wednesday, 1 July 1964 at 1800. I was in France and could not attend. I was awarded my degree *in absentia*.

At the end of this holiday in early August there were tears from Lucienne's mother. I think she liked me well enough but was sad about the prospect of losing her daughter to a distant land. Lucienne and I returned together by train and ferry to England where we went our separate ways, Lucienne to Bristol and me to Southend. I soon returned to Bristol to meet Lucienne and bring her back to Southend. My father very generously lent me the use of his trusted Ford Anglia for this purpose. At Clifton College we had a semi-formal meeting with Mr Lane, Lucienne's employer and the Head of School House, and his wife Mrs Lane. They congratulated us on our engagement and on my exam results. We then drove back to Southend to begin the next phase of our lives.

My parents organised my belated joint 21$^{st}$ birthday and engagement party in mid-August 1964. It took place at our house in Henley Crescent. It was attended by myself and Lucienne, my paternal grandparents (my mother's parents had died some years earlier), my Auntie Margaret and Uncle Frank, my mother's brother Harold and his wife Jennie, my cousin John and his girlfriend Sheila whom he later married, and Auntie Haidée and family. It was a happy occasion.

# My First Job

Lucienne's arrival at Henley Crescent necessitated some adjustments to the household arrangements. Basically, Nick, then aged 12, moved into my bedroom to share with me in two single beds, while Lucienne took Nick's small corner bedroom at the front. My mother made a big effort to help Lucienne feel at home and I think she was generally successful. There were inevitable tensions, notably between Lucienne and my father, but nothing too serious. We all settled down to the new routine.

Lucienne met our neighbours, relatives and friends, and was a great hit. I was proud of her. She was not too enamoured with my friends Richard Coakes, Dave Hall and Pete Wyss who were not very tactful. She got on particularly well with my paternal grandmother who lived with her husband Tom, daughter Margaret and son Frank, at their bungalow in Hobleythick Lane. Lucienne always had a close affinity with my grandmother, who confided things to her that were otherwise unspoken within the family and about which I have subsequently learned. Lucienne has always got on well with elderly people and young children.

At the start of September 1964 I started work at the National Provident Institution, known as NPI. I was assigned to their head office at 48 Gracechurch Street in the City of London. It was a short stone's throw from the Tower of London, the Stock Exchange, the Bank of England and other well-known monuments. It was also close to Fenchurch Street mainline station. For this reason I chose to travel to London on the railway line from Westcliff to Fenchurch Street rather than on the alternative line from Prittlewell to Liverpool Street.

Working in the City was exciting. Part of the uniform of a City Gent was a dark suit and tie, but I never wore a bowler hat or carried the customary rolled umbrella. My starting salary was £720 per annum out of which I paid £10-6-0 per month to British Rail for my season ticket and £16 per month to my mother towards my keep. After income tax and National

Insurance I was left with about £25 a month. I was still better off than when I was a student.

NPI had a distinguished history of selling life assurance and pensions. When it was established in 1835, an advertisement in The Times newspaper offered nine types of insurance and invited readers to apply. The NPI prospectus reminded readers that: "The man who daily expends upon himself more than is necessary for his proper sustenance, and takes no thought for the period of age and decrepitude which is likely sooner or later to befall him, is deficient in that prudence which is the handmaid of every virtue". In 1842 NPI moved its offices to Gracechurch Street, London, where I was now working. In 1966, the year after I left, NPI moved to a new head office in Tunbridge Wells.

On my first day at NPI I was given a short induction tour of the various departments, together with two other actuarial students who were starting at the same time. It was all rather daunting and I can't say I was overwhelmed with interest by what I saw. It seemed to be a lot of tedious paperwork and record keeping. I was assigned to a section dealing with group pension schemes set up by companies for their employees, as opposed to personal pension policies for individuals.

The section head instructed me on the work I would be doing. He also explained my actuarial training. Basically, it was "on the job" training supplemented by a great deal of private study for the actuarial exams. There were four levels of examination: preliminary, intermediate, associate, and fellowship. My degree in maths exempted me from the preliminary exams, so I started at the intermediate level. I was told it took an average of 7 to 10 years to fully qualify as an actuary at fellowship level. The record was 4 years but it had been known to take as long as 30 years. Provided you or your company paid the fees you could keep taking the exams every year until you passed. Some people never qualified but still managed to carve out a useful career for themselves. My section head was a case in point.

I worked in an open plan office where I was given a desk opposite someone called Roger Friend who commuted each day from Guildford. He was good company and knew the ropes. I had my own telephone that scared me a bit since I was not used to taking calls. Fortunately my phone was fairly silent.

Each company pension scheme managed by NPI had its own drawer of files. They consisted entirely of paper records because computers were only just coming in and there was much resistance to the new technology. The paper records included card indexes with one card for each employee in the scheme giving details of their age, annual salary and pension contributions. Most of my work was about keeping these records up to date when employees had salary increases, retired or died, or when they left the company's employment or new people joined. Out of interest it may be noted that pensions in those days were not readily transportable. There was therefore a disincentive for people to change employers. On the other hand, most workers did not expect to change employers. Rather, they expected a job for life.

My favourite task was preparing quotes for new pension schemes. Today there is a plethora of pension products available whereas in those days there were only two basic options: with-profits or without-profits. With-profits policies had lower guaranteed returns but shared in NPI's profits with the prospect of generous bonuses. Without-profits policies had higher guaranteed returns but did not benefit from any NPI profits. The most important part of the quote was the determination of the annual premium that the company would be required to pay to fund the scheme. I was taught to calculate this by an iterative process. You first made a guess at the premium, then worked through the consequences of this for each employee. This determined whether the scheme would be over- or under-funded, leading to a refinement to the first guess of the annual premium. You continued this process until the answer did not appreciably change. I did all this using a clumsy mechanical calculator operated by hand. It took about one day to complete the operation, which then had to be checked by my section head. The complete calculation would now be done in a few seconds on a computer.

In October 1964, shortly after I started work, my father was inaugurated as Worshipful Master of the Thundersley Masonic Lodge. This was a moment of great pride for him, the pinnacle of his Masonic career and far more important to him than any achievements in his day job. The celebrations included a dinner-dance to which wives and guests were invited. It took place in the banqueting hall and ballroom of Garons in Southend High Street. In view of the importance of the occasion he invited Lucienne and me to attend, and also my Uncle Frank. My mother and father were of course hosts of honour and presided at the centre of the top table. After the dinner my father made a speech, something that came to him quite easily.

It was customary for the wife of the incumbent Master also to say a few words. My mother was paralysed by the prospect. Several weeks in advance she went to the doctor who prescribed her a course of tranquillisers. She practised in earnest what she wanted to say. On the evening of the event she must have been a bit comatose. I have no recollection of what she said but it went down a treat with much applause. She was greatly relieved. In her own way Lucienne also found the evening a bit stressful. She was not used to this sort of showy event. My mother and father danced the first waltz. Everyone said what a fine dancer my father was, but at the time I was not able to judge.

Meanwhile, my work at NPI and my actuarial studies seemed like unrelated activities. I had a syllabus to work through and several set books to read, together with sets of exercise questions and past examination papers. I did most of this work on the train journeys between Southend and London, an hour in the morning and an hour in the evening. It made the journey pass quickly. I was given Wednesday afternoons off each week to visit the Institute of Actuaries in High Holborn where we could do private study and resolve any problems with a qualified person who was usually available.

I thoroughly enjoyed my studies. They opened up new avenues of interesting maths in the fields of statistics, probability and game theory. Here are two typical problems in game theory:

1. What is the expected number of spins to achieve a sequence HTTHT with an unbiased coin? (The answer is 36).
2. Two players X and Y take turns to throw a die. At each throw they note the score obtained, between 1 and 6. The winner is the first player to gain a lead of 20 or more. What is the probability that X wins if Y throws first and scores 6? (This is a more difficult problem).

I found the methods for solving these problems intriguing and quite useful in later life.

It was becoming clear to me that actuaries, of whom there were about 10 at NPI, were treated rather like gods. They were far superior to accountants. They all had senior positions and were reputedly on massive salaries. They were also unapproachable. There was one exception, a recently qualified actuary called Malcolm Taylor who joined us for the weekly visit to the pub on Friday lunch times. We usually went to a pub across the road quite close to The Monument and Pudding Lane, where the Great Fire of London is supposed to have started.

Occasionally I met up with my father for lunch, and sometimes with my friend Richard Coakes who was still working in The Strand. But usually I just bought a sandwich to eat in the Staff Common Room during the obligatory lunch hour. My favourite sandwich bar was just down the road towards the Tower. NPI did not have a staff canteen, which was not unusual for offices in the City of London because there were plenty of food outlets in the vicinity.

Every morning and evening I walked to and from Westcliff station, a distance each way of about two miles. Some evenings Lucienne walked to meet me, at least until she too started work. She was anxious not to waste her time and wanted to earn some money. She found a job soldering electric circuit boards at E K Cole Ltd, a well-known TV and radio manufacturing company. She walked every day to the factory located in Thornford Gardens, about a mile away from Henley Crescent. Lucienne did this job full-time for several months whilst she lived with us. It was in the factory canteen that she was first introduced to jam roly-poly with custard.

Lucienne spent Christmas with us and was surprised by our curious English customs and fayre. At the New Year she returned to France to spend a few days with her family. She made the journey via Haidée's house in Kenton where we both stayed overnight on the eve of her departure, me sleeping on the expansive sofa in their front room. Early the next morning Uncle Billie drove us both to Heathrow where I saw Lucienne off.

Early in the New Year of 1965 I had some correspondence with my old friend Robbers who was back at Bristol University and staying in Baldock Hall, one of the university halls of residence. In those days the internet and emails did not exist. Telephone calls were possible but usually as a last resort. The normal way of corresponding was by letter, and the short delay never seemed to be a problem. I exchanged letters with Robbers and arranged to meet him in Bristol. In early February Lucienne and I travelled there by train. Robbers met us with his car at Bristol Temple Meads Station. We joined up with Dave Patrick and his girlfriend. Dave was in his first year as a PhD student. We visited the Rummer for sherries and continued for the inevitable curry. Robbers seemed to be a bit more like his old self but there was still a barrier between us, no doubt because there remained so many unspoken words about his attempted suicide. That was the last time I was to see Robbers. Lucienne and I stayed the night at Dave Patrick's house that he was renting with some other post-grads. We listened to Ella Fitzgerald before turning in.

I continued with my job at NPI and with my actuarial studies. In March I was transferred from the Pensions Department to the Investment Department. This was where all NPI's money was invested to generate the company profits and fund the bonuses for its with-profits customers. I never understood at the time what stocks, shares and debentures were all about and the section head could not have been too impressed with me. But my studies were going well and in early May I sat the intermediate exams of the Institute of Actuaries in High Holborn. On the last day of the exams Lucienne met me outside the examination hall and we went for a pub meal together to celebrate, returning to Southend in the evening on the late train from Liverpool Street.

By this time we had decided that we would marry in July in Lucienne's home town of Roanne. Her mother had already initiated things. It was understood that after our marriage we would live in England, but I don't think our plans were concrete. In fact, I was already wondering if I had chosen the right career.

Lucienne announced that she would be returning to France to be with her family for the last two months before our marriage. I was a bit devastated at the prospect of such a long separation, but I understood that this could well be her last opportunity to spend any length of time with her parents.

Lucienne left for France from Southend pier in mid-May. I took her in my father's car down Southend High Street and parked in Royal Mews at the top of Pier Hill. I carried Lucienne's suitcase the short distance to the pier where we caught the train. At the end of the pier we said goodbye and Lucienne climbed on board the ship. The moment of separation was very emotional. I walked all the way back down the pier in a daze. I walked up the High Street and caught a bus from Victoria Circus. When I got home I realised I had forgotten the car.

I immersed myself in my work routine. With my actuarial exams behind me I used the time on the train to study French. I bought a useful French primer that helped turn my school French into a more practical subject. I was also helped by having a strong incentive.

At about this time I had a phone call one evening from the Bristol police. I was shocked to be told that Robbers had disappeared from Hall. He must have cracked up again under the strain of impending exams. They asked me if I had any idea where he might be. I was not much help. Some time later I learned that Robbers had eventually been found unharmed, but he was never to complete his veterinary course. I had no idea what became of him until recently when, in August 2015, I attended a Very Old Boys reunion at Southend High School where I met up with many old friends whom I had not seen for 50 years or more. One of them was Andrew Sargent, a contemporary of mine at Bristol University. He had learned from a third party that Robbers had killed himself by jumping in front of a train at Islington Underground Station as a result of depression. I was shocked and looked into it further, to discover that he died on 14 February 1973 in Royal Holloway Hospital, London, as a result of multiple injuries from impact with the train. At the time Robert had been working as a Laboratory Technician and living at 50 Drapers Road, Enfield. The coroner recorded an open verdict. Robert had been a good friend at university and I wished that I had made a greater effort to keep in touch.

My continued unease with my job made me think about the possibility of returning to Bristol to do postgraduate study, like my friend Dave Patrick. As I turned the idea over it appealed to me more and more. I took a day off work and travelled by train to Bristol to discuss the options with the head of the Maths Department, Professor Howarth. He looked up the marks I had obtained in my Finals and told me I had been close to a First. This news encouraged me. Professor Howarth said that I was good enough to do research, but in view of my year's absence from university I would be a bit rusty and should first do a one-year MSc course. If that went well I could transfer to a PhD which I would then have to complete in just two years. All this was dependent on me obtaining funds via a Science Research Council (SRC) grant. The university would apply on my behalf. I would not know the outcome until August. I agreed to go ahead on this basis.

I communicated with Lucienne by letter and occasionally by phone. Her parents did not have a phone but I think Lucienne used her neighbour's. Our wedding date was fixed for

Saturday, 17 July 1965. The preparations were well underway for the venue and reception in Roanne, as well as the wedding dress. Richard Coakes had agreed to be my best man and Lucienne's friend Liliane Carfantan had agreed to be chief bridesmaid.

After consulting with me and my parents, Lucienne sent out wedding invitations to my relatives and closest friends. As was the French custom, she also sent out cards announcing the wedding (in French) to many of my other friends and their parents. These people mistook the announcement for an invitation. Fortunately, none of them felt able to make the long journey to Roanne and so declined. It turned out alright.

I told NPI that I would be getting married and asked if they would give me three weeks off. They said that since I had not even been there a year and I had already taken some leave it was out of the question. This galvanised me into action. I gave them notice that I would be leaving at the beginning of July. On the same day that I gave notice I received the results of my actuarial exams. I had passed with marks in the top 20% and was now a qualified actuary at intermediate level. NPI tried to persuade me to stay saying that I had a bright future. I said that insurance work was not for me and my mind was made up.

My last day at the National Provident Institution was Friday, 2 July 1965. I was to be married with no job and no certainty of any funding as a post-grad in Bristol. I consoled myself with the thought that I could always live with Lucienne in France in her house in Rue Branly. I was not too worried. The important thing was to be with Lucienne.

# Marriage

The day I left the National Provident Institution there was just two weeks to go to our wedding. I had a stag night with several of my school friends. It was a relatively sober affair but we had a good time. We went to a few pubs in Southend and finished up at a curry house in Southchurch Road. One of my friends smuggled a plate out of the restaurant and they all signed their names on it. I still have this plate somewhere, the signatures legible and stained with curry. It would be another 50 years before I was to see these school friends again.

Meanwhile, my father was planning the trip to Roanne. Without motorways it was a two-day journey by car. My parents had never been abroad before let alone experienced driving on the right. My father had booked the car ferry from Dover to Calais well in advance and had obtained from the AA a recommended route through France. It was rather a long-winded route designed to avoid Paris, taking us from Calais to St Omer, Béthune, Arras, Bapaume, Peronne, Soisson, Chateau-Thierry (where we were to stay overnight), Provins, Nemours, Montargis, Briare, La Charité, Nevers, Moulins, Lapalisse, Roanne. This was a veritable expedition for my parents and they were both anxious. My father also had to obtain additional insurance for driving in France. This entailed applying for the mysterious "Green Card" for which he had to pay an extra premium.

Our departure was planned for Saturday, 10 July, a week in advance of the wedding. We set off from Southend in the Ford Anglia at about 4:30 in the morning, my mother in the front with my father driving, and my brother and I squashed in the back. My brother Nick was about thirteen at the time. The small car was heavily loaded with a full boot and roof rack. We drove up the Old Southend Road to the Blackwell Tunnel to take us under the Thames, then down the old A2 through Rochester towards Canterbury and Dover.

Somewhere near Rochester my father began complaining of a pain in his abdomen. It got worse and we stopped. I assumed it was his ulcer playing up again brought on by the worry of the journey. I don't think he had slept the night before. My mother began saying that we should return home and I would have to make my own way to France. My father quickly rejected this. In the end I took over the driving and my father seemed to improve.

We arrived in Dover in good time for the 8:30 ferry. The two-hour crossing went without incident and we set off through France taking it in turns to drive. It was quite a marathon as we slowly ticked off the kilometres and the towns along mostly quiet rural roads. My parents were greatly relieved when around 5 o'clock we arrived at our hotel in Chateau-Thierry. It was a warm evening and we installed ourselves, quite relaxed, on the sunny terrace with pre-dinner drinks and the menu. I translated for my parents and they had their first taste of French food. We all enjoyed it and my father seemed to be much better.

The next day we set off on the second leg of our journey, which went much more smoothly. We arrived in Roanne at 4:30 in the afternoon. I was overjoyed to see Lucienne again after such a long separation. There was a chemistry between us that has never waned.

Our respective parents met each other and exchanged polite conversation through our interpreting. Madame Page was keen to honour my parents and brother and put them up in the two rooms over the shop on the first floor. Lucienne was sleeping in the back bedroom and I was assigned to the basement room in the other house at Rue Branly.

Lucienne and I had a busy schedule over the next few days. We had to make visits to the Mairie to complete certain formalities about our marriage, complicated by the fact that I was a foreigner. Then we had to meet the priest, l'Abbé Baroyer, who was to marry us at the church of St Etienne near the town centre of Roanne, this time complicated by the fact that I was not a Catholic.

We also both had to have medicals. The law in France was that each partner in an intended marriage had to be checked over to ensure they were aware of their physical state. But you were not obliged to show your partner the results. We had our medicals with Dr Boël, the family doctor.

Lucienne and her mother were also busy with last minute preparations for the big day. All this time the shop remained open requiring constant attention. I was swept along and just went with the flow.

Meanwhile my parents were finding it difficult to adjust and fit in. They were suddenly immersed in a completely different culture with a language barrier, and no doubt felt helpless and surplus to requirements. They also had trouble coping with the domestic arrangements with only a sink to wash in and no flushing toilet.

The three other family guests then arrived from UK; my Aunt Margaret, Uncle Frank, and Haidée. Mme Page had booked them into a hotel somewhere in town but they ate most of their meals at Rue de Matel. There were a lot of people at table for Mme Page to cater for and she had to pull out all the stops. I think my parents would have been happier in a hotel, but as guests of honour of M and Mme Page that was out of the question. I suppose they might have stayed in the new house at Rue Branly, but I have the impression that Mme Page viewed the principal bedroom as a sort of shrine reserved for the use of her daughter after she was married.

I was not really aware of the many tensions that were building up. They reached an explosive peak on 14 July, Bastille Day, one of the most important public holidays in France. It was a Wednesday. The shop was closed for the day and Mme Page was busy preparing a sumptuous evening meal at Rue Branly to which everyone was invited. Late in the afternoon my father asked me to follow him upstairs to his room, a rather strange request. He proceeded to deliver a prepared speech to the effect that I had neglected them since our arrival and if I did not want them to be there they would go home. I was stunned and shocked by this outburst and could not imagine what had prompted it. I could think of nothing to say and walked out of the house.

Feeling totally distraught I started walking the streets. I walked for several miles first towards the station then through the back streets of the town centre. It was a hot evening. I found myself in an open area where fairground entertainments had been erected and were in full swing to celebrate Bastille Day. There was much noise and merriment. It was now past seven in the evening and I realised that the time for the big dinner at Rue Branly had been and gone. I felt guilty about messing things up. I wanted to marry Lucienne but did not want to go through all this fuss. I continued walking through the town, along the canal, and eventually at dusk I returned to Rue Branly to find the house deserted. Soon after this Lucienne arrived at the house. She had been worried and had been searching for me everywhere. I tried to explain what had happened and I think she understood. She said that the meal had gone ahead but was a bit subdued.

The next day I saw my parents. My mother was upset and my father full of remorse. No doubt Aunt Haidée had helped smooth things over. We quickly resumed where we had left off with the final preparations for the wedding. I think my father was still suffering from stomach problems.

On the eve of the wedding my friends Dick Coakes and Dave Hall arrived late afternoon. I was pleased to see them because I had not been confident they would turn up. Everyone was now here and we all ate together at Rue de Matel.

Dick Coakes slept that night in a second camp bed installed in the basement room with me at Rue Branly. I think Dave Hall slept in the smaller bedroom upstairs. Before going to sleep Dick warned me that he often talked loudly in his sleep and not to be worried. I heard nothing and slept soundly.

The schedule on the wedding day had all been carefully worked out by Mme Page. I was oblivious to the arrangements and just did as I was told. At about 8 a.m. I drove Lucienne in the Ami 6 to her hairdressers in Rue Anatole France. I returned to Rue de Matel to await instructions and put on my suit.

The first main event was the civic marriage ceremony that was quite separate from the church ceremony. Lucienne's brother acted as chauffeur in his luxurious Citroën DS and drove me to the Mairie. I don't know how Lucienne got there. Everyone else was transported in a coach hired by Mme Page for the day. The French guests were André, Janine and Philippe who was a page boy, Tonton Charly and Tatan Hélène with their daughters Christiane and Nicole as bridesmaids, plus several family friends and neighbours. Another neighbour looked after baby Catherine, Philippe's sister.

When I saw Lucienne in her wedding dress at the Mairie I was stunned. I knew I was making the right choice. The civil wedding took place at around 11 a.m. I had to repeat several sentences in French most of which I did not understand. We then signed some documents. After the ceremony we all walked out of the Mairie and down the steps to the main square. My mother asked Lucienne if we were now married. Perhaps she was not sure whether this was the right moment to congratulate us.

The next phase was the church ceremony. Everyone piled into the coach except Lucienne and me who travelled in the back of André's DS. He took us on a strange tour of the back streets of Roanne. We passed the church at least once and then continued on another grand tour. I did not ask any questions and probably assumed that we were just a bit early. A year later I learned that the reason for the delay was a funeral taking place at the church, arranged at the last minute. Unknown to Lucienne and me it was the funeral of Janine's uncle. Nobody but the close family was told about the funeral for fear of putting a damper on our wedding ceremony.

Eventually we stopped outside the church of St Etienne. Lucienne went in before all the other guests. Curiously, it is the custom in France for the groom to arrive last. Because I was not Catholic we were not entitled to any organ music, but I did not really notice. I was being swept along on a tidal wave. The ceremony went according to plan with all the attendants playing their parts admirably. We processed out of the church to encounter another of Mme Page's surprises, an impressive "guard of honour" of people holding French and English flags (Plate 29).

We drove to Rue Branly for champagne, canapés, ham rolls, and more official photos. I learned many years later that the caterers Mme Page had booked did not turn up, so she had to improvise at the last minute making her own canapés and ham rolls. I did not notice anything amiss. In the afternoon we all drove to Le Barrage de la Tache to take refreshments and relax, admire the views, and pose for yet more photos.

The next big event in the day's programme was the evening banquet. This was held at the Grand Hotel over the road from the Troisgros hotel and restaurant. By this time everything was blurred. We sat down to a formal meal. Speeches are not an important part of French wedding protocol, but at the end of the meal I stood up and managed to make a short speech in French and in English.

There followed a ritual in which the best man had to disappear under the table to extract the bride's garter. Later in the evening the garter was sold in a Dutch auction, the garter going to the last person to pay something. The money raised was to go towards our honeymoon. Unfortunately, the auction was a bit of a disaster because the English participants did not understand what was going on. Only a few francs were raised, which were generously supplemented by Lucienne's brother André and my father. The garter episode was later reported by Dave Hall in the Southend Standard.

After the dinner there was a live band with dancing and we managed to do the first waltz, probably a Viennese. We disappeared before the end of the festivities and stayed that night in Hotel Troisgros in a room on the top floor.

The next morning we did not want to eat breakfast in the hotel or at any café where we might be recognised. By prior arrangement the Ami 6 was parked outside our hotel. So we loaded all our stuff and drove off to a rather seedy bar near the cemetery where we had coffee and croissants.

Rather mysteriously, we found Haidée's slippers on the back seat of the car. We drove round to Rue de Matel and dropped them off outside the back door. We managed to do this without being seen by anyone. Then we set off on our honeymoon.

We had no particular plan and not much money. The general idea was simply to head south. After we had driven for about an hour Lucienne happened to look in the glove compartment to find that all the car papers were missing. This was serious because it was illegal in France to travel without them. We stopped in a village wondering where they had gone and debated what to do next.

In the end Lucienne went to a public phone box and rang Mme Pommet, her parents' next-door neighbour. She was able to get Lucienne's father to the phone. He said he had broken into the car the night before and hidden the papers under the back seat for security. He must have climbed in through the boot because the car was locked. This all seemed to us a most peculiar stunt, hardly believable and totally out of character. Perhaps he had had too much to drink, again unusual. We found the papers and asked no more questions about them or the slippers. We continued our journey.

It was a hot day and as we drove further south it became even hotter. We were carefree not really knowing where we were going. We drove through Lyon, Vienne, Valence, Montélimar, eventually arriving in Avignon late afternoon. We parked on the main square beneath the Palais des Papes, something that is now impossible to do. The atmosphere of the square was animated with street entertainers, musicians and restaurants plying for business, their tables extending onto the main square. We selected one of the restaurants and ate outside in the warmth of the evening. There were fixed price menus at 7 Fr and 10 Fr (about £1). We chose the cheaper one. We found a garret room for the night in a small hotel on the corner of the square.

The next day we left the hotel and drove a few miles outside Avignon to the Pont du Gard, a notable ancient Roman aqueduct bridge crossing the river Gard. Something happened that almost put an end to our short-lived marriage. The water looked inviting so we decided to go for a swim. We walked across the bridge and explored the river banks where we found a cave suitable for changing into our swimming costumes. The water was cool, refreshing and still. We struck out across the river aiming for a beach on the other side. But on reaching the middle we encountered strong currents that swept us downstream away from our intended landfall. When we eventually reached the other side there was no longer any beach, only a shear rock face. We were out of our depths with nowhere to rest. The only option was to swim back to where we had come from. Exhausted in mid-stream we really thought we might drown. We finally reached safety feeling rather shaken.

We continued driving southwards bearing right towards Nîmes rather than left towards the more expensive Côtes d'Azur. In the evening we arrived at the Mediterranean town of Le Grau-du-Roi on the edge of the Camargues. We found a room in a hotel on the waterfront and celebrated that evening with a shellfish meal. We were both violently sick all night and part of the next day. We stayed there one more night but did not eat in the restaurant. In fact, I don't think we ate anything that day. In the afternoon Lucienne went sunbathing on the beach. I stayed in the shade. This was the first time we realised we had different attitudes towards the sun.

We drove on to nearby Aigues-Mortes, an ancient walled city with history going back to Roman times. We did some sightseeing and may have stayed one night there.

By this time we were running short of money and started heading back north. We visited some distant cousins on Mme Page's side of the family and stayed with them at least one night. They lived in a rambling farmhouse near Salindres, north of Nîmes. The stairs to the first floor were located outside on the edge of a central courtyard. There were several generations living in the house including two elderly women. They disapproved of the modern style of Lucienne's engagement and wedding rings.

We now had only just enough money for the petrol back to Roanne. We arrived at Rue de Matel on the Saturday having been away for a week. We had not thought to take a camera with us so there is no photographic record of our honeymoon.

M and Mme Page welcomed us back. We installed ourselves in the new house in Rue Branly where we slept in the reserved principal bedroom upstairs. We ate most of our meals together at Rue de Matel.

Meanwhile, my parents and the other English guests had left the week before. I think that my parents, Margaret and Haidée went straight home. Frank spent a weekend in Paris. Dick Coakes and Dave Hall went to visit Lucienne's best friend Liliane. The night before my parents left, my father wanted to tell M Page that they would be leaving at six in the morning. He said in his best French something that sounded like "six eux". In the morning M Page presented my parents with six hard-boiled eggs.

We were happy in our new married life but had little money. Our financial horizon was no more than a few weeks ahead. We did not want to be dependent on Lucienne's parents so we both needed to find work, at least until I knew what my future would be. Lucienne found employment fairly quickly at Prisunic, a nationwide department store that bordered on being a supermarket but never quite made it. There were two Prisunic shops in Roanne, the main one in the centre of town in Rue du Lycée (later to be renamed Rue Charles de Gaulle when it was pedestrianised), and the other smaller one in the suburb of Le Côteau. Lucienne worked at the latter. She travelled there every day on her bicycle, a distance of some two miles, working from 0800 to 1200 and 1400 to 1800, returning home for lunch. She worked on the meat counter.

Through her job at Prisunic Lucienne found a possible opening for me. The "chauffeur de camion" (van driver) was due to take his month's holiday in August and they needed someone to fill in. I was given an appointment to see the director, M Michau, who was based in the main store at Rue du Lycée. I found my way to the store and was ushered to his office on the top floor. M Michau was a tall man of military bearing, meticulously dressed in a suit, tie and waistcoat. He had great presence. I sensed immediately that all the staff were terrified of him. He asked me a few relevant questions, like whether I could drive, and I got the job.

For the first week I shadowed the driver before he disappeared on his annual holiday. I soon learned the routine. Most of the day was spent making frequent journeys to the rubbish dump behind the abattoir with van-loads of flattened cardboard boxes. There was always a mountain of boxes in the back yard of the store, dumped by all the various departments when they restocked. We had to keep on top of it. We gave ourselves breaks mid-morning and mid-afternoon for refreshments. This comprised a "petit canon" in the form of a glass or two of dry white wine taken in the rural bar behind the abattoir. The other main job during the day was ferrying the director, M Michau, and his deputy, M Andouche, to and fro between the two stores in Roanne and Le Côteau. At the end of each day I drove to the big warehouse at Faubourg Clermont, a suburb of Roanne, to load up the van with all the orders for the next day that had been telephoned through from the various departments in the store. The very last job was to drive back to Roanne and park the van in the back yard of the store ready for unloading the next morning. I usually got back to Rue de Matel on my bike about 6:30 p.m. where I met up with Lucienne for our evening meal.

The next morning the first job was always to unload the van, carrying the boxes to the correct "reassorts" (reassortiments) in the store. I had to start this at 0730 in order to be finished by 0800 when the store opened. The unloading was the most tiring part of the job. There was plenty of it and it was mostly heavy. At about 0800 I rewarded myself for breakfast with a slice of custard tart, beautifully fresh, cool and creamy, that I bought from the store's cake department. Both Lucienne and I worked six days a week for not much money.

To perform this job I had to get used to driving the van, a large Citroën Tube. The most difficult part was reversing into the narrow alleyway beside the store in Rue du Lycée. This road was the central thoroughfare through the town and always busy with cars and pedestrians. It required skill, confidence and gesticulations on my part to obtain space to reverse and complete the manoeuvre successfully.

During this time there were several visitors to Rue de Matel. They would usually turn up unannounced. One day I arrived back to meet La Tante de Calais (M Page's younger sister Lucie) and her husband who was severely handicapped following a stroke. They were accompanied by their daughter Margaret-Marie, a rather twisted young woman who was irreverent towards her father and was soon to become a nun in Madagascar. M Page was always close to his younger sister whom he had looked after in his early teens when they were orphaned. Mme Page asked privately why they could not have turned up a few days earlier in time for our wedding. Soon after, one of Lucie's sons Claude arrived for the night. He was a lorry driver and parked his massive articulated vehicle outside. He was a simple and amiable man without graces and complications, unlike his brother Bernard.

One morning I received a letter forwarded to me from my parents in Southend. It was from Bristol University. It said that I had been awarded a research grant for the MSc course. This was wonderful news and removed a major uncertainty in our lives.

At about this time Lucienne began suffering from morning sickness. She went to see Dr Boël who pronounced her pregnant. We were both shocked and at the same time elated. But it was not quite what we had planned. The brief thoughts of returning to our carefree student life in Bristol were dashed. We were going to have to come to terms with the responsibilities of parenthood.

# Return to Bristol

We left Roanne mid-September in the Citroën Ami 6 loaded with our luggage and several boxes of provisions from the shop. These included such mundane things as pasta, arachide cooking oil, tins of pâté, petits pois and French beans, saucisson, wine vinegar, garlic, and bottles of Kiravi red wine, none of which you could readily buy in UK. We stopped the night in a cheap hotel somewhere north of Paris. The next day we drove to Calais and made a short visit to La Tante de Calais and her family. They lived in a modest terraced house in a cobbled backstreet of the town. I also met her jovial, but profligate son Bernard and his wife in their butcher's shop, a business which was soon to go bankrupt.

We then caught the ferry from Calais and arrived in Dover early evening. We presented ourselves at Customs saying we wanted to import the car. The most difficult part of the process was obtaining insurance. The only way we could do this and drive off in the car that same evening was to join the AA on the spot, which we did. We then drove to Southend to stay with my parents for a couple of days.

We returned to Bristol to find somewhere to live. We settled for a flat in Hampton Road, Cotham, a short walk from the University. We rented it from the landlady, a Mrs Mary Lythgoe. The flat was in a Victorian terraced house like so many in Bristol. It was on the top floor up two flights of stairs and comprised a living room, a kitchen and a bedroom. The toilet and bathroom were one floor down and shared with the Lythgoes. As our flat was located under the eaves, all the rooms had sloping ceilings. Mrs Lythgoe did not have a husband or partner. She lived on the ground and first floors with her spoilt son Peter of similar age to us. Peter worked for an insurance company in Bristol. We did not tell Mrs Lythgoe that Lucienne was pregnant.

We settled into our new environment and I registered at the University. Lucienne soon found a job working in a factory located in Malago Road, Bedminster, a suburb of Bristol the other side of the city centre. She did general office work including stock taking, earning about £6 a week. She was worried that if they discovered she was pregnant she might lose her job.

Our daily routine started with me driving Lucienne to work early in the morning. She was still suffering badly from morning sickness and on the way we frequently had to stop, holding up the rush-hour traffic in the city centre, while Lucienne was sick at the kerbside. After dropping her off I would return to the university where I worked until collecting Lucienne in the evening. We ate a simple meal after which I usually worked until the early hours. On reflection, it must all have been much harder for Lucienne than for me. My role was easy compared with hers. She stuck at her job until the baby was imminent.

My MSc course entailed lectures on a syllabus dealing mainly with topics in fluid dynamics. I also had to write a mathematical thesis on a research project. The degree of MSc was to be judged partly by examination and partly by the thesis. We could choose the topic from amongst several ideas suggested by different lecturers. I chose a topic on the wind-driven ocean circulation suggested by Professor Mike Rogers who became my supervisor. Prof Rogers was a fluid dynamicist with a keen interest in computers. He was a man of immense capability and influence that I did not recognise at the time. He was ambitious and far-sighted, and one of the founders of computational fluid dynamics. At that time he was one of very few people who appreciated the incredible importance that computers would have in our everyday lives. He founded the Computing Department which he headed as a sub-division of the Maths Department. I believe that he met with much resistance in trying to make Computer Science a subject in its own right. Eventually, in 1972, the group was formally recognised as a separate entity by the University, and in 1975 Prof Rogers became Director of the Computer Science Centre. Soon after this he also became Head of the Department of Mathematics.

To return to my MSc studies, the idea for the research topic was vague. It was to investigate the ocean circulation using spherical coordinates rather than using the local "beta-plane" approximation in Cartesian coordinates as was commonly done. With the spherical approach it was natural to consider the circulation as a small perturbation about a solid body rotation. I immersed myself fully into my investigation. On top of this, the lectures covered a syllabus that required plenty of hard study. I had little spare time.

It was the first year of this particular MSc course so the lecturers were enthusiastic and keen to make it a success. Two key lecturers were Dr Philip Drazin and Dr Howell Peregrine, both to become eminent professors later in their careers. There were only four of us doing the MSc course, although the PhD students also attended the lectures. They were on leading-edge topics giving us an up-to-date knowledge of current research on water waves, fluid flow, turbulence and the then fashionable subject of singular perturbations. I can honestly say that I never worked harder than for my MSc. It was a good experience for me. What I learned has been invaluable ever after.

During the first term my supervisor, Professor Rogers, invited Lucienne and me to a dinner party at his house. He and his wife lived in a rambling farmhouse somewhere on the outskirts of Bristol. We were presented with an enormous saddle of lamb which he brought to the table to carve. There were several other guests at the dinner, all university academics. Lucienne and I felt a bit out of place amongst the lofty conversation. One guest, Dr Howell Peregrine, was particularly galling with his many boasts, including that of having achieved his ambition of attending both Oxford and Cambridge Universities for his first and second degrees, respectively. Dr Peregrine had recently joined the Maths Department as a lecturer. In fact, he turned out to be very good at his job.

Meanwhile, our landlady Mrs Lythgoe was friendly and helpful. She knew by now that Lucienne was pregnant and became quite motherly towards her. She and her son Peter went away every weekend to stay with her sister in Highbridge, Somerset, where they helped out in her sister's pub. So at weekends we had the house to ourselves. We took advantage of this on Friday and Saturday nights by going downstairs to watch her television. The highlight was "The Avengers" starring Patrick MacNee. We always tried to leave everything just as we had found it, but Mrs Lythgoe became aware that we watched her TV. She did not mind, but asked us to pay a small amount each week towards her gas fire that we used in cold weather.

My student grant from the Science Research Council and Lucienne's low wages did not provide us with much money. My grant was £500 a year. This was a bit more than when I had been an undergraduate but was reduced when I told them that Lucienne was earning. This was hard for us because we were short of money and the means testing nullified all the effort of Lucienne's work. Altogether we had £11 a week to live on. Out of this we paid £5-10-00 for rent and about £3 for food. We made savings wherever possible. For example, we bought the cheapest cuts of meat, the only fish we ate was coley, we ate a lot of potatoes, and we bought our eggs from Mrs Lythgoe who collected them weekly from a Somerset farm. The eggs were cheap because they were rejects, mostly irregular shapes. For a special occasion we bought a half-shoulder of lamb costing 2s/10d (14p).

There was not much in the way of heating in our flat, just an electric fire in the living room which we could not afford to run. There was a grate in the bedroom but when we tried to burn some coal it filled the flat with dense smoke. So we abandoned that form of heating. The only way to heat the kitchen was by turning on the gas rings on the stove, but that was expensive. In winter we were bitterly cold. Grandma Parsons came to the rescue by giving us her paraffin heater. This was an efficient device that quickly threw out a massive amount of heat. It was also very unsafe.

Our main luxury was the car, Lucienne's Ami 6. Now that she was settled in the UK Lucienne had to take a driving test because her French licence did not count. She passed her test easily.

We did not go out much for reasons of money shortage, work, and Lucienne not feeling on top form from her pregnancy. But fairly early on we did make contact with my cousin Donald Hudson who lived with his wife Maureen and three young children in a smart house the other side of Bristol Downs. Donald was about ten years older than me and had a good job doing something in the oil and gas business. Lucienne and I would often baby-sit for them and they once invited us for a meal. We returned the compliment by inviting them to our flat. We made room for them to eat in our bedroom because it was cosier. I cannot remember what we ate for main course, but Lucienne prepared crêpes suzettes for pudding.

When Lucienne was well into her pregnancy she suddenly had a craving for cucumber. It was 10 o'clock in the evening. This craving was strange because she did not like cucumber. She asked me to find some urgently. This posed a problem because it was not the season for cucumber. In those days you could buy things only when they were in season, like fruit, tomatoes, lettuce, new potatoes in the summer months; Brussel sprouts, parsnips, cabbage, main crop potatoes in the winter months. This shortage of "all the year round" fruit and vegetables in the 1960s would have made it hard to keep up the "five a day", but that was not necessary because our basic diet was healthy. Hamburgers and pizzas were a rarity. The situation began to change in the early 90s with fast food outlets and imported produce from around the world. To return to the cucumber, all the shops were closed so I thought my best chance would be an Indian Restaurant where I had heard of raita, though never eaten it. I tried several curry shops in the city centre with no luck. I visited the posher hotels in the hope they might do cucumber sandwiches, but again no luck. The last hotel I tried was The Hawthorns. All they could offer me was an apple. It was now approaching midnight. I took the apple back to Lucienne rather disappointed that I had not done better. We later joked that the failure to satisfy this craving was the reason for Isabelle's birthmark on her thigh, shaped like a cucumber slice.

We spent Christmas in Southend and then travelled to France for the New Year. We actually went by car in spite of the risk of snow and ice. It was bitterly cold in Roanne, quite difficult to reconcile with the hot summers I had spent there previously. On New Year's Eve we were all invited to a friend of Mme Page, a certain Lucienne Bordat. She was married with a young child. We thought the invitation was just for drinks but we had got the message wrong. We found awaiting us champagne and a mountain of fresh oysters. This was most embarrassing because Lucienne and I had already been invited by André and Janine to accompany them to a restaurant later that evening. They were waiting for us outside in their car. After many apologies we left M and Mme Page with the Bordat family and went with André and Janine to the restaurant where they treated us to a fine meal. My main course was pintade (guinea fowl) sitting on a bed of paté on toast that soaked up all the juices. I had never eaten guinea fowl before. After the meal there was dancing (smooching) until midnight when we saw in the New Year. We felt indebted to André for the many restaurant meals he had bought us and wondered if we would ever be able to repay him.

One Saturday in mid-February of 1966 Mrs Lythgoe invited us to visit her sister's pub in Highbridge to celebrate Lucienne's birthday. We arrived to find they had organised a party with Lucienne as guest of honour. In order to please her they had prepared snails, a French delicacy which the English had heard of but never seen or tasted. They must have gone to great trouble to find a supply from somewhere. But not knowing that the snails needed to be cooked they were served up cold straight from the tin. As the birthday girl, Lucienne was presented with this dish with great pride by Mrs Lythgoe's sister. Lucienne had to eat it in front of everyone. She pretended how much she liked it.

I was completely ignorant of the process of childbirth, as was Lucienne. The baby was not yet due, so it came as a surprise when Lucienne's GP sent her to hospital after she experienced early labour pains. She was admitted to the labour ward in the hospital on Blackboy Hill at the top of

Whiteladies Road near Bristol Downs. She stayed there for five days during which time she was not allowed to move. That was the regime in those days. When things settled down she was discharged.

In that same month of February we drove to Southend to see my parents. On the way back we visited Haidée in Kenton. She announced that Paul Hudson, my cousin and the younger brother of the aforementioned Donald, had recently bought a small maisonette just a few miles away. Haidée was anxious for us to visit Paul and meet his French wife Michèle whom he had recently married.

Paul and Michèle were disorganised. Paul gave the impression of being so extremely busy that we were lucky to catch him, an annoying trait that has persisted throughout his life. On this particular occasion he was making meticulous checks of his bank statement against cheque stubs to the now iconic background music of "Michèle Ma Belle" by the Beatles. I asked him how big the discrepancy had to be before he made deeper investigations. He said a few pence. Paul and Michèle had a baby, Suzanne, and divorced soon after.

Lucienne was now becoming big and it was hard work for her to go up and down the stairs to our top-floor flat. We had been making preparations for the arrival of the baby with some financial help from our parents. We had a proper pram and Lucienne had worked hard to prepare a fairy-tale cot.

When her time came Lucienne was admitted to the same hospital on Blackboy Hill at the top of Whiteladies Road. I took her there, then went back to the flat and waited, I can't remember how long. I received a call on Mrs Lythgoe's phone to say we had a baby girl, everything was alright and I could see my wife and daughter at visiting time. The next slot was from 6 p.m. that evening. I drove to the hospital and parked outside in good time. I was keen to go in but somehow felt it bad manners to be too early. So I was actually a few minutes late after the doors opened. I think I had a bunch of flowers. Lucienne was disappointed that I was the last father to arrive. I was quite overwhelmed to see this new baby that somehow we had produced together. She was beautiful and bright red.

I later returned to the flat and told Mrs Lythgoe the good news. It was Friday, 15 April, 1966, just after Easter.

Lucienne must have spent about 10 days in hospital, the norm at that time. I spent my days visiting her and working hard at my studies. Lucienne brought our baby back to the flat and we readjusted. We were proud of Isabelle Christine. I think we qualified for a child allowance because otherwise we could not have managed financially.

The doctors recommended that Lucienne drink a pint of Guinness each day to give her strong milk. We dutifully bought bottles of Guinness at great expense on our small budget. Lucienne drank it although she hated it. This is yet another example of how the medical profession does an about-turn on key topics.

One week after the birth we had a visit from Mme Page. As always she gave no advance notice of her arrival, but Lucienne had sensed it. It was evident to us that the best route was by plane from Lyon to Bristol airport. There was only one plane a week that she could conveniently catch, so on the appointed day I drove to Bristol Airport in the hope that she would be there. She was indeed on the plane and was astonished to find me waiting for her. Mme Page bought lots of things for the baby while she was with us. Soon after we had a visit from my parents to see their first grandchild.

When Isabelle was young we always spoke French to her. Of course, this came naturally to Lucienne but was more of an effort for me. But I learned fast and became an expert in "Potty French". We had no television and few acquaintances or social outlets, so there were no distractions from giving Isabelle a French upbringing, albeit a bit pseudo. Isabelle learned French as her first language and spoke without inhibition or accent. Her first words were "à boire", not "Maman". This was because we did not realise that babies needed to drink water. My parents were initially frustrated that they could not easily communicate with her, but this did not last long. Isabelle had clearly absorbed the English she had heard around her. As soon as she came in contact with other children she began to speak English without difficulty.

Shortly after her first visit Mme Page made a second visit, this time accompanied by M Page. They were both very supportive. We took them sightseeing round Bristol and visited the zoo.

Meanwhile my MSc course was going well. I had done some good work towards my thesis and my studies on the course content were thorough. I felt I was in control, unlike for my first degree. I wrote up my thesis and had it typed and bound. At the beginning of June I sat my series of MSc examinations. The results were announced soon after.

At this same time my cousin John Besent married Sheila Miller on Saturday, 11 June 1966, at Union Church, Totteridge, the same church where his sister Jean had been married. Lucienne and I did not attend the ceremony because it clashed with my MSc Exams.

I obtained the degree of MSc with commendation, an unusual accolade and the first to be awarded in the Maths Department at Bristol. Apparently the external examiner, Professor Keith Stewartson[14] of University College, London, had been impressed by my thesis and the predictions I had made about the deep oceanic equatorial current. I was awarded my degree at the July Congregation. After this success I was accepted for a course of research towards a PhD.

Soon after Isabelle was born Lucienne resumed work, this time for Bristol County Council as a teacher of French to adults. She enjoyed this work and had great rapport with her students. Her hours were unsociable, alternating between day and evening. When she was working I arranged to work at home in the flat to look after Isabelle.

---

[14] Keith Stewartson FRS (1925-1983) received many honours for his pioneering work on rotating fluid flows, shear layers, magneto-hydrodynamics, and flow at both high and low Reynolds numbers. In 1981 I was to meet him again when he was the Cabinet Office Chairman of my Individual Merit promotion panel. In April 1983 he suffered a heart attack while on a lecture tour of the west coast of the USA. He was flown home and taken to University College Hospital where he died two weeks later.

That summer we drove to France to stay with Lucienne's parents. Isabelle travelled on the back seat in a banana box. My parents and brother Nick also visited Roanne that summer. They arrived independently and stayed at Rue de Matel for about two weeks. My parents made several day trips to surrounding places of interest like St Etienne, Charlieu and Ambierle. One day I took them to Vichy where we had a go on the pédalos on the river Allier. Once again my father found the ordeal of travelling to France stressful and he was taken ill with stomach pains. He was attended by Dr Boël and recovered in a couple of days.

My parents' visit was timed to coincide with Isabelle's christening which took place on 28 August 1966 at the church of St Etienne where we had been married. We had to make the arrangements with the priest, Abbé Baroyer, whom we met in the vestry after mass one Sunday morning. He plied us with Kir, a mixture of white wine and Crème de Cassis, which he himself drank in some quantity. It was potent and went straight to our heads. The day of the christening was a happy occasion with both sets of grandparents present. Isabelle's godmother was her aunt Janine, Lucienne's sister-in-law. Her godfather was her cousin Philippe, Janine's son.

My parents and Nick returned to England having made their second and last visit to Roanne. Lucienne, Isabelle and I stayed on for a few more weeks. After we left that summer the shop closed for good. It was converted into a dining room to be used for special occasions, and furnished with pieces taken from Rue Branly including the piano and dining table. The house at Rue Branly was then rented out.

On return to Bristol I settled in as a PhD student, with just two years to do my research and complete my thesis. I received a small increase in my SRC grant.

I can remember several of my contemporaries. There was Eric Towers, he who was in the same year as me as an undergraduate and the only one to achieve a First. He was now starting his third year as a PhD student. His topic was something obscure about existence theorems for a class of solutions of the Navier-Stokes equations. As far as I know he never managed to solve his problem and failed to complete his thesis. Then there was Ken Morgan who spent his time writing computer programs. He had a distinguished career eventually becoming Professor of Computational Fluid Dynamics at Swansea University. Dave Burridge was interested in weather forecasting and was to become Director of the European Met Office in Reading. There was another applied maths PhD student called Peter Taylor who worked on modelling the turbulent flow round power station chimneys. Like me, he was married with a young daughter. I believe he eventually emigrated to Canada. John Byatt-Smith made a name for himself with his PhD thesis by solving the long-standing problem of finding an explicit solution of the integral equation for solitary water waves. After a short spell at the University of Berkeley, California, he spent the rest of his career in the Maths Department at Edinburgh University where he became a Senior Lecturer. John and I invented the game of infinite noughts and crosses in which you had to get five in a row on a board that could be extended indefinitely in any direction. We spent many hours playing this game on the blackboard and trying to develop winning strategies. Then there was Ron Smith who developed a successful ansatz for a particular class of fluid flow problems. On completing his PhD he went to Cambridge as a Royal Society Research Fellow, then became Professor and eventually Head of the Maths Department at Loughborough University. There were also several PhD students on the pure maths side working on problems in the field of logic, the liveliest of whom was Judith Hart.

There were always a few PhD student drop-outs, no doubt because we were left largely to our own devices with minimal supervision. The thinking seemed to be that if you could not make it by your own efforts you were not worth a PhD. There was no production line process like now.

In choosing an area for my own research it was natural for me to continue in the field of the wind-driven ocean circulation and to stay with my supervisor, Professor Rogers. He was busy in his new role as Head of the Computing Department and was not able to give me much time. I was unaware of his many commitments and ambitions, except that I had to make appointments to see him through his secretary at least a week in advance. I saw him for about 30 minutes once a fortnight to tell him what I had been doing. He listened and gave me general encouragement, but that was about all.

I spent most of my first term reading the literature and trying to home in on a problem that I felt was meaty enough and at the same time tractable. But I was drifting around in a directionless manner unable to find anything suitable. This was mainly due to my lack of experience. My supervisor did not help with any ideas. I wondered if I would ever produce anything novel for my thesis.

I became interested in the recent theories to explain the western boundary currents in the world oceans, like the Kuroshio in the Pacific Ocean and the Gulf Stream in the North Atlantic. I soon understood the explanations in the literature in terms of vorticity and the importance of the variation of the Coriolis force with latitude. But none of this explained why the Gulf Stream separated from the east coast of the USA at Cape Hatteras (North Carolina) to cut across the Atlantic, leaving a cold water gyre to the north (the Labrador Sea).

Several theoretical models and large scale numerical simulations had recently been published in the literature to try and explain Gulf Stream separation, but none had been successful in predicting it. In all these models the western boundary current stuck resolutely to the western coast. I certainly had no clue as to why this was.

I then came across a recently published 1966 paper by a well-known Swedish oceanographer, Pierre Welander, who examined yet another model of the wind-driven ocean circulation. There was nothing particularly interesting about the paper except that it incorporated density stratification very simply by means of just two layers, the lighter layer sitting on top of the heavier layer.

I began playing with this model and soon realised that if I increased the amplitude of the wind stress pattern I could make the lower layer surface in the NW corner of the ocean basin. This forced the western boundary current to separate from the coast and follow the surfacing line of the lower layer. I worked out how to calculate the path of the separated current and found that with a realistic wind-stress distribution it looked something like the real path. I spent several weeks tying up loose ends but I had done the essential work in just a couple of days. It was now the end of my second term as a PhD student.

At Easter 1967 I attended the Ninth British Theoretical Mechanics Colloquium held at the University of Strathclyde in Glasgow, 10-13 April 1967. There were over 200 attendees. We stayed in one of the University Halls of Residence. Some 60 contributed papers were divided into three groups, corresponding roughly to the areas of solid mechanics, incompressible fluid mechanics, and gas dynamics. There were also invited papers from eminent mathematicians like Brooke-Benjamin and Lighthill, whom I knew through their published work but had never met before.

Isabelle made her very first steps one evening while Lucienne was out teaching. She walked staggeringly from the chair to the spare divan bed that we had in our sitting room. This was a proud moment. When Lucienne was free she took Isabelle in her pram to Bristol zoo, an event that took place at least once a week. In the Spring Isabelle was able to play in Mrs Lythgoe's back garden. We made regular visits to Southend to see my parents and spent Christmas with them. On one occasion we took Isabelle on Southend Pier with my brother Nick.

A rare event was one afternoon in Spring 1967 when we managed to arrange for Mrs Lythgoe to look after Isabelle while we went to the local cinema to see the film "Bonnie and Clyde". This was a landmark film because it broke many taboos. It was a violent film in which death was for the first time depicted realistically. The film was surrounded by much publicity. At each showing St John's Ambulance were in attendance to stretcher people out as they fainted. By today's standards it was no more violent than Dr Who. We have become anaesthetised to violence and, like entropy, there is no going back.

On 5 July 1967 my Aunt Margaret married Malcolm Edwards at St Mary's Church, Prittlewell. Frank was now the only unmarried sibling of my father's.

In that same month Lucienne's young nephew Philippe came to stay with us in our flat in Bristol for a few days. He travelled alone arriving by plane at Bristol airport, a brave thing to do for a nine year old. Philippe's visit was timed so that we could take him back to France by car when we made our usual summer pilgrimage to Roanne.

The four of us drove to France via Southend to see my parents and break the journey. Something happened which is difficult to understand. We stayed overnight in Southend and got up early next morning for the drive to Dover. It was Monday, 7 August. We packed our luggage into the car and were about to leave when my father took me to one side. He told me that my grandfather (his father) had died during the night. They had all been up all night but had decided not to wake us. He went on to say that my grandmother had expressly wished us to continue on our journey to France as planned. My father said he was of the same opinion and urged me to be on my way. Stunned by the news I did as he suggested.

In retrospect this was the wrong thing to do, but this was how my parents dealt with such things. I had known that my grandfather was old and frail but my parents never told me about his illness. They were doing what they always did when a relative died. They kept me in the dark and out of the way. There may also have been the practical issue of not wanting to put the four of us up for what would typically have been at least another two weeks before the funeral (unlike two or three days in France). I broke the news to Lucienne on the way to Dover and she was horrified. My grandfather was buried while we were on holiday in Roanne.

Back in Bristol I continued with my research investigating various oceanographic problems. I spent a lot of time writing an enormous computer program that attempted to model numerically my mechanism for Gulf Stream separation with greater accuracy and fewer approximations than in my analytical model. My program took a couple of hours to compile and run on the University computer, a PDP-11, so I had to run it by myself late at night, with special permission. The program never worked properly and I could not stop my ocean basin from leaking. I eventually abandoned that line of research, perhaps prematurely, because the problems were certainly soluble. I went on to do some work on the problem of Rossby wave propagation over topographical features like the mid-Atlantic Ridge, but nothing I did was as good as my earlier work on Gulf Stream separation.

At Easter 1968 I attended the Tenth British Theoretical Mechanics Colloquium held in Oxford, 2-5 April. We stayed at St Catherine's College.

I was now beginning to think about jobs. I considered applying for a university post. With the rapid expansion of the universities at that time there were many posts available. But I homed in on the Scientific Civil Service, which I saw as a halfway house between university and industry.

My first job application was at the Admiralty Research Laboratory (ARL) in Teddington, next door to the National Physical Laboratory and close to Hampton Court. I was called for interview. In the morning I visited Maths Group located on the main site in Queens Road (now completely erased to make way for a smart housing estate). I was hosted by the Group Leader, a Mr (Fred) Steel. He interviewed me informally and introduced me to various people including the manager of the KDF9 mainframe computer Bill D'eath, pronounced "Deeth", and his assistant Barry Boothman. I also met Dr Ian Yuille whose pet project was computer-aided ship design, a mathematician Dr E M Wilson, and several computer programmers, two of whom were a husband and wife partnership. I also met George Heselden, an eccentric mathematician of whom more later.

My formal interview was scheduled for 2 o'clock that afternoon at Upper Lodge, an annex to ARL close by in Bushy Park. I was driven by Fred the "tilly" driver to Upper Lodge for my appointment. The panel comprised Dr (Peter) Crane, Mr (Phil) Lindop, and Dr (John) Gill who was the chairman. Dr

Gill was head of the Oceanography Group. They asked me several technical questions about my thesis. Mr Lindop, obviously concerned about whether I had my feet on the ground, asked me how I would explain the Coriolis force to a layman. I managed to divert this question by describing how, in a survey made of derailments on the Union Pacific Railroad, the 19$^{th}$ century American scientist Maury had found a tendency for engines to leave the track on the left. Mr Lindop appeared mystified but Dr Gill was pleased with the answer.

I later discovered that Dr Gill was at that time in negotiations with the US setting up an exchange agreement with the Maury Centre of Oceanographic Sciences just outside the Navy Research Labs in Washington DC, an establishment I had no knowledge of at the time. I obviously touched the right buttons in mentioning Maury and was made a generous offer to start work in October as a Senior Scientific Officer. It was not even conditional on my obtaining a PhD.

Soon after this I had a job interview at the Royal Aircraft Establishment, Farnborough. They told me just before the interview started that however well I performed it would be impossible for them to offer me a job at Senior Scientific Officer level. This persuaded me that the job offer from ARL was too good to refuse, so I wrote to accept.

It was the custom for PhD students finishing their studies to treat the other students to a meal. We invited several of my contemporaries to our flat for a meal of roast chicken. John Byatt-Smith took it upon himself to serve the chicken. He cut delicate razor-thin slices off the breast and legs. I was intrigued by this pedantic approach having learned in France how to cut a chicken into eight portions, a much more practical solution.

By May 1968 I felt I had done enough towards my PhD. I wrote up my work in long-hand. The original still exists. Before going on our summer holiday to Roanne in July/August I had to think quickly of a title for my thesis. I chose the rather dull title "Two Problems in Oceanography". I should have called it "Gulf stream Separation and Other Problems". I left the completed manuscript with a freelance typist who was able to do the mathematical symbols on a special typewriter.

With my thesis out of the way we spent the summer weeks in Roanne. One day we were approached by a man in the local park. He said he owned a shop selling children's clothes and he thought that Isabelle would make an ideal model. She was certainly an attractive child, chubby, well-mannered and always smiling. The next day we returned to the park to meet the man who took photos of Isabelle dressed in various outfits. In today's climate we might have been more apprehensive.

One Sunday Abbé Baroyer invited himself to lunch, something he did frequently. He had no qualms about imposing himself. Mme Page prepared the table in the new front room, formerly the shop, with the best linen, china and cutlery. Abbé Baroyer arrived after Mass on his bicycle and seated himself at the head of the table. He talked incessantly while Mme Page obediently brought in course after course. I understood little of what he said except that he was critical of the food. For example, he said that in the best traditions of fine dining the green salad was not served at the beginning of the meal, as Mme Page did, but after the meat course in order to cleanse the palate. None of this stopped him eating and drinking in large quantities.

Lucienne's brother André and his family no longer lived in their small house, which they now rented out. They lived in a much larger two-storey house they had had built on part of the adjoining land. We often went there for meals and to relax on their large verandah.

One day in France I removed the floor mat in Lucienne's Ami 6 car to find to my surprise that there was nothing beneath it. The metal floor had rusted away to reveal the road. I bought some sheet metal and a pop-rivet gun and made some basic repairs. There were no MOTs in those days.

On return to Bristol I proof-read my typed thesis and had it bound in three copies. I submitted my thesis to the University in September 1968. My *viva* took place soon after. My internal examiner was my supervisor, Professor Rogers. My external examiner was Dr Francis Bretherton[15], an oceanographer and fluid dynamicist then based at Cambridge. It all went well and I was told informally I would receive my PhD. Dr Bretherton recommended that I try and publish my Gulf Stream work. They then invited me to lunch at the Berkeley restaurant opposite the main university buildings. Lucienne and Isabelle met us afterwards.

Some weeks later, after starting work in Teddington, I wrote up the part of my PhD thesis dealing with Gulf Stream separation and submitted it to the Journal of Fluid Mechanics. It is customary for a PhD student publishing his work to acknowledge the support of his supervisor. Although I got on well with my supervisor, Professor Rogers, I did not feel I had received much support from him. So I did the rather dishonourable thing of omitting any acknowledgement. My Gulf Stream paper was accepted for publication without changes. It turned out to be the best and most cited piece of work I ever did. Try Googling "Gulf Stream Separation" to find near the top of the list my seminal paper: "A two-layer model of Gulf Stream separation", Journal of Fluid Mechanics (1969), 39: 511-528 © Cambridge University Press.*

---

[15] Francis Bretherton obtained his PhD at Cambridge University. His research areas were atmospheric dynamics and ocean currents. From 1974-1981 he was Director of the National Center for Atmospheric Research at Boulder, Colorado. From 1982-1987 he was Chair of the NASA Earth System Science Committee. From 1988-1999 he was Director of the Space Science and Engineering Center at the University of Wisconsin-Madison, where he became an Emeritus Professor in the Department of Atmospheric and Oceanic Sciences.

# BOOK FIVE
# Work (1968 – 2018)

# Starting Work and the 1970s

On Monday, 7 October 1968 I started work at the Admiralty Research Laboratory (ARL) in Queen's Road, Teddington. My starting salary was £1760, more than three times my grant as a PhD student.

The main activities at ARL were oceanography, submarine hydrodynamics and propulsion, electromagnetics and degaussing, underwater acoustics, submarine sonar and detection, infrared radiation, military photography, operational assessment, computer science, a range of mathematical research topics, plus engineering and draughtsman support. Most of this was a mystery to me. The location of ARL was chosen to benefit from the expertise of the National Physical Laboratory (NPL) next door, whose canteen Glazebrook Hall we shared for our mid-day meals.

ARL Teddington was established shortly after WW1. Many famous scientists had worked there during and after WW2, including Francis Crick FRS and Nobel Laureate (1940-1947), Martin Beale (1951-1960) one of the pioneers of mathematical programming, Edward Lee (1951-1955 and 1971-1974) who built the first infrared spectrometer, Jack Good (1959-1962) the inventor of the Fast Fourier Transform (FFT), R.V. Jones FRS (1938-1939) the father of Military Intelligence, and Cyril Hilsum FRS (1947-1950) who invented the basic technologies for semiconductors and liquid-crystal flat-panel displays. The last two were later to be my mentors and examiners on promotion panels. Some brilliant scientists had also worked next door at NPL, including the mathematicians Alan Turing and James Wilkinson, the latter still there in my time at ARL. Some of this atmosphere rubbed off on to me, but generally I was unaware and ignorant of the significance of all these people.

The Superintendent at ARL when I joined was Mr Burrows, soon replaced by Edward Lee, and later from 1974 by Alex Mitchell. I was told that Mr Burrows kept flasks of neat ethyl alcohol in his office for his private consumption, prepared by the on-site chemist, Dr Soole.

Two weeks before starting work I made a trip to Teddington to look for accommodation, leaving Lucienne and baby Isabelle in our Bristol flat. I found a temporary B&B with Mr & Mrs (Len and Vera) Rockcliffe who lived in a rambling three-storey house in Gloucester Road quite close to my place of work. Mr Rockcliffe had spent his entire career with the Admiralty, retiring from "O" Group the year before I joined. He had left his address with the Personnel Department, offering temporary accommodation in his house for new recruits like me. I took up the offer and was made most welcome.

Mr and Mrs Rockcliffe, both devout Christians, were kind and hospitable. They were staunch members of their local church of St James in Hampton Hill where they were dedicated to the local community. They lived comfortably but meagrely making their own bread every day, which was unusual in that era. Mr Rockcliffe was a qualified and experienced mechanical engineer. He spent his spare time refurbishing a large vintage Austin in his garage. When I told them I had left my wife and daughter behind in Bristol they promptly invited them to stay with me at their house until we could find something more permanent. We accepted the offer and Lucienne and Isabelle joined me at Mr and Mrs Rockliffe's. We must have stayed there for a week or two.

We left all our belongings in Bristol while we searched for a suitable apartment to rent in Teddington. We found a ground floor flat that we liked in Waldegrave Park in nearby Strawberry Hill. The houses were large of the Victorian era, and mostly converted into flats. The landlords, a Mr and Mrs Shack, lived in the house opposite. Mrs Shack referred to her husband simply as Shack. We came to an agreement about the rent and Shack said he wanted payment up-front to secure the flat. The three of us returned the same afternoon with a cheque. To our surprise and dismay Shack declared that somebody else had just offered a higher rent and, hard luck, we had lost out on the flat. Mrs Shack felt sorry for us and intervened, telling her husband that he should honour his agreement with us. Reluctantly he relented and we got the flat.

That weekend at the beginning of October 1968 we drove back to Bristol in the Ami 6 to finalise our departure from Hampton Road. We rented a "white van" from a hire company off Whiteladies Road to shift our few belongings, which were still too much to fit into our car. On the Sunday we drove with a van-load to the new flat in Teddington. There were no motorways and the journey was quite long and tiring. After unloading we made the journey back to Bristol. We were both exhausted and I was not driving the big van to the best of my ability. With only a half-mile to go I was driving in darkness up the steep incline of St Michaels Hill when I clipped the wing mirror of a parked car. I stopped the van and we got out to inspect the damage. The parked car had a small scratch whereas there was no discernible damage to our van. We tried to find the owner of the parked car and even went into the nearby pub, but without success. We left a note on the windscreen.

The next morning I returned the van to the hire company and reported the incident. My honesty cost me my £30 deposit, a massive sum in those days. I should have kept quiet. We cleared our few remaining things from the flat and said goodbye to Mrs Lythgoe.

We moved into our flat in Teddington. It was comfortable and relatively spacious comprising a hallway, a kitchen, a bathroom/toilet, a large living room that was also our bedroom, and a long back garden to which we had direct access. Our young neighbour upstairs was an engaging character, a British Airways steward. He boasted how he bought his shoes in Italy and his clothes in Paris. He told us about the goings-on of flight attendants - an eye-opener.

I started work at ARL and soon settled into the new routine. On my first day I was surprised to be given the choice of joining either Maths Group or "O" Group (the Oceanography Group), both of which I had visited at my interview. I chose the safer option of Maths Group. I was given a large office on my own on the first floor of Main Block overlooking Queens Road. The road was partially obscured by the 10ft perimeter wall which had glass shards embedded along the top as a security measure.

I was assigned the task of making improvements to a complex computer program that calculated the inviscid fluid flow round submerged bodies, such as submarines. I did the theory and Barry Boothman, a programmer and Deputy Computer Manager, implemented the changes in software. I then devised various tests to check that the amendments were working properly. It was not stimulating work. I walked the mile to work each day, sometimes collected in the evening by Lucienne and Isabelle in the Ami 6.

Everybody at work wore suits and all the men wore ties. Those in higher positions were always addressed by their title and surname, never by their Christian name. It was a matter of respect. Most people smoked in their offices. It was considered quite normal and those few who did not smoke never complained. Smoking was just not an issue.

The office next to mine was occupied by Dr Ian Yuille, later heralded as the father of computer-aided ship design. Dr Yuille always complained about the lack of recognition he received for his work. I think he was right. Dr Yuille had a long and distinguished career in naval architecture, recognising the huge potential of digital computers for solving structural problems in the design of ships. He later wrote up his experiences in a book entitled "I fathered a Goddess - Autobiography of a Naval Scientist", published in 2012.

The discontent in Maths Group was just part of a general feeling of malaise that I sensed. The root cause was the advent of the electronic desktop calculator, and later the personal computer, making the cumbersome Maths Group mainframe computer redundant except for the biggest computations. It was an English Electric KDF9 that kept many people in work. Few people recognised the key role that personal computers were to play in our work and everyday lives. The new technology was resented because it upset the status quo.

I sensed jealousy towards me from other members of Maths Group, notably from two of the women. Like me, they were both SSOs (Senior Scientific Officers) but they had taken half their careers to reach that level whereas I had walked straight in. I realised I had been fortunate to have been offered SSO. It was much better to be at the bottom of the SSO pay scale than at the top of the lower SO scale, the normal entry point for PhDs. It meant I avoided a gruelling promotion board with all the associated hassle and uncertainty. I was starting off higher up the ladder by one large rung. The Scientific Civil Service was very hierarchical.

I was later to appreciate that my Civil Service rank at the beginning of my career was higher than my father's at the end of his. This never upset him, in fact, quite the opposite. He boasted to his friends about my modest achievements often to my embarrassment.

I met many interesting characters at work. This was an era when eccentrics were tolerated in organisations. In fact they flourished and were often a source of great innovation. The atmosphere was relaxed compared with today and no doubt several people were not pulling their weight. This was mainly because there were no rigorous tasking procedures in place. At the time, we always felt under pressure but in retrospect the work environment was relatively stress-free and the management was slack. Although the various projects had to be completed within budget, there never appeared to be any real shortage of money. The freedom we were given encouraged original thinking of which there was plenty.

A memorable and eccentric character was George Heselden, a fellow member of Maths Group and a talented mathematician, although I was never quite sure what he did to earn his salary. The first thing you noticed on entering his office was the array of assorted pill bottles on his desk and the overpowering smell of medicines and potions. George was a hypochondriac. He kept his nails long and had the disturbing habit of cleaning them with his pen-knife while we were talking. In spite of all this we had many interesting exchanges and conversations. He gave me several useful leads in my work and I found him most supportive.

George was a bachelor who lived in a flat in Hampton Hill, Teddington. One dark winter's night when walking home he was mugged. He was shocked rather than seriously hurt, but the experience made him feel vulnerable. George addressed the problem with his usual rigour. He decided that to deter future muggers he had to drastically change his appearance. He bought shoes with built-up heels to compensate for his lack of height. He grew a full beard and bought a homburg hat and full-length leather overcoat. The transformation made the diminutive George look quite sinister and menacing.

A few years later when George was near retirement he actually married, much to everyone's astonishment. His wife was a glamorous buxom woman, a former film star and notorious sex symbol of the 1940s. Her stage name was Christine Norden. She was almost a foot taller than George. They married in March 1980 and George was her fifth husband. One of the craters of the planet Venus was named after her. George later developed and named a mathematical formula in her honour. How on earth they met given their different backgrounds is a mystery. It seemed a most unlikely match, but a happy one that lasted until her untimely death eight years later.

Another strange character was Pete Foster. He was about 40 years old and a chain smoker. Pete was scruffy and had the lowliest of positions in the organisation. But when you got to know Pete you realised he was clever. He just had no ambition, at least not at work. He was an expert on Medieval Latin and spent his spare time translating ancient documents, specialising in maps. His talents were somehow wasted at ARL.

Shortly after joining ARL I was sent on a "Tour of Establishments" for new recruits to the Scientific Civil Service, together with several other new entrants. These included Ian Roebuck who was to become a good friend, magistrate, leading Lib Dem councillor in Weymouth, a wise negotiator, a good scientist, and an accomplished speaker even about subjects of which he knew little. The tour lasted about ten days. We visited research sites around the country including the Admiralty Underwater Weapons Establishment (AUWE) in Dorset where I was later to work. It was a useful tour that gave me an appreciation of the breadth of government research work, somewhere between the universities and industry.

This relationship with the universities was reflected in the pension arrangements. At that age I had little interest in pensions. My retirement seemed so remote that it could be disregarded. But we had to choose between joining either the Civil Service Pension Scheme or the Federated Superannuation Scheme for Universities (FSSU). I chose the latter because I still believed I would finish up working in a university. At the same time I had to choose between retiring at age 60 or 65, a meaningless question to someone who was only 25. Fortunately, I was later able to transfer my benefits accrued under FSSU to the more generous Civil Service pension scheme.

I was not alone for long in my large office. In early 1969 a fluid dynamicist called Mike Murray joined me on transfer from the hydrodynamics group. He was good company and bright. Mike was a confirmed bachelor who lived with his mother. Over the next 30 years our two paths were to cross frequently.

At about this time I was summoned to a meeting with the leader of "O" Group, Dr J S Gill, he who had chaired my recruitment interview panel and no doubt had been behind my job offer. I think he was disappointed that I had opted to join Maths Group.

Dr Gill asked if I could produce a mathematical model to describe the discharge of the hot coolant water from a submerged nuclear submarine. There was concern that our nuclear boats might be vulnerable to infrared detection of their coolant water at the sea surface from an aircraft or satellite. I knew how to tackle this problem having studied the flow of turbulent plumes in my MSc course. Dr Gill introduced me to a young member of his team, Roger Panter, who was doing scaled-down tank experiments to simulate the coolant water discharge process. My mathematical modelling went well. I extended the existing theory to deal with trajectories that began at angles to the vertical. I was able to predict the mean trajectory and spread of the coolant water and its temperature at the sea surface as a function of the efflux rate. My results agreed with Roger's tank experiments. On the basis of this a sea trial was planned for October 1969 to obtain some real data with a nuclear submarine. I quickly appreciated that Dr Gill's Group was much nearer the "coal-face" than Maths Group. We were dealing directly with in-service navy vessels and sea officers. I began to realise what was at stake and that we had a heavy burden of responsibility to get it right.

In January 1969 I took time off work to travel to Bristol with Lucienne and Isabelle to receive my PhD degree, a happy occasion. My parents and Lucienne's mother also attended the ceremony.

Around Easter 1969 I attended the 11th BTMC (British Theoretical Mechanics Colloquium) at Nottingham University. I was trying to hold on to my university contacts because I still thought I might return to university, maybe as a lecturer. I was to attend only one more of these annual events, the 12th BTMC held the following year at the University of East Anglia.

Close to our Teddington flat were Pinewood TV and film studios, so it was not unusual to see well-known actors in the street. One such character living in Teddington whom we often saw was Patrick Newell, the fat man who played "Mother" in the original TV series of The Avengers. Just down the road from us in a beautiful detached corner house lived the actress Mary Holland who sprang to fame as "Katie OXO" in her long-running series of TV adverts, well before Lynda Bellingham. Her house, like many others in the area, including the flat where we were currently living, was later demolished to make way for an apartment block.

Near to Katie OXO's house on the opposite corner was a derelict tennis court. We often went for walks this way and Isabelle once spotted an old tennis shoe in the undergrowth on the edge of the court. It was always there and for this reason became iconic.

We liked our Teddington flat and enjoyed the freedom of the garden. There was a patio and lawn, and a dense wooded area at the bottom of the garden. This made us think about buying our own house, something Lucienne had always been uncomfortable with in view of the large debt incurred by a mortgage.

I had been invited by Dr Gill to go on the afore-mentioned sea trial. The trials vessel collecting the data was to be HMS *Hecla*, a Royal Navy oceanographic research ship, in consort with an SSN (nuclear attack submarine).

One grey day in October 1969 I set off from Teddington mid-afternoon with Dr Gill and a young Scientific Officer called Tony White with whom I was to have a lasting friendship. We were driven in the ARL "Tilly" to Heathrow where we caught the evening BEA flight to Belfast, Dr Gill travelling first class as befitted his rank. I think this was my first experience of flying. During the journey I mentioned to Dr Gill that Lucienne was pregnant and expecting in five or six weeks. He said that maybe it was not a good idea for me to go on this sea trial. I thought it was a bit late to change my mind. In any case I was still naive about the risks of childbirth.

On arrival at Belfast airport a navy driver met us at the exit of the terminal building. He insisted that we sit in the staff car while he took care of the luggage. It was now early evening, getting dark and raining. He drove us to Londonderry to join the ship, HMS *Hecla*, a distance of some 100 km. On entering Londonderry we were stopped and searched by heavily armed troops at several barriers and checkpoints. This was the start of "The Troubles" in Northern Ireland. I was shocked and astonished because, after all, this was part of the United Kingdom. Two months previously in August 1969, after episodes of extreme violence in Londonderry and neighbouring provinces, the British Government had sent in troops in what was described as a "limited operation" to restore law and order. The conflict was only just beginning, although it had existed in one form or another for centuries.

Eventually, we reached the Londonderry dockside in late evening. We climbed the Hecla's gangway and waited for our luggage but mine did not appear. The driver had forgotten it on the pavement at Belfast airport. This was a bad start to my first sea trial. We were due to sail before dawn in just a few hours. I was assured that the driver would fetch my suitcase and have it on board before we sailed. When I awoke early next morning we were already underway. I was relieved to be reunited with my suitcase.

We sailed to an exercise area in deep Atlantic waters NW of Ireland. I shared a cabin with a young Sub-Lieutenant. This was my first experience of the disciplines and pleasures of navy life at sea. The work routine got underway and everything seemed to be going well.

One morning a few days into the trial Dr Gill took me to one side and asked me to follow him to his cabin. His attitude was sombre. He sat me down and said he had some bad news. He handed me a cryptic signal just received by the radio operator. It said:

"WIFE IN HOSPITAL. BABY BORN. SINCE DIED."

I was stunned. Did it mean that my wife had died or had the baby died? Whatever it meant I could not take it in.

The Captain arranged to get me home. This is when the Navy is most impressive, responding to a crisis. A frigate, HMS *Naiad* was in the area. We sailed on a converging course to within helicopter range. I left *Hecla* in the ship's Wasp helicopter and landed on *Naiad*, a vessel capable of 35-40 knots and much faster than *Hecla*. I was made welcome, and after a late evening meal was given a temporary bunk by a companionway somewhere in the bowels of the ship. I did not sleep much. I could feel from the ship's vibrations that we were moving fast through the water.

Early the next morning we were within helicopter range of land. I was flown in the ship's Wasp to Prestwick Airport near Glasgow and transferred straight onto the 8 a.m. BEA flight to Heathrow. Somehow I made my way to Middlesex Hospital.

I was totally unprepared for what had happened and I do not think I was a great support to Lucienne who was upset beyond words. I really had no idea how to deal with death having never even attended a funeral.

We named our baby Alan Philip, born on Tuesday, 21 October 1969. He lived 23 hours. A week later we buried him in Hounslow Cemetery, not far from our house. The Catholic priest led the ceremony. My love for Lucienne was not sufficient to comfort her in her grief. I could not reach the depths of her despair. I was dreadfully sad but tended to push things under the carpet. Lucienne would have preferred me to show more emotion which was not in my character. This was to be the first of many family bereavements occurring whilst I was away at sea.

We began negotiating to buy a property in Ashley Drive, a small three-bedroomed semi-detached house in Whitton, a sub-district of Twickenham about three miles from my place of work. Lucienne's initial apprehension about a mortgage was overcome. We bought the house for £5700. The previous owner was an elderly gentleman who had died. As it turned out, with the massive inflation and pay rises of the 1970s we were never to make a better investment.

We moved into our house in early 1970. It seemed the height of luxury compared with our previous dwellings. It was located on a small estate of similar houses all built in the mid-1930s. Our neighbours in Ashley Drive were sociable and we quickly made friends.

The exception was the couple in the house adjoining ours. They were the Burnhams. She was demented and had dreadful shouting scenes with her husband that the whole street could hear. Sometimes she locked him out of the house forcing him to sleep at night in the garage. If we made the smallest noise in our house she would bang on the partition wall. She once held her baby screaming from an upstairs window to demonstrate how we had woken him up. Fortunately, the Burnhams soon moved out to be replaced by Marjorie and Arthur and their two young children Sharon and Mark. We became good friends and kept in touch with Marjorie, even when much later she was a widow living in Somerset.

In the next house were Mr and Mrs Smith, a friendly middle-aged couple. Mr Smith had a gammy leg, possibly the result of an accident. Their daughter was married to André, a young Frenchman. They had a baby daughter named Fleur. André was a waiter at the restaurant in the Greyhound Inn at the exit to Bushy Park opposite the side entrance to Hampton Court. He was charming but a bit of a waster, and possibly a lothario. Their marriage did not last long.

On the other side of our house were John and Thelma. At that time they had no children and both went out to work. Consequently we saw them only occasionally, although they were neighbourly.

Next door to them were Mr and Mrs Baisey, an elderly retired couple. They were friendly but tended to keep to themselves being of the older generation.

In one house further down lived John and Laurel Fadden with their young sons Craig and Paul (later to die of cancer in his twenties). John worked for Walls in their experimental laboratory making and testing innovative sausages and pies, some of which we sampled with mixed results. Walls frequently did a special run of pork sausages for the Queen's Royal Household. They always made enough for the workers to take some home, a few of which John occasionally supplied to us. They were bigger than the average sausage and exceedingly good and meaty. John was a bit of a "groper", and when the occasion presented itself tried to press himself on other women in the street, including Lucienne.

Then there were the Sleggs who lived a few doors down. Mrs Slegg had a severe heart condition. This meant that Mr Slegg had to do most of the housework while holding down his job and looking after their three young boys. I can't think how they all fitted into that small house. Mrs Slegg was to die young. Mr Slegg later remarried.

Mr Gentry was a retired primary school teacher who lived with his mother opposite the corner to Bryanston Avenue. He later gave Isabelle coaching lessons to prepare her for secondary school entry examinations at Putney High School, an expensive independent school that we thought would be best for her. As it turned out she was unhappy there, but her suffering was not to last too long.

On 30 May 1970 Lucienne lost another baby boy. We named him Charles. After this she began seeing a top Harley Street gynaecologist called Mr Geoffrey Chamberlain. To think that our two sons would now be in their late 40s.

We were keen to make improvements to our new house. Lucienne took the initiative by demolishing the front garden wall, the fireplace in the lounge, and later the pantry in the kitchen with help from a 5lb hammer and John Fadden. These activities usually took place clandestinely, so it seemed to me, while I was at work, but they provided the impetus for the rebuilding process most of which we did ourselves.

I sanded and revarnished all the downstairs woodblock floors, we repainted the house interior and wall-papered everywhere, we rebuilt the bathroom and kitchen, bought a second-hand, convection gas oven for £5 which was immediately renovated free-of-charge on conversion to natural gas (and is the only oven in which I ever made successful Yorkshires). I then repainted the outside of the house in white and yellow, a colour scheme that prevailed long after we left until circa 2010.

Soon after moving into our house Isabelle started school at St Edmund's Roman Catholic Primary in Nelson Road, Whitton. We developed a routine where I took Isabelle to school in the car, usually with Lucienne, so that she could keep the car during the day after driving me to work in Teddington.

On 30 November 1971 my cousin John came to stay with us. He was starting a new job with Surrey County Council, but he and his wife Sheila had not yet completed the sale of their house in Sutton Coldfield or the purchase of their new house in Claygate. John stayed with us, except over Christmas, until the second week of January 1972. He was good company. John was a qualified estate manager and an expert in all things rural. His visit is remembered for his sage advice that once the leaves of a tree have turned yellow they are unlikely to turn back to green.

We had several visits from Mme Page in the early 70s, sometimes accompanied by M Page. Their visits were always unannounced. On one occasion while I was at work and Lucienne was collecting Isabelle from some after-school activity, Mme Page locked herself out of our house. She broke one of the small leaded window panes next to the front door so that she could put her hand through to open it from the inside. I replaced the glass and did a botch-up job of repairing the lead using putty. My repair lasted a long time, and until recently was still visible on Google's Streetview.

We took M and Mme Page on a sightseeing tour of London by car. We drove into Downing Street and parked outside No 10 to take a photo. At that time Downing Street was open to the public just like any other road, with no security except for a policeman outside the front door. Soon after this it was sealed off because of Irish terrorist threats.

Lucienne set her mind on obtaining a Teacher's Certificate. In order to qualify for entry she first had to study "O" Level English Language, English Literature and Latin through correspondence courses, all of which she passed with top grades. Her French Baccalauréat qualification did not seem to count enough. Her exam results enabled her to enrol in 1973 as a mature student at Digby Stuart Teachers Training College in nearby Roehampton, now a University. After successfully completing the two-year course she did a one-year Bachelor of Education course at London University gaining an Honours 2:1 degree. All this was a considerable achievement.

I also became involved in teaching. Mr Rockcliffe told us he had developed cancer in his leg. It had to be amputated above the knee, which was terrible news. He had been taking an evening class at Twickenham Technical College teaching basic maths to young catering students. He asked me to take over from him while he recovered from the operation. I did this job for two terms and did not enjoy the experience. The catering students were in their late teens with no academic ability and absolutely no interest in mathematics. I found them difficult to teach and impossible to control.

A positive outcome of this experience was that it put me in contact with the Open University. To earn extra money I became a Tutor for the Mathematics Foundation Course entailing five or six hours of my time each week. I diligently marked the monthly exercises of about 20 students, responding to their queries by letter and phone. About once a fortnight I met them in the evening at Kingston College for a two-hour tutorial. The students were all well-motivated adults, quite the opposite of the caterers. The work was poorly paid but interesting and satisfying. I was still toying with the idea of one day returning to a university environment.

Meanwhile, Lucienne found a part-time job teaching French at Whitton Secondary School in Percy Road. She did not particularly enjoy the experience and she found her fellow teachers apathetic and uninspiring. This put her off becoming a school teacher. Instead she started teaching French to adults, something she did successfully for many years and always enjoyed.

The back garden of our house was totally overgrown when we moved in. One of my first jobs was to dig it over. We were amazed to discover just below the surface a plastic sheet covering an enormous pit full of rubbish including masonry, old mattresses and bedsteads. After getting rid of it I was left with a large crater. At this time I was planning a garage at the bottom of the back garden so, fortuitously, I was able to use most of the earth that I dug out for the concrete base of the garage to fill the hole left by the rubbish. I then planted potatoes in the back garden to break up the heavy soil. When I had finished building the garage I laid the back garden to lawn and built a concrete path across it leading to a new gate and wooden fence at the bottom. Vehicular access to the garage was round the back alleyway.

Following John Fadden's example I rented a 5-rod allotment (about 25m) on a large plot of council land nearby, adjacent to a cemetery. My allotment had been neglected and the soil was heavy clay. I found the work back-breaking and not very fruitful. I had some success growing potatoes, parsnips, sprouts, spinach, lettuce and onions, but the crops were meagre and much of the produce was either eaten by pests or became diseased. I put in considerable effort over three seasons with limited return. Nevertheless, I learned a lot about growing vegetables and the never-ending problems of weeds and pests. It was while working on the allotment one Sunday morning that I first met my brother's girlfriend Chris, later to become his wife. Nick and Chris stayed with us for Sunday lunch.

Soon after our house move in 1970 Dr Gill invited me to join his team to continue the work I had been doing for him in Maths Group. I was pleased to accept because it opened up paths to more interesting work. I moved offices and started sharing a room with Roger Panter who was to remain a good friend for many years.

Dr Gill gave me the task of providing a theoretical framework for the many possible non-acoustic ways of detecting submerged submarines, for example, electro-magnetic effects, infrared effects, and hydrodynamic effects such as pressure changes, turbulent wakes, internal waves and surface waves generated by the submarine's passage. Whilst acoustics remained the principal means of detection there was always (and still is) the concern that maybe there are other ways, often referred to by sceptics as "unsound" methods. There was always a fear that the Soviets might develop a new means of detection that we did not know about, thereby putting us at a disadvantage. Our budget for the work was small. My job was to provide the underpinning theory to assist in the interpretation of results from our small programme of sea experiments. The aim was to enable predictions to be made of the risk of non-acoustic detection of our submarines in a wider world context.

I applied myself to the new task with enthusiasm and made good progress. There were many interesting mathematical problems to address: for example, the generation of internal waves in stratified media by the motion of submerged bodies, the effect of the internal waves on modifying the spectrum of sea surface waves thus providing possibilities for satellite detection, the hydrodynamics of isotropic turbulence, and the propagation of electromagnetic waves at the air-sea interface. At about this time I went on my second sea trial on RMAS *St Margarets* out of Plymouth. The trial was organised by my colleague Roger Panter. One of the aims was to assess the false alarm rate by collecting and analysing data on infrared background noise when viewing the sea surface in different sea states, possibly a rather extravagant use of ship's time.

Dr Gill seemed satisfied with my theoretical work. It gave him a small amount of leverage for obtaining information from the US through our exchange agreements. The US had a much bigger research programme in this area. He asked me to join him on his next trip to the States and Canada. This was my first foreign business trip and my first visit to North America, so I will give some details.

It used to be a requirement for all official travel to be conducted using government service facilities wherever possible. So all travel to North America, no matter what the final destination, had to be made by RAF VC10 from Brize Norton in Oxfordshire to Washington Dulles. The RAF VC10s were fast but noisy aircraft. They contravened the US airport noise abatement laws but were exempted because of their military status. After they ceased to fly commercially the RAF bought up all the retired aircraft for spares, and continued to fly them until September 2013.

Itineraries in the US had to be planned around the outward-bound flight to Dulles that left UK at midday every Friday, and the return flight that left Dulles also on a Friday at around 6 p.m. This meant that every trip to North America had to be at least one week's duration. The arrival in Washington at around 3 p.m. local time on a Friday had the benefit that most of the weekend was free to sightsee, after taking out the time for any onward travel on US commercial airlines. This state of affairs continued until the early 90s when we started paying for all travel and accommodation out of our research budgets, and commercial airlines offered better deals.

So mid-afternoon on Thursday, 29 April 1971, Dr Gill and I were driven in a staff car from Teddington to RAF Brize Norton in Oxfordshire. The armed forces hospitality budget used to be exceedingly generous. On arrival at Brize we were escorted to the Officers' Mess for a meal. I was astonished to see whole sides of poached salmon (an expensive fish in those days), massive joints of roast beef and pork, and many other extravagant offerings to tempt us. In addition we had the freedom of the bar with wine, beer and spirits. Dr Gill took all this in his stride. We were then accommodated overnight in the officers' quarters.

The reason we had to arrive at Brize the night preceding the mid-day flight was that the RAF had a complex countdown procedure, rather like astronauts flying to the moon. At 6 a.m. a loudspeaker in my room, connected to the base tannoy, announced it was time to get up for breakfast. The arduous procedure of checking and rechecking the passenger list began. After breakfast we were shuttled in a bus to the main terminal for more queueing and more checks. VIPs travelled separately and avoided much of the hassle. As a Senior Principal Scientific Officer, Dr Gill was entitled to first class travel on commercial airlines, but his rank was insufficient to qualify as a VIP for military travel. None of this was of any concern to me because the whole experience was so exciting.

We eventually boarded the VC10. The flight-master conducted further checks to verify we were all in the right seats, facing backwards for safety reasons. The mini-VIPs then boarded to be seated by the emergency exits with extra leg-room, followed by the more important VIPs who quickly disappeared behind curtained seating. Eventually we took off.

The flight itself was without incident. We ate a standard aircraft meal with no TV and no alcohol, in case we had to ditch in the North Atlantic, but I did hear glasses clinking behind the VIP curtains. We landed at Dulles airport mid-afternoon local time.

Dr Gill and I took a taxi for the 20-minute drive from Dulles to downtown Washington. I was astonished by the sheer size and complexity of the road layouts. We were booked into the Roger Smith Hotel (since pulled down) on Pennsylvania Avenue just one block away from the White House. As we walked up the steps into the hotel foyer a man behind us was mugged for his wallet. This was my first taste of the USA.

We checked into our rooms on the sixth floor to unpack. On the way up Dr Gill asked me to call in at his room in 20 minutes time for a strategy talk. I was impressed with the size of my room with its centrally heated radiators and en suite facilities. At the appointed time I knocked on Dr Gill's door and was admitted. Before saying anything he drew the curtains, turned up the TV to maximum volume, and turned on all the taps in the bathroom. He explained that security procedures were necessary to avoid the risk of being overheard. Dr Gill was always paranoid about spies and security, no doubt because in his early career he had worked on eaves-dropping techniques at the secretive Post Office Research Station at Dollis Hill.

After these preparations Dr Gill began talking about the plans for our forthcoming meetings. He finished by telling me about the event we were to attend that evening. One of Dr Gill's projects in Teddington had been on "helium speech". Tanks of air traditionally used by deep-sea divers contain 20% oxygen and 80% nitrogen. The trouble with nitrogen is its solubility in the blood at high pressure causing nitrogen narcosis (disorientation) and "the bends" if the diver surfaces too quickly. One way to avoid these problems is to replace the nitrogen in the breathing mixture with helium which does not dissolve in the blood. But because helium is a light gas it causes the vocal cords to vibrate at a higher frequency resulting in a "Mickey Mouse" type of sound known as helium speech. Dr Gill was the first person to devise and build a real-time processor that made helium speech sound normal. This would be relatively easy nowadays with digital technology, but with the analogue technology then available it was quite a feat. The donkey work was done by Ron Morris, but Dr Gill was the brains behind it. For his achievement the Helium Speech Society of America had invited Dr Gill as guest of honour to their annual banquet in Washington. I was going along as his colleague.

Having explained all this, Dr Gill added that a mishap had occurred on the flight. A bottle of malt whisky he had packed in his suitcase as a gift for someone had broken in transit. All his clothes were now saturated in whisky, including the suit and shirt he was due to wear that evening. I commiserated with him.

We set off for the special evening dinner. I joined Dr Gill in the crowded hotel lift, a powerful smell of whisky emanating from his person. People looked at him sideways no doubt thinking he was some kind of "lush".

At the banquet we were seated at a table with several notable people. One of these was Alan Shepard who in 1961 had been the second person, and the first American, to travel into space, the first being the Russian cosmonaut Yuri Gagarin. In 1971, just three months before our meeting, Shepard had been appointed commander of Apollo 14 and became the fifth person to walk on the Moon. Somewhere I have his autograph. At the conclusion of the after-dinner speeches Dr Gill and several other attendees, including the renowned high flyer Ralph Goodman, decided to visit the bars in nearby Georgetown. But it was now 11 p.m. (4 a.m. UK time) and I declined feeling pretty shattered. I took a taxi back to the hotel and crashed out.

The next two days being Saturday and Sunday were free days. We could not have chosen a more inappropriate weekend to be in Washington. It was the weekend of what became known as the May Day riots, part of the US Vietnam anti-war movement. On Saturday 1 May 35,000 protesters set up camp in the parks near the Washington Monument in central Washington. On Sunday 2 May riot police raided the encampment firing tear gas and knocking down tents. The campers scattered towards neighbouring parks and college campuses. Early on Monday 3 May, while most protesters slept, 10,000 federal troops moved in backed up by 7000 police and security forces. They protected the perimeter of every monument, park and traffic circle in the city. Paratroopers and marines were deployed via helicopter. The protesters employed hit and run tactics disrupting traffic and causing chaos in the streets. While the troops secured the major intersections and bridges, the police roamed through the city with tear gas making sweeping arrests, over 12,000 in total.

I spent that weekend wandering round Washington on foot, oblivious to most of this. I visited the museums, the Washington Monument, Lincoln Memorial, Arlington Cemetery and other tourist attractions. The cherry trees lining the avenues were in full bloom and spectacular. I was aware that some sort of demonstration was going on because I came across large crowds of hippies and squads of riot police in the central parks with helicopters clattering overhead. But none of this interfered with me. I do not know how Dr Gill spent his weekend. He said later that I had been a bit reckless to venture outside.

I met up with Dr Gill in the hotel breakfast room on Monday morning. Our first meeting was scheduled for 10 a.m. at the Naval Research Laboratory (NRL) a few miles south of central Washington. Our plan was to take a taxi, but there weren't any. We were advised that any form of travel was dangerous and that all routes out of Washington were blocked. So we retired to our hotel rooms to await developments.

I looked out of the open window of my room at the streets below. People were running riot throwing trash cans and overturning parked cars. But the police seemed to be getting the situation under control. My throat was tickled by tear gas wafting up. At around mid-day we managed to find a taxi driver willing to run the gauntlet.

We had meetings and exchanges during the week at NRL and Johns Hopkins University outside Baltimore. I gave a presentation at Johns Hopkins on my theory of submarine-generated internal waves. It went down well with the US guru Marshall Turin (they did have strange names these Americans) whose many papers I had read.

We visited the Maury Centre for Oceanographic Sciences. The Director of the Centre, Brackett Hersey, invited us to his house for dinner one evening. We ate a magnificent dish of spiced beef with an oriental seasoning that I have never been able to reproduce. I think it must have contained Chinese 5-spice, something unknown to me at the time.

On Friday night we flew out of Washington for the Canadian phase of our trip. The first destination was Victoria on

Vancouver Island, British Columbia. We stopped overnight in Montreal and ate in a restaurant where they spoke Canadian French. It seemed like a sort of pigeon French to me.

We resumed our journey to Victoria the next morning. After changing flights in Vancouver we arrived in Victoria on Saturday afternoon in a propeller aircraft. I liked Victoria with its corner sweet shops, rather like an English town. It was here I had my first sighting of the Pacific Ocean. It was also where I had my first live sighting of our Queen who was doing a Royal Tour of Canada at the time[16]. My view of her was limited to seeing her sweep past in a motorcade.

On the Monday we had the first of several meetings at the Defence Research Establishment Pacific (DREP) in nearby Esquimalt. One meeting was with Blythe Hughes, a young scientist doing interesting work on the spectral properties of ocean surface waves as measured by observations of the sun's glitter pattern. It had applications to submarine detection, but who could rely on the sun shining all the time?

After our visit to DREP we set off on Wednesday 12 May on the long journey to Halifax, Nova Scotia, on the Eastern seaboard. This entailed a short flight from Victoria to Vancouver to change planes where we were met with a considerable delay. It turned out this was because the Queen was flying out. There was at that time a recognised protocol that no flight could leave or depart from an airport within 30 minutes of the Queen's Flight.

We eventually boarded a big jet to take us to Halifax. It was my first flight in the new style of wide-body Jumbo that had only recently come into service. I enjoyed the flight and admired the superb views of the Canadian Rockies. Dr Gill travelled first class and emerged bleary-eyed from the baggage hall in Halifax.

---

[16] Queen Elizabeth and Princess Anne visited Canada from 3-12 May 1971 to mark the centennial of British Columbia's entry into the Canadian Confederation (the birth of Canada as a nation on July 1, 1867). They visited Victoria, Vancouver, Tofino, Kelowna, Vernon, Penticton, William Lake and Comox, B.C.

For the next two days we visited Canada's Defence Research Establishment Atlantic (DREA). The first evening we were invited for dinner to the house of John Stockhausen and his wife. John was a Canadian acoustician working at DREA and a great friend of Dr Gill. The second evening we went to a lager festival at the DREA lab where Barrie Franklin (who some years later I was to know well) gave a talk about the differences between beer and lager. We flew back to Washington mid-day Friday to debrief Dr Valentine Flint at the British Embassy and collect our gratis duty-free bottle of spirits. We then drove to Dulles for the RAF flight home - altogether an exciting trip.

Soon after joining "O" Group I had become close friends with Tony White who was such a supportive colleague during my sea trial on HMS *Hecla*. He was learning classical Spanish guitar and practised during the lunch hour. These were the days when one-hour lunch breaks were still sacrosanct with a proper meal in the canteen. I found myself attracted to guitar music, particularly after hearing records of Julian Bream whom I found much more attuned to the emotion of the music than the technically perfect John Williams.

So I accompanied Tony to his guitar class one Monday evening in nearby Isleworth. The teacher was a Mr Birdett. He lent me a guitar and I took to it immediately. I then bought my own guitar from Mr Birdett for £15, together with a rigid case for £5. The guitar turned out to be a good one. I began to play the basic repertoire of Sor and Carcassi and made rapid progress, soon moving on to later composers like Tárrega, Lauro and Villa-Lobos, learning some of the latter's easier choros, preludes, and studies. I practised in my office at lunch times until a woman upstairs complained that it sounded too funereal and depressed her. I often wonder which piece that was. The only person at work who took any interest in my playing, other than Tony, was Yvonne Sale who actually sat in my office occasionally to listen. Phil Lindop opposite me used to slam his door, or perhaps it was just a very fierce spring.

One evening I went to the Fairfield Hall in Croydon, possibly with Tony, to attend an inspirational guitar concert given by Andrés Segovia. He had short stubby fingers like bananas, but this did not seem to affect the brilliance of his playing. This must have been one of his last public concerts.

I later had lessons in Richmond with John Mills (the guitarist not the film actor) before he became famous as a recitalist. I continued playing until the late 80s when I dropped it to take up the organ. Long guitar nails in the right hand are not compatible with keyboard instruments. My abandonment of the guitar was perhaps a mistake because I was never able to achieve the same level of accomplishment on the organ. I have since reverted to the guitar and it all came back quickly. If it had not been for Tony White I would never have picked up a guitar.

Meanwhile, I was tasked at work to carry out a detailed assessment of the risk of submarine detection by non-acoustic means. As usual, the drive was intelligence that the Soviets might be ahead of us. I calculated the quantitative probabilities of detection in a range of scenarios for all the non-acoustic processes I could think of and model, and wrote everything up in a report. The management liked the report but made me rewrite the conclusions, which they thought over-stated the risk. There were politics involved that I was not aware of.

Dr Gill was committed to his work and always full of ideas. A favourite habit of his was to call people to his office about five minutes before five o'clock to discuss his latest brainwave or his latest unlikely solution to some outstanding problem. These meetings often lasted a couple of hours, poring over data records spread out on tables and floor, and discussing hypotheses ad nauseam. This upset some people anxious to get home.

Sometimes Dr Gill phoned me at home. He once called me on a Sunday morning with an important idea that could not wait. He asked me to meet him at a pub somewhere in Erith. I drove there to find him standing by his car. He told me to get back in my car and follow him to a different pub in case our telephone conversation had been tapped. We went to a pub with a large beer garden so that we could sit outside with less chance of being bugged. This sort of clandestine operation was typical of Dr Gill, but it was not without good reason.

Many people, Lucienne in particular, felt somewhat alienated by my reluctance to discuss my work. This was a handicap for me, at least initially after the openness of university life, but I had been indoctrinated about security. When I first joined ARL I was briefed by the security officer who told me that statistically speaking there was always at least one Soviet spy working in our midst. Experience showed this to be true.

For example, in 1934 three undergraduates at Trinity College, Cambridge, were recruited as spies by the KGB, the USSR intelligence service. They were Guy Burgess, Donald Maclean, and Kim Philby, soon to be joined by Anthony Blunt and John Cairncross. These five became arguably the greatest spy ring the West has ever seen, penetrating deep into British Intelligence Services. Additional alleged KGB spies recruited at Cambridge were Leo Long, Michael Witney Straight, Dennis Proctor and Alister Watson. The latter was born in Southend in 1908 and probably went to my old school. At its peak these spies had all achieved influential positions in government departments. In particular, Alister Watson became head of Submarine Detection Research at the Admiralty Research Laboratory, Teddington, the group in which I was shortly to work. Watson was suspected of espionage in 1965 and grilled for six weeks. He admitted meeting senior KGB officers in London but denied passing secrets. MI5 remained convinced he was a spy and had him removed from secret work at Teddington. He took up a less sensitive post at the Institute of Oceanography in Surrey. He died in 1981.

A second example of espionage in our midst was the notorious Portland spy ring that operated within the Admiralty Underwater Weapons Establishment (AUWE) in the late 1950s. Ethel Gee, a records clerk at AUWE, worked with Harry Houghton, an employee in the naval dockyard, passing secrets to a KGB Colonel who worked under cover as a Canadian businessman called Gordon Lonsdale. All three were captured and imprisoned in 1961. An AUWE colleague, Sam Mason, told me how one cold morning in winter on the way to work, he stopped on Chesil Beach to help Harry Houghton whose car was on fire. Harry had forgotten to remove a rug that he had draped over the engine the night before to stop it freezing up. Sam Mason later regretted the help he had given.

Lonsdale was later exchanged for the British spy Greville Wynn. Houghton and Gee served only 9 years of their 15 year sentences. They were released in 1970 and married, living in Poole. Of course, we only know about the spies who were discovered. Others must have gone undetected. We were taught to treat everyone at work as potentially suspect.

Meanwhile, Dr Gill thought it would be good for my career to become involved in the more mainstream topic of underwater acoustics and sonar. I began in the summer of 1973 by participating in a sea trial code-named Square Deal. It was a collaborative trial with the Americans involving three ships: a US ship towing an acoustic source, a second US ship towing an experimental V-shaped acoustic towed array receiver designed by Hank Aurand of the Naval Ocean Systems Center (NOSC) in San Diego, and our own ship RMAS *Bullfinch* towing a single array. The work was to be carried out in the North Atlantic sea areas of Bailey and Rockall. There were several objectives: to use the towed arrays to obtain data on horizontal noise directionality, to obtain data on acoustic propagation loss in the area of the Rockall Trough, and to measure the noise and detection performance of our arrays. It was my job to analyse the UK data after the trial.

We set off from Plymouth on 26 August 1973. Crew members included Dr Gill and Bert Allwood. Bert was a delightful man with no pretensions and whom everyone liked. I had a lot of interaction with him. He had so many interests spanning music, mathematics, physics and philosophy. Bert had two sons, Ralph and Peter, who at the time were both choral scholars at Kings College, Cambridge. Ralph went on to become Director of Music at Eton College and later often appeared on TV, lately as a judge in Gareth Malone's BBC programme "The Choir". Peter became Headmaster of Lichfield Cathedral School before moving to the Dragon School, Oxford, and has since retired. Peter and Bert made a brief visit to our house in Weymouth in 2009. Bert died in September 2012.

I believe that senior management disapproved of Dr Gill attending sea trials because he was a Group Leader with responsibilities and management duties back at base. But he was a "hands-on" scientist and was never happier than when he was messing about with electronics and pieces of equipment or examining data records. On one occasion Dr Gill was refused permission by the Director to take part in a submarine trial. So he drove up to Faslane in his mini at his own expense and took holiday leave to join the boat for the duration of the trial. Maybe this is why he never achieved further promotion.

To return to Square Deal, the weather during the whole exercise was appalling, never less than Sea State 6 and deteriorating to Severe Storm Force 11 with rollers the height of houses. Part of the trials plan required us to run for several days on a fixed course down the Rockall Trough parallel to the course of the transmitting ship that was towing the sound source. This put us broadside to the heavy seas making it hazardous and impossible to eat or sleep. Sometimes the ship rolled so far I thought she would never recover. Dr Gill seemed unaffected by all this. I can picture him sitting on the floor of the bridge lab trying to fix pieces of equipment that were in pieces and rolling round the deck. He and I were the only scientists not seasick. The ship's crew suffered, the cook never emerging from his bunk after the first day. It did not matter because cooking was impossible in those conditions. We ate sandwiches and thick soup. In spite of all this the trial was a success in achieving all its objectives.

Our Group at ARL had a rolling contract with the Department of Applied Mathematics and Theoretical Physics

(DAMTP) at Cambridge University. The contract supported an oceanographer, Dr Adrian Gill [17], who was free to study any subject he wished provided it had some remote connection with underwater acoustic propagation. MoD in London were questioning the relevance of this funding to our defence work. So in the autumn of 1973 I was sent on a three-week liaison visit to Cambridge.

I stayed in digs near the station and walked each day down Trumpington Street to DAMPT in Silver Street. I was given a desk in an office with the Australian hydrodynamicist J S (Stewart) Turner, an expert on fluid mixing processes. Stewart was busy completing his now classic book "Buoyancy Effects in Fluids". I was flattered that he asked me to proof-read two of the chapters. He allowed me to carry out some laboratory experiments on double diffusion processes, in this case oceanic "salt fingers".

There were many well-known scientists at DAMPT who up to that point had just been names to me. I met Sir Geoffrey Taylor OM, the father of modern fluid dynamics, then in his late eighties, sitting and chatting in the common room. I regularly saw Michael Longuet-Higgins arriving at work on his bicycle. I met the young Herbert Huppert later to rise to fame for his work on volcanoes, and whose son Julian was to become the eccentric Liberal Democrat Member of Parliament for Cambridge. Stephen Hawking was there but relatively unknown at that time. Julian Hunt was a young Lecturer, later to become a Tory minister and Lord Hunt of Chesterton. And of course there was the Head of DAMPT, the brilliant George Bachelor. I enjoyed my time at Cambridge but doubt if my visit changed anything. At least the research contracts with Adrian Gill continued.

---

[17] Adrian Edmund Gill FRS (1937 - 1986) was an Australian meteorologist and oceanographer best known for his textbook *Atmosphere-Ocean Dynamics*. Born in Melbourne, Australia, he worked in Cambridge from 1963 to 1984. He was elected a Fellow of the Royal Society a month before his early death from cancer.

Whilst in Cambridge, Adrian Gill brought to my attention the interest of some Russian oceanographers in my 1969 paper on the Gulf Stream. Several papers examining my model had been published in the Russian literature. A visiting scientist from Scripps in San Diego, Pearn P Niiler, complimented me on my work but said it was a pity I had not been trained as an oceanographer, a comment that hurt me.

It was not until the mid-70s that western oceanographers became interested in my Gulf Stream model. Several papers then appeared dealing with extensions and generalisations. After a publication by George Veronis (one of my heroes) at Yale, the "Parsons Model" became known as the "Parsons-Veronis Model" of ocean circulation. My name became further diluted with more contributions leading to the "Parsons–Veronis–Huang–Flierl Model". In the early 90s a group at Imperial College revisited my work, spawning yet more papers. Interest has continued with a recent paper in 2009 by a group at MIT entitled "An Eddifying Parsons Model". Much of this can be Googled.

At about this time a young US exchange scientist joined us at Teddington. His name was Dr William (Bill) Moseley. He was given a desk in my office. Bill had charisma and was good company. He was also talented and ambitious. I liked him and so did Lucienne who found him handsome. We got to know his wife Sandra and their five boisterous children. Sandra was a bit like Jerry Hall complete with Southern accent. They rented a large house in the Hampton Hill area of Teddington. Bill and Sandra both liked a drink and enjoyed loud parties, some of which we attended. They made monthly trips to the US Air Force base at Upper Heyford, Oxfordshire, where they stocked up on massive American rib-eye steaks and duty-free gallon flagons of gin and white bacardi rum.

Very soon after Bill's arrival we suffered the Three-Day Week, the so-called "Winter of Discontent". The Three-Day Week was one of several measures introduced by the Conservative Government under Edward Heath to conserve electricity, the production of which was severely limited by the industrial action of the coal miners. From 1 January to 7

March 1974 commercial users of electricity were limited to three days of consumption each week and prohibited from working longer hours on those days to make up the shortfall. Essential services such as hospitals were exempt. Television companies were required to cease broadcasting at 10.30 p.m. Consequently, for the first three months of 1974, immediately after Christmas, we had no electrical power at work for two days each week. There were no PCs in those days so that was not an issue, but by 3 p.m. it would begin to get dark and cold with no hot water in the radiators. We were issued with candles and instructions on how to use them efficiently and safely. We sat huddled in semi-darkness at our desks wearing overcoats and writing by candlelight like scenes out of Dickens. Bill Moseley was totally astonished at our British aplomb and sheepish acceptance of these conditions.

My analysis of the Square Deal data went well. In Spring 1974 I travelled to the US with Mike McCann (of integral p-squared-dt notoriety) to discuss and exchange results. We flew first class on a commercial airline. In those days there was no such thing as business class. Lucienne and Isabelle accompanied me to the first class lounge at Heathrow where we helped ourselves to canapés and champagne before take-off. Mike and I flew on what must have been one of the last BEA Comet flights, taking us over the pole to Los Angeles. It was all very decadent. I think this VIP treatment was all fixed up by our man in Washington, Dr Valentine Flint, who somehow managed to circumvent the strict requirement to fly to the US by Royal Air Force VC10. From LA we flew the short hop to San Diego where we arrived after midnight local time, staying in a hotel close to beautiful Shelter Island.

We had a free weekend to see the sights in San Diego. On the Saturday evening we had dinner at the house of Rodney Bown and his wife Gloria. Rodney was at that time an exchange scientist at NOSC from AUWE Portland. He had cultivated an American accent while in the US which he retained on his eventual return to Portland. Both he and his wife were always very conscious of rank, yet he became a Weymouth taxi driver on retirement.

On Monday we had technical meetings with Hank Aurand and a young graduate student named Burley Brunson at NOSC. We were invited to dinner that evening at Hank's house where we ate roast beef, exceedingly tender. On another night Burley Brunson took us in his car to see the night life in Tijuana just over the border in Mexico, an eye-opener. The six-lane highway on the US side of the border became a dirt track on the other side. Tijuana itself was seedy and impoverished. It was not surprising there were so many illegal Mexican immigrants in Southern California. I bought Lucienne a small articulated fish made of Mexican silver. This trip was the first of many to San Diego and California. It is a wonderful place of optimism and innovation where nothing seems impossible, all helped by a warm climate the year round.

At the end of our week in San Diego, Mike McCann and I checked in at the airport for our flight to Washington. The receptionist informed us that the first class seats we had booked were no longer available. She said that the whole of first class had been block-booked a few days earlier by the San Diego Padres (pronounced Pardrays). This was clearly supposed to be explanation enough. When we asked why a bunch of seminaries had taken our seats the response was one of shock. Had we never heard of the best major league baseball team in the US? We said we could not care less about the San Diego Padres. We just wanted our first class seats that had been booked and paid for weeks ago. Eventually we settled for the two rear-facing air hostess seats in first class. During the flight we ate steak and lobster and watched the Padres team manager handing out wadges of banknotes to the players, probably pocket money.

On arrival in Washington we checked in at the Americana Hotel, Crystal City, not the best of hotels but well within our *per diem* rate. It was close to the British Embassy, before it moved much later to Massachusetts Avenue. Immediately outside my hotel room was a massive ice machine, something I had never seen before. It woke me during the night with its regular discharge of great quantities of ice cubes.

Our man in Washington, Dr Valentine Flint, had once again done us proud, fixing us up with a huge rental car. We spent the weekend driving round Washington and made a long sightseeing trek to the Luray Caverns and Shenandoah National Park. We had several meetings the following week, one with a Lt Cdr Pete Tatra at the Maury Centre who had written a fast-running acoustic ray tracing program that the US had agreed to give us as part of an exchange agreement.

One evening we drove to a Steak House in Rosslyn that Dr Flint had recommended. This was my first taste of real American steak, an enormous slab of prime sirloin that was cut in front of us from one of several carcasses cooking over red-hot grills around the restaurant. It cut like soft camembert and was probably pumped up with steroids, but very tasty.

Basically, all our work was aimed at keeping one step ahead of the Soviets, so it was vitally important they did not know what we were doing. This called for tight security that prevented us from telling friends and family about our work. We were even discouraged from finding out too much about the work of colleagues. So I was never really sure what my US colleague Bill Moseley was doing at ARL even though we shared the same office. It became a bit clearer when in late September 1973 Dr Gill asked me to accompany Bill to the remote island of Unst in the Shetlands. This turned out to be some assignment.

I set off with Bill on a cold day in late September on a flight from Heathrow to Aberdeen in one of BEA's Vickers Viscount turbo-props. At Aberdeen we sat in a dismal transit lounge waiting for our connecting Loganair flight to the Shetlands. It was cold and raining heavily. I had just bought a book of the championship chess games recently played between the Russian world champion Boris Spassky and the volatile but brilliant American Robert Fischer who controversially took the title. This was the Cold War in miniature. Bill and I passed the time discussing this east-west encounter.

The landing at Sumburgh at the southern end of the Shetlands was hair-raising because the pilot had to avoid the lighthouse on the precipitous cliffs of Sumburgh Head before lining up with the runway. From here we flew in a small 6-seater island hopper landing in fields on several islands to drop off or pick up locals. Each time the pilot landed he sounded a loud claxon to disperse the sheep. The journey gave us a fine view of the islands. Eventually we landed at the RAF base at Saxa Vord on Unst, the most northerly of the inhabited Shetland Islands, where we were put up in the officers' mess.

To give an idea of its remoteness, Unst is further north than Bergen in Norway, and further north from Inverness than Manchester is from London. The island is largely boggy grassland with some coastal cliffs. The main village is Baltasound, formerly the second largest herring fishing port after Lerwick.

We met up with Norman Field who comprised the advance party. Norman, a frequent visitor to Unst, worked at Teddington in the Sonar Group. Over the years I was to work closely with him. He was a qualified chemist but was now working as a "jack-of-all-trades". Norman got on well with military personnel and I later found him to be indispensable on any sea trial.

Norman welcomed us to the site and explained the set-up. RAF *Saxa Vord*, at the northern end of Unst, was a radar station operated by the RAF during the Cold War, so named because it never featured direct military action. This was because both East and West possessed nuclear weapons in sufficient quantities to guarantee their mutual annihilation. The era of the Cold war was relatively stable and was not to end until the early 90s, after which our future would become more precarious. The radar station on Unst provided a long-range early warning capability for the airspace north of Scotland. For a fascinating insight into the history and purpose of this secretive site, see "A History of RAF *Saxa Vord*" on the internet.

Of particular relevance to me was the naval presence at *Saxa Vord*. After WW2 the United States developed and laid around the world several deep-water sound surveillance systems (SOSUS). These were large, low frequency passive listening devices equipped with arrays of hydrophones and processing facilities that could detect submarine positions by triangulation over hundreds of miles. Their purpose was to protect the US from the threat of Soviet ballistic missile submarines. One of these SOSUS systems covered the Iceland-Faroes Gap to detect the passage of Soviet submarines between the Norwegian Sea and the North Atlantic. The underwater cables from this SOSUS system ran ashore to a receiving station on Unst. Bill had access to the data because he was a US scientist, whereas I was denied access. The job of Norman and me was to monitor the recording equipment and keep it going day and night so that Bill could later have at his disposal a complete data set.

Unst in autumn was a most desolate place. Norman would drive me in the jeep along the steep track from the RAF base to the receiving station on the cliff top of Saxa Vord where I did my night shift. It was bitterly cold and spooky with the wind whistling round the building. I worked there on shifts with Norman and Bill for about three weeks. I have never experienced such a lonely and isolated time.

Norman had been there a week already to set things up and he knew his way round the island. On spare afternoons we drove in the jeep to view derelict crofts and deserted beaches where Vikings had once landed. I met up with a local man, Bertie Henderson, who had a part-time business connection with the site at Saxa Vord. He sold sheep skins and meat which I duly bought to take home.

At the end of my period of duty on Unst I travelled back with Norman leaving Bill to manage on his own for a couple of weeks. For some reason Norman and I did not fly back to Aberdeen. Instead, we travelled by overnight ferry in rough seas from the Shetlands to Aberdeen where we caught the BEA regular flight to London.

On arrival at Heathrow my suitcase failed to appear on the carousel. I reported it missing at the BEA desk. They asked me to describe my suitcase and its contents. When I told them it contained a leg of lamb they quipped that I should not worry because my case would walk home. I was not amused. At about 23:00 that night a taxi arrived at our house in Ashley Drive with my case. The leg of lamb and sheepskin were both intact. I was impressed. They told me that my case had mistakenly been put on the Aberdeen flight to Dublin.

In late 1974 Dr Gill put my name forward to be considered for promotion, thus beginning the long Civil Service annual process of filtering and selecting candidates for interview. I passed the initial hurdles and went to London for my interview in April 1975. At about the same time Dr Gill made a sideways move to take over "L" Group that specialised in sonar development. His post as head of "O" Group was filled by Dr Valentine Flint on completion of his two-year stint at the Washington Embassy. I was successful at my interview and was promoted to PSO (Principal Scientific Officer) with effect from 1 July 1975. This was the "Career Grade" that any aspiring scientist might expect to reach before retirement. Dr Flint, my new boss, congratulated me on "joining the club".

Dr Flint was a man whose name and appearance were perfectly matched. He was short, stocky and slightly rotund because of his liking for good food and fine wine. He had a short, thick neck supporting a head that looked like it had been chiselled from a block of granite. His voice was fine and rich, and he spoke eloquently. He was a wonderful "bullshitter". In a conference or seminar setting he could answer any technical question even if he knew nothing about the subject. At the end of his monologue we would all think how wise he was, only reflecting later on the emptiness of the content. This was a useful skill. I found him a difficult but likeable man. They do not make people like this any more.

Dr Flint's leadership was marked by several new technologies he had picked up from his experiences in the States. One of these was the dictaphone, a relatively new tool. I would hear him through the wall dictating his letters into this new-fangled device that generated a tape to be passed on to the typing pool. These sheltered women were terrified of the new technology but were seduced by Val's fruity voice. Dr Flint's other hobby-horse was the conduct of sea trials. His vision was that the technical report of any sea trial should be completed before the scientists left the ship rather than three months later. This was to be achieved using the latest computer technology for on-board data analysis and word processing. It was an admirable ambition that we in his team failed to grasp at the time.

In addition to the women in the typing pool there were women who ran the telephone switchboard. All incoming and outgoing calls had to go through the switchboard. If we wanted to make a private call, for example, to home or to the bank, we had to declare this to the switchboard operator who made a note of it. At the end of each month we received a bill for our private calls. It was a wonderful relief when some years later everyone had direct lines and the switchboard became obsolete.

In 1975 Chris Harrison joined our group. He was a young PhD from the Scott Polar Institute in Cambridge with a good theoretical background. In addition to his scientific abilities Chris was a virtuoso flautist, playing in both professional and local amateur orchestra. He once gave a lunchtime performance at work, the highlight being the "Flight of the Bumblebee". On and off I kept in touch with Chris over many years. He was confident and outspoken, a typical "angry young man". When he ate an apple he ate the lot, core and all. I got on well with Chris because of his mathematical background. In fact, I got on well with most people. Chris's fractious personality resulted in many confrontations with Dr Flint, and particularly with Mr Phil Lindop.

Mr Lindop was the other PSO in the group, but much senior to me. He seemed to have a permanent chip on his shoulder because he thought his work was not sufficiently recognised. Even if his contribution to a project was remote he would insist that his name appear on the report. He was protective of his data and constantly suspected others of plagiarising his results. He even confronted me once when I wrote a report that included some results I had obtained on the Square Deal trial, about the amplification of signals in the sound channel when a source vessel crosses the edge of the continental shelf. Mr Lindop declared this to be his own domain and I was trespassing. I published the report anyway.

Another scientist in the group was a young man called David Pugh, affectionately known as Pip. He was an oceanographer from Cambridge. Pip was quiet, friendly, intelligent, and popular with everyone. One day he failed to turn up for work. We heard nothing for a few days. He was then found dead in his flat with a plastic bag over his head and an orange stuck in his mouth. It was very sad. The manner of his death remains a great mystery to me.

The "O" Group clerk at that time was Margaret Bowditch. Her job was to look after all the files and administration of the Group, and to act as Dr Flint's secretary. Margaret was a delightful woman of great charm and with a caring disposition. She took a particular liking to me. Lucienne and I met with her socially on several occasions. Her husband John was a Sergeant Major at the nearby Royal Military School of Music at Kneller Hall, Twickenham, the training ground for British Army bands. She often gave us free tickets to concerts given by the massed bands. On one occasion we attended an open-air evening concert that finished with Beethoven's 1812 symphony complete with real cannons and fireworks. Sadly, both John and Margaret died relatively young of cancer.

On 6 April 1975 our second daughter Caroline Estelle was born. This had been a long and stressful process to achieve success, but we were delighted with the result. Lucienne's labour was brought on by laughing at a Morecambe & Wise programme. At 11 p.m. in a state of controlled panic I drove Lucienne to Hammersmith Hospital to meet her private consultant, Mr Chamberlain. He turned out to be on holiday so his deputy took charge. When we brought baby Caroline home we were so concerned that nothing should go wrong that we made all visitors wear face-masks. I think they were astonished.

Caroline was christened at St Edmund's Church, Whitton in the early summer of 1975. Her godparents were my brother Nick and our French niece Nicole who was not able to attend.

In Spring 1975 we had an unlikely recruit at ARL, a Mr Rod Litchfield who had written to us asking about possible part-time work. He was a tall, gangly character about my age and a lecturer in Chemical Engineering at Surrey University in Guildford. He was looking for work during vacations and other spare times. Rod offered specialist programming skills which were rare at that time, so we took him on as a part-time consultant.

Rod was given a desk in our big office along with Roger Panter, Chris Harrison, the American Bill Moseley, and me. We had a jolly time together. I liked Rod and he fitted in well. The only strange thing about Rod was his eagerness to relate to us his sexual exploits with his wife. When I eventually met his wife, Theresa, I felt embarrassed to look her in the eye.

I visited Rod at his university department in Guildford. He specialised in optimising industrial chemical processes. He gave me a tour of the labs. We had an enjoyable lunch and a bottle of Piesporter in a restaurant on the campus run by catering students and trainee chefs. He told me he was bored with his university job. It remained a mystery to me how he had so much spare time and I wondered if the university knew, or even cared about his moonlighting.

Rod's job at our lab was to write software to assist in Phil Lindop's analysis of his acoustic propagation data. Phil was organising a sea trial for late September 1975. He was short of scientific staff to man the trial so it was arranged that Rod and I would join the team. This was the only trial I ever went on where the weather was so bad that we never did a stroke of work.

The plan was for Rod and me to join the trials ship, RMAS *St Margarets*, off South Uist in the Outer Hebrides. We flew from Heathrow to Glasgow and took a Loganair flight to the island of Benbecula. From here we hired a taxi, a 4x4 Jeep, to Lochboisdale on the island of South Uist, a journey of some 45 minutes. The landscape was flat and desolate, not unlike the Shetlands. We checked in at the local inn on the waterfront at the head of an inlet looking out to sea, except that the poor weather obscured the view. It being a Sunday and under

Scottish jurisdiction there was no alcohol for us that day. There was also minimal heating. We settled down to a miserable evening meal listening to the wind whistling and the rain lashing against the window panes. What with this and the inn sign outside creaking on its hinges I was reminded of the Admiral Benbow in Treasure Island.

The ship was supposed to rendez-vous with us at noon the next day, but I was unsure exactly how the transfer process was to be accomplished. The next day dawned with more rain and a full gale. We sat at the window of the inn overlooking the inlet awaiting the appointed hour. There were no mobile phones in those days. Noon came and passed with no sign of the ship. By five o'clock it was getting dark with the weather worsening. We resigned ourselves to another cold night at the Admiral Benbow.

The next day was again cold and wet but the wind had eased. There was still no sign of our ship. Then around lunchtime we spotted a small black dot on the horizon. It slowly grew bigger occasionally disappearing in the troughs of the waves. After twenty minutes we identified it as a small boat. It turned out to be the *St Margarets* launch coming to collect us. The ship's captain had decided it was too dangerous in rough seas for the ship to negotiate the inlet.

We were pleased to be finally on board. We set off for the trials area. The bad weather did not relent and there were repeated postponements of activities. It was just too dangerous to deploy the heavy equipment overboard. I remember taking Rod up to the highest deck on the ship above the bridge. It was an open deck overlooking the foredeck with a spectacular view in rough weather. We held on tightly to the rails as we headed into mountainous seas. The noise of the wind was so strong it was impossible to talk. We watched mesmerised as the ship's bow slammed and buried itself into each approaching wave sending a shudder of vibration through the ship that lasted several seconds. The foredeck was awash for what seemed an eternity before slowly rising and shedding tons of water in readiness for its encounter with the next wall of water. Our emotions were a mixture of fear and wonder at the forces of nature.

We later found relative shelter in the lea of one of the Hebridean Islands. I had a go at fishing to pass the time. Rod caught a Red Gurney that one of the crew filleted with two quick flicks of his sheath knife. All I caught was a large herring gull that got its wings caught in my line as it tried to eat the bait.

I really cannot recall the purpose of this sea trial. I just remember that after a week of inactivity and no change in the weather Phil Lindop decided to call it a day. I arranged for Rod and me to be dropped off with Phil at Campbeltown. The rest of the team returned with the ship to Plymouth. Phil Lindop went straight home but Rod was keen to stay overnight in Campbeltown. I did not mind too much. I knew that if I returned to Teddington immediately I would be called upon the next morning to give a presentation to a group of "Perishers" who were scheduled to visit. It was many years before I gained the confidence to talk easily to large groups of people.

For the uninitiated, the Perisher course, officially known as the Submarine Command Course, is a 24-week course that any aspiring officer must pass before taking command of a Royal Navy submarine. It has a high failure rate and is widely regarded as one of the toughest command courses in the world. An unsuccessful candidate is not permitted to return to sea in the Submarine Service. However, he is permitted to remain in the Royal Navy by moving into the surface fleet.

So Rod and I checked into a hotel in Campbeltown. It was still raining heavily and the town was cold and dismal. Before darkness fell we walked the few streets and noted the large number of bonded warehouses full of whisky casks. We then climbed the water-logged hill behind the town for a view of the harbour. We returned drenched to our hotel to be welcomed with typical Scottish hospitality to cold rooms, no heating and a restaurant that was closed. I tried to explain to Rod that this sea trial was not typical. The next morning we caught the bus for the depressing four-hour journey to Glasgow and our flight home. It rained all the way.

Shortly after this trip, Rod and his wife suffered a terrible loss with the cot death of their second child Benjamin, aged 3 months. I did not see much of Rod after that. I believe he left Surrey University to run his own business. Rod died young in July 1999 from cancer, aged 56.

Bill Moseley made return visits to Unst during that year to collect more data. I believe he made these visits by himself. At the end of the year he had some sort of breakdown that was hushed up by him taking extended leave. Possibly he was on medication to overcome the loneliness of Unst. I do not really know and I never asked him. Shortly afterwards his term at ARL ended and he returned to the States, but we remained in touch as close friends, even to this day.

After Bill Moseley left we were joined by two other US exchange scientists, both from the Naval Research Laboratory, Washington. One was John de Santo, a theoretician working on a new method of modelling acoustic propagation called the parabolic method, now a standard technique with several advantages over the more standard ray tracing and normal mode methods. The other was Burton (Burt) Hurdle, a venerable and veteran acoustician who was using his time in the UK to write a book. Even at that time Burt was close to retirement, but to my certain knowledge he was still working at the Naval Research Laboratory (NRL), Washington, in 2004. My recollection of Burt is of an amiable, bumbling character with a vast knowledge. He was a man who always had time for everyone and could surprise you by turning up unexpectedly anywhere in the world that you might be. After he eventually retired Burt continued to work in Washington for nothing under a "dollar-a-day" contract that allowed him access to NRL and its facilities in return for his considerable expertise and scientific output. He died age 97 on 4 March 2015.

In August 1976 we celebrated Isabelle's first communion with a meal and "croque-en-bouche" at the Greyhound Inn close to Hampton Court. The "maitre d" was André, the young Frenchman who had married the daughter of our Ashley Drive neighbours, Mr and Mrs Smith. Most family members were present including my uncle and aunt, Harold and Jennie Besent.

One day Dr Gill told me that his teenage son William had been arrested on a charge of the attempted rape of a girl late at night on her way home from Weybridge station. William had been positively identified in a line-up. He protested his innocence and Dr Gill never doubted it. William claimed to have an alibi and Dr Gill was sure the police would soon release him, but they decided to continue with the investigation and held William in custody. They questioned him and his friends and relatives for several days. They seemed determined to press charges. Dr Gill was distraught at the treatment of his son. He became convinced that the police were more interested in obtaining a conviction than in seeing justice done. Eventually, after a long process William was released without charge. It was established as a case of mistaken identity. After this experience Dr Gill, always an upright and law-abiding man, lost faith in the police and vowed never to cooperate with them again.

As early as 1974 there were rumours that the Admiralty Research Laboratory would soon be closed as part of a government rationalisation programme, the various projects transferring to other establishments. By 1976 the plans were firmed up. We were told that the sonar and oceanography work with which I was involved would be in the vanguard, the first to move. We were to join the Admiralty Underwater Weapons Establishment (AUWE) on the Isle of Portland in Dorset.

Of course, nobody was happy about being forced to relocate and uproot their family, myself included. To stop people "deserting the ship" and applying for other MoD jobs in the area, for example in London, an embargo was placed on all transfers except for those people close to retirement. The only other option was to resign.

I started looking again at university posts but saw nothing that attracted a comparable salary. In June 1976 I applied for a well-paid job of geophysicist at Schlumberger in their oil and gas exploration division, with offices close by in Twickenham.

I was interviewed and offered the job on the spot. They wanted me to start immediately. I said I was committed to an important sea trial in two weeks time and felt honour-bound to complete it before starting work with Schlumberger. This was not the right response and the job offer was withdrawn.

In August 1976 Lucienne and I decided to visit Dorset, particularly Weymouth, in order to see what the area was like. We drove down for a weekend visit with Isabelle and baby Caroline, booking into the Queens Hotel by Weymouth Town Bridge at the end of St Thomas Street. That summer was exceedingly hot with record temperatures. The town was animated and the beaches packed with happy holidaymakers. Even late at night we could walk the streets in shirt sleeves. In those days there were no drunks on the streets because the pubs closed at 22:30 and there were no nightclubs. We were impressed with Weymouth and its beautiful bay. What with all the sandy beaches and palm trees along the sea front it reminded Lucienne of the Côte d'Azur. We had an enjoyable evening meal together at the Sea Cow Bistro on the waterfront. This trip made us warm to the idea of moving to Dorset.

I was now becoming occupied with regular sea trials and frequent trips to the States and Canada. On one of these trips I was invited out to dinner in Mystic Seaport, Connecticut, by US department head Howie Schloemer. He offered me a job as an exchange scientist at the Naval Underwater Systems Center (NUSC), New London. This was the first of several invitations to work abroad all of which I turned down. I liked my job in UK and did not want to go through the family upheaval.

I was still in Dr Flint's Group but was actually working for Dr Gill. He asked me to take part in the first of a series of sea trials at the US Atlantic Undersea Test & Evaluation Center, known as AUTEC. This is an American base on the tropical island of Andros in the Bahamas, located some 150 miles south-east of Southern Florida, and north of Cuba. As they say on their website, *"AUTEC provides instrumented operational areas in a real world environment to satisfy research, development, test and evaluation requirements and operational performance assessment of warfighter readiness in support of the full spectrum of maritime warfare."*

UK submarines often use this site, by arrangement with the US, to be put through their paces and assessed on their performance in simulated operational situations. In February 1977 the recently commissioned UK nuclear attack submarine HMS *Sovereign* was due to be working at AUTEC. ARL was involved because *Sovereign* was to tow the first complete UK towed array sonar system that our group had developed and to which I had contributed. Over the years we were to forge a close working relationship with the submarine squadrons leading to the installation and testing on RN submarines of several experimental equipments we had developed.

I travelled to AUTEC with a colleague from Teddington, a young hydrodynamicist called Ray Bartlett. The plan was that after spending a few days at the AUTEC site the two of us would transfer to *Sovereign* to record and analyse data with the new sonar system on the passage back to the UK.

The route by plane to AUTEC was quite long, involving a car drive to Brize Norton RAF base in Oxfordshire the night before, and after the usual count-down, leaving Brize about midday in an RAF VC10 heading for Washington. There were strong headwinds forcing the VC10 to land and refuel in Gander, Newfoundland, in heavy snow, where we were treated to a free hamburger in the canteen. We arrived at Dulles airport, Washington, around 1600 local time. On these trips we often stayed the night in Washington in order to visit the UK Embassy. On this occasion we flew directly on a commercial flight to West Palm Beach in Florida. As we landed in a steep descent the pilot made the usual passenger announcement: "Welcome to beautiful West Palm". After spending the night in this idyllic resort we had time for a quick swim before boarding a small AUTEC plane to take us to the Bahamian island of Andros. The airfield was a swathe cut out of the jungle. From here we were collected by an RN driver who took us through the swamps and bush to the AUTEC base.

Ray Bartlett and I arrived at AUTEC to be met by Norman Field who was already there. Once again he was doing his logistics job. Norman showed us round the base which was like a small village with a white sandy beach nearby where there was a cocktail bar. I drank pina coladas there with Norman in moments of relaxation. The white beach, blue sea, and all-enveloping heat were like a scene from a James Bond film.

During the first night on the base my colleague, Ray Bartlett, began suffering violent pains in his groin. On the advice of the navy medic an emergency transfer was arranged to the mainland hospital in West Palm Beach. It turned out that Ray had kidney stones so he did not return with me across the Atlantic on HMS *Sovereign* as planned.

A few days later I joined HMS *Sovereign* out at sea by boat transfer late at night. In choppy conditions we located the submarine and I scrambled up its steep, black casing in darkness to be grabbed by helping hands. I climbed down through a hatch to the confines of the submarine where it was warm and welcoming, a completely different world to the harsh conditions of a few moments ago. Space was at a premium and I was assigned a sleeping berth in an empty torpedo rack deep in the bows of the vessel. Even though nuclear boats are relatively spacious compared with diesel-electric boats, they are still claustrophobic.

The plan was to cross the Atlantic, stopping somewhere in the middle for an exercise with surface ships. For the whole of this time we would be testing the new sonar array being towed behind the boat. We were also testing the new towed array processor designed by John Bouffler. It was named JASOND after the first letters of the months during which it had been designed and built, July to December.

Altogether, the time at sea on this trip was three weeks. I soon got into the routine of submarine life. There are several differences between life on RN submarines and surface ships. Firstly, because everyone lives so close together the officers are more informal with the crew than on surface ships, although discipline is actually more rigorous. Secondly, there is no consumption of alcohol whilst at sea, except for a glass of sherry in the Officers' Mess before lunch on a Sunday. This is a safety precaution in case in extreme circumstances a submarine escape is necessary whilst submerged. The other notable difference is that you soon lose track of whether it is night or day, the only clue being whether you are eating breakfast or dinner.

It is interesting to note that both the UK and French navies are "wet", except in their submarines services. In contrast the US Navy is totally dry, which may account for the mayhem caused ashore by their sailors when a US naval ship visits a foreign port.

Some days into the Atlantic crossing we began the naval exercise. Our objective was to "sink" the fleet tanker RFA *Gold Rover* without being detected and destroyed by the defending frigates. The exercise lasted some 24 hours after which we claimed to have sunk *Gold Rover* several times, later to be confirmed by analysis of all the logs. At the end of the exercise we surfaced to take on board the Admiral who congratulated the crew. Our Captain, like most submarine commanders, was a bit of a cowboy. He wanted to impress the Admiral so took the boat into a fast dive to its deepest depth. (The precise maximum depth capability remains classified). I happened to be in the toilets at the time and found I was unable to open the door as a result of the compression. I was stuck there until we rose to a lesser depth, missing my opportunity to talk with the Admiral.

My time on board was exciting and our equipment worked well. Near Plymouth Sound the Captain let me look through the periscope at the approaching land. I later described the experience to Lucienne. Her first reaction was that this was an out-of-date technology in need of replacement. I did not see this at the time but in retrospect she was spot-on.

Soon after this trial I moved from "O" Group, still headed by Dr Flint, to join Dr Gill's team in "L" Group. I had a great respect for Dr Gill and valued his opinions. But he was not sufficiently valued by his superiors because he had the wrong priorities. Whereas some senior people were motivated by personal ambition, Dr Gill was motivated by an old-fashioned sense of duty. This attitude was shared by many of his contemporaries, all of whom were patriotic and put their country and the armed forces first in all their dealings. Dr Gill once said to me that we owed it to the officers and men in the front line to provide them with the best equipment possible. Everything he did was with this one aim in mind, and with the greatest urgency. I began to share some of this motivation, which is difficult to appreciate if you have never experienced front-line conflict.

# Early Sea Trials

In mid-1976 I transferred to "L" Group, headed by Dr John Gill. The group comprised some 30 scientists plus support staff working on research and development on new systems of SONAR (SOund Navigation And Ranging). There were several section leaders including Bert Allwood, John Bouffler, Ross Stamp, David Weston, John Long, and now me.

The team for which I was responsible comprised Colin Hammond, Alan Morrell, Roy Baker (later), Derek White, Jim Cumber, Geoff Sands, and of course Norman Field. The core of our work was to develop and improve the performance of our towed array sonar systems. A key part of this was to test prior theories against the data collected at sea with our experimental equipments, and to use the results as a basis for further improvements. It was an iterative cycle. In the early days we did not pay for ships' time from our budget. Ships were provided out of the Navy budget of which I was oblivious. As a result we were less than economical in the use of resources and the design of experiments. But without this freedom we would not have made some of the breakthroughs that helped us and the US to keep ahead of the Soviets. Over the years, particularly between 1975 and 2000, I organised and took part in a great many scientific experiments at sea. This was during the heart of the Cold War.

More recently, with the huge cost of sea trials being paid for directly out of research budgets, senior management decided they could be replaced by computer simulations. What they failed to understand is that most scientific advances are made by the anomalies observed in live experiments, in other words, things that go wrong or fail to agree with theoretical prediction. This has certainly been my experience and is how progress is made. Computers are unable to predict what they are not programmed for. (This point might be argued nowadays with the progress in artificial intelligence).

I should explain now a bit about underwater sonars, whose purpose is to detect sound. Radio waves do not propagate well underwater. They are heavily attenuated. So radar is not an option underwater. The only waves that do transmit in the oceans over long distances are sound waves. So sonar systems are the "eyes and ears" of ships, and particularly submarines. They employ sensors called hydrophones that detect sound, the underwater equivalent of microphones. Hydrophones convert sound waves in the form of pressure signals to electrical signals that can be readily recorded and processed. Our work was to develop sonar systems that could detect other vessels, particularly Soviet submarines, at the longest possible ranges.

There are two types of sonar. There are active sonars that send out acoustic pulses (pings) and listen for reflections off other vessels, and there are passive sonars that detect the sounds emitted by other vessels. The advantage of active sonars is that they provide information about range by measuring the time interval between the transmission of a ping and its return. Their disadvantage is that they give themselves away because other vessels can detect the ping. The advantage of passive sonars is that they are covert. Their disadvantage is that they rely on the vessel they are trying to detect making some sort of noise. During an encounter a submarine commander will use his passive sonar for as long as possible in order to remain covert. He will use his active sonar only at the last minute in order, for example, to obtain a more accurate firing solution for his torpedoes.

The Cold War at sea was concerned on the one hand with making bigger and better sonar systems for the detection of enemy vessels, and on the other hand with making our own vessels less vulnerable to active and passive sonar detection by reducing their acoustic reflection coefficients and reducing their radiated noise levels.

Until the 1960s all UK sonar systems, active or passive, were mounted on the hull of the vessel. These systems were quite successful in detecting and tracking noisy Soviet attack boats and missile boats like the Yankee Class. But the new Delta classes of Soviet submarines were quieter and more difficult to detect. To counter this, the US began developing a new type of passive sonar system that was towed behind the vessel. It was called a towed array.

The advantage of a towed array is that its length is not limited by the size of the vessel. This allows it to be made very long, up to several hundred metres, thus providing it with better directivity, an improved in signal-to-noise, and a detection capability to longer ranges particularly at the lower frequencies. It is also physically removed from the main source of interfering noise, namely, the mother vessel. On the other hand, a towed system is more cumbersome because it has to be reeled in and out. Towed arrays provided a giant leap forward in the West's capability against the Soviets. Suddenly, sonar operators were swamped with detections from hundreds of miles.

John Long at ARL Teddington was largely responsible in the early seventies for the UK initiative to develop its own towed array system. This is when I became involved, and I had some success. My name subsequently became associated exclusively with towed arrays even though they represented less than 50% of my work.

The performance of a towed array is usually limited by self-noise, that is, noise detected at the hydrophones resulting from vibrations of the structure and turbulent pressure fluctuations generated by its passage through the water. This noise can swamp the real signals from other vessels that the system is trying to detect. So a key objective of my work was to minimise the levels of self-noise. Success meant longer detection ranges and better performance at higher tow speeds.

Part of my job on transferring to "L" Group in early 1977 was to gain an understanding of the noise mechanisms in towed arrays and to use this knowledge to design quieter arrays with better performance. The previous incumbent of the post told me that the mechanisms were so numerous and complex that a quantitative understanding was impossible. So I was presented with a challenge.

Colin Hammond and Jim Cumber in my team were both talented engineers. Derek White was an ex-draughtsman who assisted with array design. Norman Field did a bit of everything. He was qualified as a chemist but his work took him into every area including electronics, materials, engineering and data recording. His greatest strength was in organising sea trials, which he did for me over many years. Altogether, we were a strong team.

Our work involved building prototype equipments, testing them at sea, returning to the lab to analyse the data, then implementing modifications and improvements to the equipment. All the time I was trying to gain a better theoretical understanding of what was going on. This entailed modelling the noise mechanisms and the vibrational modes of the array structure and comparing predictions against measured results. The cycle was tied to the availability of suitable ships for carrying out the sea trials.

If a particular equipment or modification was successful it would eventually go into production with a contractor, and then into service on RN vessels. But the time-cycle was long, far too long. In the 1990s the MoD made a great effort to reduce this cycle from decades to a few years with limited success.

In the 1970s the prototype equipments we designed were built by a contractor called Ameeco Hydrospace based in Erith, Kent, prior to their move to Andover in the 1980s. I made frequent visits to the Ameeco factories with Norman Field and Colin Hammond to discuss the plans and progress of the equipment builds for the next round of sea trials. Norman's contribution to the meetings was always practical. He had a good feeling for what was feasible and what was not. People turned to him for advice and he was good at thinking on his feet. He had a phenomenal memory for facts and for who had said what at previous meetings. Colin on the other hand had a great feel for what was good design and what was bad. We made several advances through Colin's insight for which unfortunately he never obtained official credit in terms of promotion. He was just too vague and off-hand at interviews.

These were the days of extended lunches at contractors. Most of the work was conducted in the mornings. Ameeco then treated us to a fine lunch at a smart pub or restaurant. It seemed the right way to do business. Workers these days rarely even have a lunch break. The norm for visitors is now a working lunch of bites and limp sandwiches. I suppose we could have been accused of bribery and corruption, especially when we were given bottles of sherry at Christmas. Some years later, a ruling was introduced that we should not accept hospitality from contractors unless we felt that offence would be caused by declining. Any hospitality we did accept had to be entered in a "Hospitality Book" kept by one of the secretaries and freely available for anyone to inspect. This made everything transparent and above-board. Nevertheless, our rules about receiving hospitality were not nearly as strict as those of our US counterparts in American research labs.

I made many visits over the years to the Admiralty Materials Laboratory at Holton Heath in Dorset. They designed various sonar materials for me, in particular, high damping PVC hoses that housed the towed array sensors and electronics. The lab was a sleepy place with buildings and gardens spreading over several acres. I always felt that the scientists there had an easy time, but they produced many interesting materials. It was there that I first witnessed demonstrations of superglue, bouncy materials with coefficients of restitution apparently greater than one, and squidgy substances that got very hot when you squeezed them. Two scientists I particularly remember are Ron Long and John Clothier. Many years later these novel materials became common place in everyday applications, including children's toys. The Laboratory was closed down in the 1990s.

In the 1970s and 80s I went to sea typically every six months. Occasionally we used Royal Navy vessels, but usually we used ships of the Royal Maritime Auxiliary Service, initially RMAS *Bullfinch*, then RMAS *St Margarets*. This latter vessel was the last twin-screw, steam-powered cable-laying vessel to be active in British waters. She was a ship of great character with extensive varnished hardwood interiors and a pervading smell of cabbage.

These vessels were originally designed for laying deep-water telephone cables. They were not ideal for our work but we had to make do. We devised practical means of using them for our purposes. For example, the winches on the cable ships were on the foredeck since underwater cables are deployed over the bows. But towed arrays had to be deployed over the stern. So we installed pulley systems and sheaves to take the cable and arrays aft through the ship. The adjustments involved several visits to the ships in their home port of Turnchapel in Plymouth. Norman Field was responsible for much of this work. These vessels also had speed limitations, with a maximum of 12 knots.

The RMAS vessels were run on Navy lines but with less formality. The officers on board had various backgrounds, some making their career in the Fleet Auxiliary Service, others having worked in merchant ships, and a few ex-Navy. There would typically be 7 officers comprising the Captain, 3 deck officers one of whom was the navigating officer, 2 engineer officers, and a radio officer. This was before the days of satellite navigation and mobile phone communication. The crew comprised some 10 to 15 men. These numbers were sufficient to keep the ship operational 24 hours a day and to assist with handling the equipment.

The first towed array sea trials I went on took place in the NE Atlantic, usually in sea areas Bailey or Rockall. We went there because it was only two days sailing from Plymouth and because there was less chance of interference from commercial shipping. On the other hand there was a high probability of bad weather in these areas, often preventing us from collecting good quality data or even deploying the equipment. Sometimes at least half the period allotted to the sea trial was lost due to bad weather. It could be so bad that the cook was no longer able to prepare meals. Force 8 and 9 gales were not uncommon and I recall on more than one occasion it rose to Storm Force 10/11. There is a big difference between a gale and a storm, notably with the level of noise and the sound of the wind, which is a good octave higher in a storm. Such conditions are frightening, particularly at night. The noise is deafening and the ship lurches, pitches and rolls uncontrollably. All you can do is lie wedged into your bunk and stagger along to the wardroom for the occasional sandwich or bowl of soup. In these conditions the main objective of the bridge officer who drives the ship is to keep it heading into the swell to avoid broaching. Fortunately, I never suffered from sea sickness.

Norman Field once took a photo from the upper bridge of *St Margarets* during severe weather showing mountainous waves breaking over the bows of the ship. It was such a stunning photo that he presented a framed version of it to the ship where it was mounted on the wall of the wardroom. Whatever happened to this photo?

The food on board ship was always wholesome and plentiful. The day started at 06:00 with a cup of tea brought to your cabin by a steward. The tea was disgusting and usually went straight down the sink. Its main purpose was to serve as a wake-up call. The safest hot drink was chocolate. For breakfast there was always a choice of traditional fayre including porridge, eggs, black pudding, white pudding, kidneys, kippers and pusser's sausages, the latter not very nice. Lunch and dinner were both full meals, including soup and pudding. A favourite main course was individual steak and kidney suet puddings called "baby's heads". As the scientists were classed as officers we ate in the wardroom, served by stewards. I enjoyed all the meals and rarely missed one.

To make the most of any fair weather we often started work at first light and carried on until darkness fell. Sometimes, if other ships were involved we obtained permission to work night shifts. But night working was difficult because of the crew's union rules. After a long day's work we would all retire exhausted to the wardroom to take advantage of the duty-free drinks.

Our data was recorded on large and cumbersome Ampex tape recorders. These were top-of-the-range, expensive devices costing some £30K each. Our business was worth a great deal to the Ampex salesman who treated a group of us every Christmas to a sumptuous evening meal at Plumber Manor near Sherborne. All the data we recorded was analogue and there were no PCs at that time. I did on-line analysis using bulky spectrum analysers. I did other analysis post-trials at the lab where I spent many hours poring over the results and trying to reconcile them with my various theories.

In the mid-70s we began investigating alternative trials locations that might hold the prospect of fewer days lost by bad weather. The first promising site was one of the Scottish Lochs, Lower Loch Fine. This had the attributes of being accessible from the open sea, fairly deep, and sheltered from rough waters. We had several trials periods here and it was all very civilised. The water was calm and at night we could anchor in tranquil conditions off Tarbert. The captain sometimes set up boat transfers for evenings ashore in the local pub. Nearby we could buy top quality kippers and salmon, locally smoked, for taking home.

On one of these trials we had a near disaster. In the middle of a run we suddenly lost all incoming data. We reeled in the cable to find nothing on the end. About £100K worth of experimental towed array had parted from the cable and dropped to the bottom of the loch. The captain immediately turned the ship round and we started dragging the bottom of the loch with a grapnel on the end of a long line to try and snare the array, but without success. In those days we did not have the benefit of GPS for accurate position fixing. So I used the ship's area chart and a careful process of logic based on where I believed we had started the run and the time that had elapsed before we lost data to determine where the lost array ought to be. I illustrated this to the captain using matchsticks to measure out distances on the chart. My advice was followed and we recovered the array on the next pass with the grapnel. For some time after that the ship's officers referred to me as the "matchstick man".

But the Loch Fine location was far from ideal. The area of water that was deep enough for us to work in was relatively small. In addition, our data recordings were often contaminated by the underwater sounds of airguns that the seismic community were using nearby in their search for oil and gas. We had to coordinate our activities with theirs.

A second site we occasionally used in the late 70s and early 80s was Horsea Island, just outside Portsmouth. The facility comprised a man-made lake built around 1900 and some 1100 yards long, used until WW1 for testing torpedoes. The site continued to have various naval and commercial uses focused mainly on diving and underwater engineering, but was slowly being run down. On my first visit I drove along a rutted causeway surrounded by mud-flats and rabbits. I arrived at a large deserted office block, overgrown with weeds, where I found a lone superintending naval lieutenant and a charge-hand named Jock. The site at that time housed HMS *Phoenix*, the naval school of firefighting and damage control. The school comprised several steel structures each simulating three decks of a warship. Impressive kerosene fires were occasionally lit to enable naval trainees to practise various types of firefighting. I sometimes witnessed this from a safe distance.

With the help of Barry Pye, an electronics engineer working for me, we devised a way of using this lake facility. It centred around an old launch, not much bigger than a large rowing boat. The array to be tested was paid out over the stern of the launch, while a long hawser attached to the bows was reeled in by a powerful WW2 barrage balloon winch located at the head of the lake. The data recording equipment was stowed in the relatively dry cockpit of the launch. This enabled flow noise and vibration data to be recorded for several seconds once steady state conditions were achieved. It was very much a case of string and sealing wax, and the quality of the data was never like that achieved in open sea. I entrusted most of the data analysis to Geoff Sands, an Oxford Physics graduate who was bright but poor at writing up his work.

A third location that was much more promising and rather more exotic was an area of ocean south west of Madeira. This had the benefit of being not too distant from UK and close to the friendly Portuguese port of Funchal in Madeira. Our Foreign Office at that time got on well with the Portuguese. The area SW of Madeira was far enough from the major shipping lanes and offered good prospects of fair weather most of the year, but these advantages were partially offset by a longer transit time from UK. Madeira was sub-tropical, rather different from either Rockall or Loch Fyne.

I have clear memories of the first of these visits to Madeira and so will provide some detail. It was April 1977. *St Margarets* was fitted out at Turnchapel, Plymouth. Most of the scientists sailed down to Madeira on the ship, a passage of four to five days via the Bay of Biscay. Norman Field and I flew to Madeira to join the ship on her arrival. In the 1970s there were no direct flights from UK to Madeira. UK airlines considered the runway at Funchal to be too short. The only commercial airline to serve Funchal was the Portuguese airline TAP. So we had to fly from London to Lisbon and connect with a TAP flight to Madeira. This had the benefit of providing an opportunity to see the sights in Lisbon and visit the impressive castle.

We left UK in grey weather to arrive at Funchal airport early evening to be met by an all-enveloping warmth, with people in shirt sleeves and white shorts. Madeira is a beautiful mountainous island, volcanic and fertile with lush vegetation. We took a taxi from the airport along the tortuous coastal road to the small, capital city of Funchal. The approach is similar, but slightly more spectacular, than the night drive from Osmington to Weymouth. We were booked into a back street B&B in Funchal. We then went out for an evening meal.

The food on Madeira is simply cooked but tasty and all locally sourced. The island was still not a serious tourist resort, so everything was cheap. There was plenty of fresh fruit and vegetables, not much meat and an abundance of fish. All the restaurants had tuna fish and espada (black scabbard fish) on their menus. Espada are long thin, ugly fish with black slimy skins, ferocious looking teeth, and enormous eyes because they live in darkness at great depth, but the flesh is light and delicate. We ate espada that first night. Norman and I enjoyed spending time comparing the menus outside each restaurant before finally making a choice. The anticipation is almost as enjoyable as the event.

Before leaving UK, Norman had made contact with the local shipping agent in Funchal called Blandys. This turned out to be a wise move. We soon discovered that nothing of any commercial value entered or left the island without Blandys having a hand in it. They provided supplies to all the ships that came and went and were the founders and principal distributors of Madeira wines. They also had influence over the airport customs officials, as we were to discover to our benefit on later trips when we had to carry last-minute items of electrical equipment through Funchal airport. You could say that Blandys was the local mafia.

So the next morning we walked down towards the harbour to visit Blandys shipping agency. A signal from *St Margarets* was waiting for us announcing that due to bad weather on passage through Biscay she would be a day late. This meant we had another 24 hours to see Madeira.

Norman and I enjoyed sightseeing, so we took a bus on another hair-raising journey into the mountains to stand on the edge of the extinct volcano. On return to Funchal we had lunch in a small restaurant on a high terrace in the cliffs overlooking the harbour. We ate fish and drank a bottle of the local Vinho Verde. On strolling through one of the town squares we came across about 20 chess boards set up on tables under the plane trees. A chess master was walking along the tables playing all the boards simultaneously.

We soon realised that taxis cost little more than buses, especially for two people. Almost every taxi was a Mercedes diesel saloon, often fairly old but always in immaculate condition. So we took a taxi to a small fishing village up the coast called Camara De Lobos, then drove and walked to the top of the nearby cliff called Cabo Giao, the second highest sea cliff in the world at 580 m.

The ship arrived the next day. We had a successful and uneventful trial without a single day lost to bad weather. One night after work some crew members deployed a huge hook from the afterdeck on which they caught a massive shark. It must have been six feet long and a beautiful creature. I watched it being hauled in on the winch, twisting frantically trying to free itself. Once on deck the task of killing it was gruesome. The crew extracted all its teeth as trophies, and the flesh was filleted for the chef who fed us all a meal from it the next day. I have to say it was excellent meat. At the end of the trial the ship returned to Madeira where we disembarked with a half-day to spare before flying out the next morning.

Over the next ten years I took part in at least a dozen sea trials out of Madeira. It was always immensely pleasurable and exciting. At the end of each trip there was usually an opportunity to do some shopping before flying out. I liked to visit the large covered market in Funchal. This was full of exotic fruit and vegetables some of which I bought to take home, like massive avocados that were unheard of in UK. The adjacent fish market offered a large choice of unusual fish, all cheap. Norman often bought great slabs of tuna which he deposited in the ship's freezer for collection later in Plymouth. You could not buy fresh tuna in UK. Norman must have been eating it for months because I don't think his family liked it.

Of course, we all bought bottles of Madeira. This was ridiculously cheap, a vintage bottle 60 or 70 years old costing just a few pounds. I bought some 1915 Malmsey for my father, which he said was the most incredible fortified wine he had ever tasted. We discovered one particular wine shop that we called "the crypt" because it was in cellars beneath a disused church. One attraction of this shop was their generosity with free tastings.

Another speciality of the island was hand embroidery. There were shops in Funchal where you could buy beautiful hand-embroidered tablecloths and napkins of every shape and size. Again, these were ridiculously cheap by UK standards. You were expected to haggle and the deal was always sealed with a glass of Madeira.

Yet another local speciality was wickerware. You could buy whole suites of wicker furniture at knock-down prices. Personally I am not fond of wickerware, but one Alan Morrell would stuff vast quantities of it into the ship's hold.

Madeira is renowned for its flowers. The last thing we did before jumping into a taxi to the airport was to buy a box of flowers in the market square. The assortment always included orchids, flamingos, and birds of paradise.

Just before our second sea trial out of Madeira there was a plane crash at Funchal airport. The aircraft was a Portuguese Boeing 727. The pilot was trying to land in heavy rain and on the third attempt touched down 600m past the aiming point on the runway. For some reason he was landing in the downhill direction. The plane over-ran and plunged off a cliff onto rocks killing 131 people. This was 19 November 1977. A week later I flew to Madeira for a sea trial. The flight was tense, again on a Portuguese Boeing 727. When we landed safely a huge cheer went up from the passengers. The plane crash resulted in the building of an extension to the runway out into the sea. This persuaded other airlines to land there. You can now fly direct from UK, but it is still not the safest of landing strips.

The B&B family-run hotels that we used on these first trips to Madeira were not best-suited to our purposes. Because of the uncertainties of *St Margarets'* arrival and departure times at Funchal we were never quite sure which nights to book and for how long. This did not suit the small hotels that only accepted firm bookings paid in advance.

So we started booking ourselves into the new 5-star hotel in Funchal, the Casino Park, which accepted extensions of bookings and last minute cancellations without penalty. The Casino Park was very grand. Men had to wear a jacket and tie in the restaurant where high-class international cuisine was served. The rooms were sumptuous with balconies, fine views, complimentary chocolates, and decorated with exotic flowers. The chamber-maids visited twice a day, once in the morning to service the room and a second time in the afternoon to provide goodies and turn down the bed covers. I had never before experienced service like this.

Madeira was one of the few places in the world where our employers, the UK MoD, did not have a fixed "per diem" subsistence rate. This was because Madeira was seasonal with fluctuating prices that made it difficult to fix a single rate. So Madeira subsistence was expenses plus.

When we submitted our claims at the end of one of our trips to Madeira, the MoD Accounts Office in Bath refused to pay for the Casino Park on the grounds that we were abusing the expenses system by staying in the poshest hotel. I asked Norman Field to take up the challenge on our behalf. He successfully argued that in the long run it was cheaper to stay at the Casino Park because we did not suffer lost deposits due to unavoidable cancellations and there were always spare rooms available for last minute extensions. In any case the Casino Park was no more expensive than a smart UK B&B.

We did not confine our interest to Madeira. We tried Ponta Delgada in the Azores, also Portuguese, but found its port and hotel services limited. We also did one or two trials out of the Canary Islands, notably Santa Cruz on Teneriffe and Las Palmas on Gran Canaria. But these islands are Spanish and the port authorities never gave us the same degree of cooperation that we obtained from the Portuguese.

Generally, we could rely on good weather to conduct our experiments in these southern areas – the principal reason for going there. But this was not always the case. We once hit a storm SW of Madeira when the ship burst a rivet below the waterline in the forward hold. The ship took on water that was stemmed by the application of a cement box. In daylight I was astonished to discover that the metal stanchions on the afterdeck had been bent in two by the force of the waves. These episodes added excitement at the expense of lost trials time.

It might be concluded from all this that we really enjoyed our trips to Madeira, and that would be true. We also worked hard and enjoyed that too. The whole experience was always exciting and exhilarating. These early trials activities provided the towed array designs that eventually went into service on RN frigates and submarines in the 1980s and 90s, and in some cases were still in service in the "noughties". They provided the groundwork for today's digital towed arrays that are employed on the latest vessels like the new nuclear Astute class of attack submarines.

I have already said that the RMAS ships we were using were not ideal platforms for our research work. They were also getting old and expensive to maintain. In the late 1960s plans were already underway to build a new, all-purpose research vessel that would eventually replace RMAS *Bullfinch* and *St Margarets*. She was to be called RMAS *Newton*.

A committee was set up to include all interested parties from various parts of MoD and elsewhere to deliberate on *Newton's* design. With his experience of ships and sea trials, Norman Field was selected to sit on this committee to represent ARL's interests.

RMAS *Newton* was commissioned in June 1976 as a dual role cable layer/oceanographic research vessel. At 3940 tons she was almost twice the weight of *Bullfinch* or *St Margarets*. She had diesel-electric propulsion giving her a top speed of about 14 knots, and in order to aid manoeuvrability at low speed was fitted with a nozzle rudder and retractable bow thruster. She was equipped with extensive laboratories for the scientists, all located on the main deck level instead of deep in the dark bowels of the ship as on *St Margarets*. The scientists' cabin quarters were comfortable and spacious with en suite facilities. There were even special cabins for females. The ship had ample gear to handle all sorts of equipment. I did several sea trials on RMAS *Newton* out of various home and foreign ports. She was eventually taken out of service in 2010, and in January 2013 was at the breaker's yard in Ghent, Belgium.

# Move to Weymouth

We had known for some time that the move to Dorset was imminent and unavoidable. After our successful 1976 reconnaissance trip to Weymouth and the discovery that they still had a Grammar School, much of our initial reluctance was dispelled. I was informed that my official start date for work at the Admiralty Underwater Weapons Establishment (AUWE) was 8 January 1979. There was considerable flexibility on when to move house and when to report for duty at AUWE, provided it was not later than this deadline.

On 5 May 1978 we put our house in Ashley Drive on the market with Snellers estate agents of Twickenham at an asking price of £29,000. At the same time we travelled to Weymouth to look at houses to buy. There was not much on the market and the prices were relatively high, no doubt inflated by the anticipated influx of people from Teddington.

We found a house in Melcombe Regis - 11a Melcombe Avenue (it should have been No 13). It was built in 1928 in an attractive area of Weymouth close to the main hospital (since reduced to a community hospital with limited services) and 200 yards from the beach. The house was large and detached with a fair-sized garden and had been empty for some time. The downside was the damage it had suffered by severe flooding from a burst pipe in the loft that had gone undetected for most of the previous winter while the house was empty. In spite of its dilapidated state we were seduced by its potential, totally underestimating the amount of work that needed doing. We offered the asking price of £35,000 which was accepted.

On Friday 7 July we made another weekend trip to Weymouth to look at the house again and to make a long list of all the things that needed doing. We also visited Weymouth Grammar School where Isabelle was interviewed by the Deputy Headmistress.

We then had a survey done on 11a Melcombe Avenue to report on all the defects. The garden was a jungle, the kitchen needed to be completely renewed, window frames replaced, rewiring throughout, complete redecoration inside and out, a new central heating boiler, and repairs to all the damage done by the burst pipe. The latter included a complete drying out followed by replacement of all the warped floorboards and doors, and replacement of the crumbling plaster and damaged ceilings. The survey report stated that the asking price of £35,000 was excessive and that £30,000 was more realistic.

Amazingly, we were not put off by the survey report and submitted a revised offer of £30,000, which was accepted. We applied to the Halifax Building Society for a mortgage and provided them with cost estimates for all the repairs. Meanwhile, we had still not found a buyer for our own house in Ashley Drive. We were persuaded to drop the asking price to £27,500.

As soon as Isabelle's school term finished in July we made our annual pilgrimage to Roanne. Lucienne's parents were now retired and more relaxed, Isabelle was able to play with her cousin Catherine, the meals were always excellent, and the weather was warm and embracing. What fond memories.

At the end of our holiday on Friday, 28 July 1978 we set off from Roanne around 06:30 for the long journey home. By that time some of the motorways around Paris had been completed and it was just possible to do the whole journey in one day, including the ferry crossing from Calais to Dover. On this occasion we were the very last car to arrive at the ferry and they just managed to squeeze us on. This had the advantage that we were first off. We arrived home in Whitton at around 22:30, totally exhausted.

On the doormat were two official looking letters that we immediately opened. The first was dated 18 July from the Headmaster of Weymouth Grammar School, Mr J Hunter. He informed us that their good impression of Isabelle had been confirmed by the report from Putney High School and he was pleased to offer her a place starting in the $2^{nd}$ year. This was good news. Isabelle did not have to take the 11-plus exam.

The second letter was not so good. It contained the devastating news that the Halifax were not prepared to offer us a mortgage on 11a Melcombe Avenue, not because of the scale of the work we had identified and costed, but because their surveyor thought the building was subsiding and was too much of a risk. This was dreadful news. Almost simultaneously Isabelle fell off the table in the kitchen while trying to reach the biscuits in the top cupboard. At the same time Fat Thelma from next door walked in with dog shit on her shoes which she spread everywhere whilst giving us important news which escapes me. We were completely overwhelmed. It was midnight and we went to bed.

I awoke early the next morning, Saturday 29 July 1978. High on adrenalin I drove by myself straight to Weymouth to restart the house search. The journey in those days was not straightforward because there were no motorways beyond Basingstoke. The M3 extension and the M27 had yet to be built. When I eventually arrived in Weymouth I visited the local estate agents in St Thomas Street where there was little of interest on the market. But Adams, Rench & Wright (no longer in business) had a property that had come on their books that same day. It was in Carlton Road North.

I went there immediately and was shown round by the owners, Mr & Mrs Ryan. I instantly fell in love with the place. It was like a castle compared with our house in Whitton and, unlike 11a Melcombe Avenue, absolutely nothing needed doing. It was close to the town centre and beach, yet had a rural feel because the road at the back, Westbourne Road, was still unmade and lined with beautiful tall elm trees, sadly to be chopped down soon after as a result of Dutch Elm disease. I drove straight back to Whitton to tell Lucienne about this house.

We knew we had to act quickly. So the very next day, Sunday, the four of us returned by car to distant Weymouth. On the way down we had a picnic lunch in a lay-by just outside Wool. When she saw the house Lucienne loved it and we decided to buy on the spot. We put a note through the door of the estate agents offering the asking price of £38,000. We returned to Whitton exhausted but elated. We were buying the house in Carlton Road North "on the rebound". We felt that maybe it was a good thing we had not become embroiled with 11a Melcombe Avenue.

On 4 August we made an application to the Halifax Building Society for a 25-year mortgage of £13,000 for the house in Carlton Road North. This sum was the maximum they were prepared to lend with my salary of £7845. After another two weeks the Halifax completed their survey and confirmed the offer.

Meanwhile, at the beginning of August we had the first serious offer for our Ashley Drive house from a Mr and Mrs Levin. They had their survey done in late August and everything seemed to be falling into place. Then on 12 September we had a letter saying that the Levins were pulling out because the survey had highlighted a lot of work that needed doing: such things as random repointing of the brickwork, one or two roof tile replacements, rust on some window frames, cracked lead flashing, and the usual queries about possible woodworm in the loft. Altogether it was nothing serious but enough to put them off. This was disappointing.

We had our survey done on the Weymouth house. Nothing bad was identified except for some damaged leadwork on the roof and some possible woodworm infestation in the roof void. On the basis of this the Ryans accepted on 25 September the reduced sale price of £37,800.

At the same time we accepted a new offer of £26,500 for our Whitton house from a Mr and Mrs Castle. Since our purchase in Weymouth was more advanced than our sale in Whitton we had to apply for a bridging loan, the interest for which the MoD agreed to pay. I fixed up the loan on 13 October through the National Westminster Bank in Teddington. On 14 October we exchanged contracts for the purchase of the house in Carlton Road North. We delayed the completion date because of my work commitments.

Immediately after the exchange of contracts I flew to the States for a week of technology meetings in Washington, New London, and San Diego. I followed this with a sea trial out of Madeira on RMAS *St Margarets*. On my return we completed the purchase and moved to Weymouth on 10 November 1978. Soon after, on 24 November we completed the sale of our house in Ashley Drive.

That winter was the coldest we were ever to see in Weymouth. It snowed heavily over the Christmas period and for several days the town was completely cut off. We had been due to spend Christmas at my parents in Southend but were forced to turn back at the Ridgeway and stay at home. All the palm trees on the promenade died in the cold, thus damaging the Mediterranean feel. We took our daughters on our homemade sledge round the icy and deserted streets.

We enjoyed our house in Weymouth. We were struck by the eerie silence at night after being used to the noise of aircraft taking off from Heathrow. We also liked the many amenities, the proximity of the town and beach, and the beautiful undulating hills and spectacular coastline on our doorstep. Isabelle liked the Grammar School and Caroline was happy at her primary school in Wyke. The shops were plentiful and there was an abundance of local amusement and entertainment. We liked seeing the many holidaymakers enjoying themselves in the summer months. It turned out to be a wonderful environment, especially with young children, and we had no regrets about our move.

Over the years we had many happy visits from our daughters and their families, particularly our granddaughters Emily, Natalie and Madeleine. When they were young, they had regular holidays with us and were able to enjoy the superb amenities of Weymouth and its sandy beaches. More recently, after our move from Weymouth, we have taken our young granddaughters Sophie and Juliette on regular summer holidays there, always during Carnival Week. Beautiful Weymouth, it is hard to find a town with so much to offer in the way of climate, shops, pleasurable activities, nature, history, and with a spectacular coastline on its doorstep.

As for ARL Teddington, it closed in 1985 as part of the government rationalisation programme. Around 1990 all ARL buildings at Queens Road were razed to the ground to make way for a smart housing estate. Of the original site there is now no sign, except for the road names, like "Admiralty Way", and the two brick pillars that used to support the large wooden entrance gates to ARL.

# AUWE and the 1980s

I officially started work at the Admiralty Underwater Weapons Establishment (AUWE) on Portland, Dorset, on Sunday, 8 January 1979. AUWE Portland was a larger establishment than ARL Teddington, and being by the sea it had a strong naval presence. AUWE was split between two sites, South and North. The South site was at Southwell near Portland Bill, and was principally responsible for the design, development and testing of underwater weapons. It was also where the Director had his offices. The North site where I worked was at Castletown and linked to a busy Navy base, HMS *Osprey*, with a dockyard and helicopter station. The North site focused on underwater detection systems, namely sonar. I had a small office on the top floor of South Block at North site.

Our proximity to the Navy Base with serving officers meant that AUWE had an Officer's Mess just like on board ship. Only scientists of rank equivalent to Lieutenant or higher were allowed in the Mess. Others had to use the base canteen. One or two scientists who were eligible for the Mess used the canteen as a form of protest for equal rights. A notable example was Martin Earwicker, a bright young scientist with leftist tendencies who never wore a tie until he was elevated to the higher ranks when he sported a bow tie. He was one of several scientists who attended Lucienne's lunchtime French classes at AUWE, supposedly in support of collaborative projects AUWE had with France [18]. Myself, I always used the canteen, simply because it was quicker and because I was not a "clubby" sort of person. The only time I used the Officer's Mess was for the Christmas lunch or for a retirement gathering.

Dr Gill was no longer our Group Head. He had made a sideways move to AUWE about a year before the rest of us

---

[18] After moving to London, Martin Earwicker had a distinguished career at MOD Headquarters. He became Director of the Science and Natural History Museums in London before being appointed in 2009 as Vice-Chancellor of London South Bank University.

moved down. Our Group Head was now Peter Clynick, quite a different character from Dr Gill. Whereas Dr Gill was intuitive, Peter was much more analytical. This is because he was originally a mathematician. He was very good at solving a problem from first principles. I formed a close bond with Peter Clynick and saw much less of Dr Gill.

Dr Gill retired from AUWE in 1984. It was only around this time that I felt able to call him John, although he had always called me Alan. Some time in 1992 I had a phone call from a colleague who said that Dr Gill had been admitted to Weymouth Hospital under observation. I visited him that evening at the hospital, just three minutes walk from our house. He seemed healthy and in good spirits and we chatted about old times. I brought him a bottle of wine. He said he would be out of hospital in a couple of days. He died the next morning.

Our new Group Head, Peter Clynick, was a keen yachtsman and kept a boat at Fareham, near Portsmouth. He sometimes took Barry Pye as a crewman on day trips in the Channel. On one occasion he invited Barry and me to join him on a weekend trip to Cherbourg. We set off early morning from Fareham on the 100-mile journey that took us about 12 hours. The weather was fair and we crossed the busy shipping lanes uneventfully, arriving in Cherbourg early evening. We went ashore and enjoyed a leisurely meal near the quay. Towards midnight we returned to the boat for a couple of hours sleep before setting off back to Fareham. Peter did not want to miss the tide. The day blossomed with sunshine and a fresh following wind. Peter unfurled the spinnaker that ballooned in front of the boat and sped us through the water. I could see the attraction of sailing but my contribution as a crewman was limited, so I retired to the galley to cook breakfast, managing to impress Peter by my skill at neatly breaking two eggs simultaneously, one in each hand, into the frying pan. This was about the limit of my experience of yachting, except for a few ventures with friends in small boats fishing for mackerel round Portland Bill.

The group in which I now worked was mostly concerned with passive sonar. It was one of several groups in the Sonar Department at AUWE. Other groups worked on active sonar, hull-mounted sonar, sonobuoys, signal processing and displays, data processing, and assessment. The head of Sonar Department was Sam Mason. I first found him to be an awesome person. After a varied and distinguished career in sonar he had become the manager of what was essentially the future technology for the *eyes and ears" of the Royal Navy. Like many of his generation, Mr Mason was totally dedicated to serving the Navy. He was loyal to his staff and aggressive to all opposition. He was an intuitive scientist who always had complete conviction that he was right, pressing home his arguments with total force. He was the sort of person that you wanted to be on your side in any confrontation. Long after his retirement he continued to give the impression that he was still running the Sonar Department. Some years later I discovered he was a devout Christian and a Methodist lay preacher with a softer side. It was only in this latter period that I felt able to call him Sam.

Most people from Teddington were happy with the move to Dorset once they had made the transition. An exception was Dave Williams who worked in my group. Dave was a clever computer buff and signal processor about the same age as me. He was responsible for developing a novel acoustic beamformer based on a two-dimensional Fourier transform. He had intended it to be a replacement for in-service beamformers based on more conventional methods of time or phase delays, but it did not catch on. However, because of the ability of Dave's beamformer to identify propagating waves outside the acoustic window, I found it to be a useful tool that I was to use a lot in my research on noise reduction. Soon after it was copied by the US people at NUSC New London, to be used extensively in the development of their hull-mounted sonars.

Dave had a large family with something like five children. He never made any secret of the fact that he was unhappy living in Dorset and would prefer a move back to London. But unable to achieve this by any of the normal routes he took more extreme measures. There were really only two ways for a Civil Servant to be removed from his post. Either you had to be a homosexual who kept it secret so making you a potential target for blackmail, or you had to commit repeated security breaches. Dave took the latter route by leaving the door of his secure cupboard open on several occasions when leaving work in the evening. These breaches were of course discovered and reported by the night watchman on his rounds. Dave was severely reprimanded and eventually moved to a government department in London where security was not an issue. I think it was the Admiralty Paint Shop. His departure was a great loss to our group.

Soon after I joined AUWE we had a visit from the Duke of Edinburgh, Prince Phillip. It was Tuesday, 6 June 1979, coincidentally my birthday. All the stops were pulled out with no expense spared in the preparations, including a specially built toilet that he never used and was dismantled immediately after the visit. I had to say a few words to the Duke about my own work as he walked round the displays accompanied by all the higher AUWE management and local civic dignitaries. I was struck by his intelligent questions and by how well-briefed he was.

At about this time a Canadian exchange scientist from the Defence Research Establishment Atlantic (DREA) in Halifax, Nova Scotia, came to join us for two years to work alongside me. His name was Barrie Franklin. I had first met Barrie some years earlier on a trip to Halifax. He was a talented engineer and physicist. Barrie came to us to learn about my work so that he could take the knowledge back to Canada to help with their national programme. He assisted me greatly in my own work at AUWE. He had a fertile mind and was innovative. I believe that Barrie was the first person to suggest using pressure gradient sensors alongside conventional pressure sensors in towed sonar systems. We even did some successful sea experiments with experimental modules. The current interest in "vector sensors" in UK and US Navy research can be traced back to this time in 1979. Barrie settled into a house in Wyke with his wife Linda and their young family.

It was during Barrie Franklin's stay that I came up with what turned out to be one of the most valuable concepts of my career. I called it "The Lower Limit of Self-Noise". I will explain what this was about. I have said earlier that a principal objective of my work was to reduce the incidence of extraneous noise in our towed arrays in order to make it easier to detect the acoustic signals of targets at longer ranges. There were several sources of extraneous noise, the most significant being vibrations and resonances in the mechanical structure of the array, and pressure fluctuations generated in the turbulent boundary layer - the region of chaotic flow generated on the outer surface of the array as it is towed through the water. The structure of the array was important in determining how and to what degree these noises reached the hydrophones inside the array. It was a complex problem.

At this point I must write an aside. A large part of my university studies was concerned with the Navier-Stokes equations of fluid motion which, together with several subsidiary equations, govern the flow and wave motions of all physical phenomena that involve liquids and gases on land, ocean and atmosphere, both subsonic and supersonic, laminar and turbulent, covering everything from weather forecasting to paint flow. These nonlinear, but innocuous-looking equations have kept thousands of people in employment for the last hundred years, and no doubt will continue to do so for another hundred years. I have continued to study different aspects of the Navier-Stokes equations during my entire working career. I can say that I have obtained a working knowledge of them and some of their properties. One fundamental property is that the equations are chaotic, that is, the minutest changes in the initial conditions can have an enormous effect on future states. This is popularly known as the "butterfly effect", where the formation of a hurricane can be contingent on whether a butterfly flapped its wings some weeks before in a distant part of the world. So it is not surprising that weather forecasts and long range predictions are unreliable.

It is instructive to look at recent medium term forecasts from the UK Meteorological Office. In April 2009 a "barbecue summer" was forecast; it turned out to be a washout. It was then forecast that the winters of 2010 to 2011 would be mild; they were the coldest for 120 years. In 2007 it was forecast that globally, the decade 2004-2014 would see warming of 0.3C. In fact, the world did not warm at all in this period. The Met Office medium and long range forecasts are sometimes right, but no more than the toss of a coin. Their current models are clearly inadequate, so the value of building bigger computers in the drive to obtain more accurate, longer term forecasts over ever-smaller mesh sizes is questionable.

Cynics will have noticed that when it became clear that the world has not been warming since 1998, climate scientists began talking instead about "climate change" and "extreme weather events". We are led to believe that such events have become more prevalent as a result of global warming. In fact, extreme weather events are no more prevalent now than at any time in the past. By way of example, consider the severe flooding in 2009 in Cockermouth, Cumbria, UK. It was widely claimed that this event was exceptional and directly linked to man-made global warming. Yet Cockermouth has a long history of flooding usually caused by heavy rainfall in the winter months. The earliest on record was in 1761, followed by 1771, 1852, 1874, 1918, 1931, 1932, 1933, 1938, 1954, 1966 and 2005. Many of these floods were just as severe as the one experienced in 2009. There are many other examples.

On a different tack, I recently studied the documented history of rainfall levels in UK for the last 100 years, which is as far back as reliable records exist. I asked questions like: "In a given year and a given UK weather region, what is the probability on the basis of pure chance of a record monthly rainfall level in at least 1 of the 6 winter months". I was astonished to find that the probabilities I was predicting using a few lines of Matlab code were similar to those predicted by the

UK Meteorological Office using their sophisticated computer models. In other words, we should not be surprised by the weather we are experiencing. It is no better or worse than it has ever been. The only difference is that extreme events are now better reported, and the rising population combined with the building of new housing estates on flood plains results in more people on the planet being affected.

The forecasters are in a difficult position. Climate change scientists cannot afford to dilute the message for fear of losing their funding. Climate change and carbon dioxide issues have become topics of big business with huge vested interests. Even our government has taken the message on board at some cost to the economy.

In summary, I would say that in the 4.5 billion years since the formation of Earth, any changes we have experienced in the last 100 years represent no more than the minutest of glitches in the record, and could sensibly be regarded as "noise". In the long term it is possible that our planet is warming, and there may even be superimposed a man-made effect. But the emphasis should be on mitigating action against the possible effects rather then trying to stop it. History shows that civilisation as a whole has always benefitted from periods of warmth. The greatest threat to humanity is not climate change but population growth. This is the root of every single one of our social, economic and environmental problems, from deforestation in South America to potholes in UK roads. There should be less emphasis on trying to save the planet, which can look after itself as it has done for billions of years. The real problem is how to save humanity from over-population.

Let me return to the more mundane problem of flow noise in towed arrays. I had made some progress in noise reduction but it was getting more difficult because I had done all the easy things. I did not have a handle on the potential for further improvement so I came up with a different approach to the problem. I imagined a hypothetical array in which the effect at the hydrophones of all these sources of noise had somehow been eliminated. We would then be left with no noise whatsoever. I then realised that in this case we would no longer be seeing the acoustic signals we were really interested in. This is because the turbulent boundary layer (TBL) contains pressure components at all wavelengths, the very longest resembling acoustic signals like the ones from vessels that we were trying to detect. So if by some means we prevented the hydrophones from seeing all the TBL pressures, including the long wavelength components, then we would also be preventing it from seeing real acoustic signals. The implication of this was that the ideal array should prevent the hydrophones from seeing every type of noise except that associated with the long wavelength TBL pressure components that resembled acoustic signals. These components therefore determined the lower limit of self-noise. It was impossible for any array to be quieter than this without reducing its ability to detect real acoustic signals.

The key follow-on questions were, what was the level of this lower limit, and how near to it were we with our current array designs? The answers to these questions would provide a measure of the potential benefits of further investment in the noise reduction programme.

As is often the case in research, an unconnected piece of work helped move things forward. AUWE had close contact with the Cambridge Professor Shaun Ffowcs-Williams, an expert on the theory of noise and vibration due to unsteady flow. Shaun was the longest serving professor at Cambridge University (1973–2002) where he was the Rank Professor of Engineering, Master of Emmanuel College, and best known for his work on trying to control Concorde's take-off noise. Shaun had done some interesting work predicting the form of the long wavelength components of TBL pressure in a flat-plate geometry. In the course of several visits to Cambridge I asked him if he could do a similar thing for the cylindrical geometry of the towed array. He produced a short paper with an astonishing result. He showed theoretically from first principles that the long wavelength components of TBL pressure in the cylindrical geometry were "wavenumber white", that is, their level depended on frequency but not on wavelength. The formula for the noise level depended on a particular function which was undetermined by his theory but which depended on frequency through a single dimensionless parameter. It was a "universal" function in the sense that its determination would enable me to calculate the lower limit of self-noise for any tow speed, frequency or array diameter.

So I set about trying to estimate this universal function. According to the theory, the noise levels from "ideal" arrays of different diameter and at different tow speeds should all collapse onto a universal curve when plotted in a certain dimensionless format. This universal curve defined the unknown function. Over the years I had collected a lot of data from arrays of different diameter and over a range of tow speeds. These arrays were certainly not "ideal", so I could not expect the data from them to collapse onto a single curve. But I thought that at least I might be able to produce an estimate of the universal function by looking at the lower envelope generated by the curves from all this data.

It turned out that I was able to make quite a confident estimate which in the course of time I refined. It became known as Parsons' Formula and represented a valuable step forward. Because it was classified I no longer have an official record of it. It became a tool for assessing the noise performance and the potential for noise reduction for any array design, and was used by contractors and researchers both in UK and abroad. You just had to plot the data according to the rules and see how close it came to the lower limit. The closer it was the better the design and the less scope for further improvement.

I had many other interesting projects at this time. One in particular concerned the development of plastic hydrophones. All sonar arrays employ hydrophones in large numbers to convert acoustic pressures in the water into electrical signals. Conventional hydrophones are made of piezoelectric ceramic. They are quite expensive and relatively heavy. In the late 1970s I had a cold call from a Dr John McGrath who worked for Thorn EMI at Hayes in Middlesex. He told me about some work they were doing to develop a new piezoelectric material made of a plastic polymer called polyvinylidene fluoride (PVDF). I obtained funding to support this work and we developed together a series of experimental plastic hydrophones for use in towed arrays and sonobuoys. The first prototypes were successfully tested at sea in 1978. However, I was never able to obtain sufficient funding to help Thorn EMI develop a production capability. I told my opposite numbers at NUSC, New London, about our capability and they had John McGrath out to visit them on the next flight from Heathrow. There are now many spin-offs from this work with commercial and military sonar applications both in the US and UK.

A related project concerned the use of optical fibres, then in their infancy. The full potential of optical fibres as carriers of digital data and as a sensing material was only just being appreciated in the 1980s. I became involved in the development of novel hydrophone devices that made use of the sensitivity of optical fibres to external pressure. (A fundamental problem is that optical fibres are sensitive to everything). The vision was of a fully fibre-optic array with no electrical connections, which would therefore be more reliable. I worked with a colleague, Robert McElany, a hyperactive Scotsman who was unable to stop talking, an extreme case of verbal diarrhoea. When he telephoned me I could put the phone down in the middle of his monologue, go and make a cup of coffee, and when I got back he would still be talking and we carried on as normal. Nevertheless, I liked Robert. We regularly visited a contractor in Harlow New Town, north of London. Robert always arranged for us to stay overnight in London so that we could see a show. The highlight was Freddie Jones and Tom Courtenay in "The Dresser" at the Queen's Theatre, circa 1982. The lowlight was Rowan Atkinson in "The Nerd" at the Aldwych Theatre in 1984. Being a Scotsman, Robert always insisted we sit in the cheapest seats. He also wanted to share hotel rooms at which I drew the line. To return to optical fibre technology, whilst it has become invaluable as a data carrier, it has not yet achieved its potential in underwater acoustics for a complete sonar system.

In late 1980 my name was put forward for promotion to SPSO (Senior Principal Scientific Officer). There were two sorts of promotion to this level. One was the management route to designated SPSO posts such as Group Head; the other was the Individual Merit route designed to give those who wanted to continue doing research a worthwhile career path with less management responsibility. Whereas there were many posts for management SPSOs, Individual Merit promotions were rare because the bar was set high. Also, whilst management promotions were within the power of the Director AUWE, Individual Merit promotions were in the sole power of the Cabinet Office in Whitehall, London. This is a government department whose main role is to provide support to the Prime Minister and Cabinet of the UK. The route involved a lengthy process of finding referees who had to be full professors at British universities, plus external assessment of a selection of published papers and a final interview with a board of academics. The whole business was lengthy and elitist and I was none too sure of my chances of success.

Professor R V Jones FRS, an elderly professor at Aberdeen, was suggested as one of my referees. Because he did not know me from Adam, he asked me to pay him a visit. On a dark and cold autumn day I travelled by train from London to Aberdeen via Edinburgh and Dundee to spend the best part of a day with him. It was an informal interview that went quite well, helped by the fact that I knew nothing at the time about the history of this great man[19], so I was not

---

[19] In WW2 Prof Jones (1911-1997) was MI6's principal scientific adviser. He was instrumental in the invention of chaff, whereby strips of metal foil were dropped from aircraft to confuse enemy radar. He was appointed CBE in 1942 for the planning of a raid on Bruneval to capture German radar equipment. Churchill had proposed that Jones be appointed CB, but the head of the Civil Service threatened to resign as Jones was only a lowly Scientific Officer, so the CBE was a compromise. He was subsequently appointed CB in 1946 and Companion of Honour in the 1994 Queen's Birthday Honours. After the war Jones was offered the Chair of Natural Philosophy at the University of Aberdeen, a position he held until his retirement in 1981, shortly after my visit.

overwhelmed. On the way back to Aberdeen railway station I walked past a fishmonger where I bought some Arbroath Smokies to take home.

The return journey to Weymouth involved an overnight sleeper that I joined in Glasgow. This was an experience I shall never forget. I arrived in Glasgow well in advance of my 23:00 night train. I was cold and hungry so had a curry in a seedy restaurant near the station, followed by a pint in a nearby pub full of drunks. I walked to the station and sat down on a bench waiting to board my train. Suddenly, the man sitting next to me was attacked from behind and throttled by a maniac shouting obscenities. I found refuge in the station bar also full of drunks, and anxiously awaited my train. When it arrived I left the bar to be confronted by another crowd of rowdy drunks apparently returning from a Glasgow Rangers football match. Some 20 police officers were trying to control them. I managed to slip through the crowd and run down the platform to join my train. I found my carriage and sleeping compartment. I went inside and locked the door, exhausted but safe. I undressed, got into my pyjamas and climbed into the bunk with a great sense of relief. The train set off and I went to sleep. Some short time later I awoke feeling ill from the curry. I left my compartment and rushed down the corridor to the toilet where I was violently sick. Just as I was leaving the toilet the train stopped in a station. A crowd of people climbed on in high spirits, including a woman dressed in a bridal gown and men in dress suits. I pushed my way through them in my pyjamas and staggered back to the safety of my sleeping compartment. After that I slept soundly all the way back to London, Euston. Ever since I have been a staunch supporter of Scottish independence.

Subsequently, when I was about halfway through this promotion process, I had a phone call from the secretary to the Director AUWE asking me to see him in his office that afternoon. The Director was Mr Ian Davies, a scientist who had made his name in radar. I walked into his office and he offered me outright an SPSO management post to head the Torpedo Group. He knew that I was already on the Individual

Merit path to SPSO but he pointed out that there was no certainty I would get it, whereas he was offering me something on a plate. He also reminded me that whereas a management promotion to SPSO was substantive (effectively permanent), an Individual Merit promotion was reviewed every three years, and if my research work was not up to the mark I could be demoted. He said he would be disappointed if I did not accept his offer because he thought I would do a good job as a manager. He was almost giving me no choice.

I went home and discussed it with Lucienne. I really did not want to be a manager so I turned down the offer. When it finally came to the big interview for my Individual Merit promotion, I was astonished to find that one of the five panellists was Professor Rogers, my PhD supervisor. There was also Professor Keith Stewartson, the external examiner for my MSc (and who was to die of a heart attack two years later). I was given a two-hour grilling. In July 1981 I was promoted to SPSO Individual Merit. I was pleased I had made the right decision. Ian Davies bore me no ill-will and sent me a note of congratulation. Lucienne and I went for a celebration meal at a French restaurant in Weybridge. We drank a memorable bottle of Nuits St Georges.

The Falklands War began on Friday, 2 April 1982 and lasted 74 days. The British government under the leadership of Margaret Thatcher dispatched a naval task force to engage the Argentine Navy and Air Force and retake the islands by amphibious assault. The nuclear attack submarine HMS *Conqueror* sank the Argentinian cruiser ARA *General Belgrano*, marking a turning point in the war. After this not a single vessel of the Argentine Navy ventured out of port. Mrs Thatcher became the darling of the armed forces. Her critics failed to understand that it was of no significance that *Belgrano* was moving away from the exclusion zone at the moment she was sunk. No competent captain would maintain a fixed course whilst entering a battle zone when under possible surveillance. Intelligence information later released confirmed that Mrs Thatcher was correct in her decision. Many people at AUWE worked day and night to produce specialist equipment at short notice in support of the conflict, particularly in helping to deal with mines. It demonstrated that in times of conflict all the red tape of equipment production could be dispensed with.

The 1980s was for me an era of exciting collaboration with the Europeans, the US, and the principal former colonial countries of the UK; Canada, Australia and New Zealand. The rationale was that collaboration saved money because resources and skills could be pooled. It also made work much more fulfilling and exciting.

I became a frequent visitor to the US Navy Labs (NUSC) at New London, Connecticut, an area of the US with an English feel to it. In winter there is thick snow and sea fog on the coast. In summer it is pleasantly warm. I usually stayed in Mystic Seaport, a former whaling village that still retained much of its historic atmosphere and character. At NUSC I made many useful friends and contacts with whom I formed a close association, including George Connolly, Bob Hauptman, Henry Bakewell, Bob Kennedy, Howie Schloemer, Wayne Strauderman, Norman Owsley, and Al Markowitz. The latter was an extrovert always bubbling with ideas, who was to die mysteriously in his hotel room during a business trip. All these scientists were fired-up and on the same wavelength as me. Whilst national security considerations hindered a free flow of information between us, we felt we were working for a common cause to maintain the sonar advantage against the Soviets.

In 1992 NUSC closed down after an amalgamation with the labs at Newport, Rhode Island, an upmarket yachting resort quite different to the old world of New London. The only advantage for me was an easier travel schedule via Boston instead of Washington and New York. On one of my last visits to New London, Bob Hauptman and friends invited me for a lunchtime meal at a local restaurant that specialised in "Combat Food". I cannot remember what we ate.

In 1983 I began a particularly successful collaboration with the Dutch. It started with a visit to Portland by a team of Dutch scientists from FELTNO in The Hague, headed by Kees van Schooneveld who was to become a close friend. He was a talented and meticulous scientist with wide experience. Kees was head of the sonar group at FELTNO and Professor of Astronomy at the University of Leiden. The Dutch were anxious to extend their own programme of work on towed arrays. We came up with an agreement involving joint studies and sea trials. The objective was to examine the feasibility of a vessel towing two arrays in parallel in order to improve directivity and detection performance. This was an innovative idea at the time. The UK would supply the array technology and some support ships and the Dutch would develop new methods of data analysis. A particular concern was whether the two arrays would tie themselves in knots. So we devised an accurate method for determining the location of every hydrophone in the twin array configuration by measuring the time delays between the many acoustic arrivals from multiple underwater shot firings. The Dutch would supply the principal trials ship, RNLS *Tydeman.*

So began a close working relationship with the Dutch scientists with regular trips between Portland and The Hague. I would fly from Southampton or Heathrow to Schiphol just outside Amsterdam then take a car or train to The Hague. We often stayed in the city centre in the Van Der Zalm hotel, or sometimes on the edge of the city in a seaside resort called Scheveningen. Our technical meetings with the Dutch were always fruitful and I found the Dutch to be most hospitable. Every nation has its distinguishing characteristics. The Dutch would take care of everything when we visited them, treating us to house parties and restaurant meals all at their expense. Equally, when the Dutch visited us they offered to pay for nothing. I wondered about our own distinguishing characteristics.

Our evenings in The Hague were often free to explore the restaurants, a favourite being Indonesian. Holland maintains a close relationship with Indonesian culture through its former colonies. The Dutch are fond of Indonesian food, which is spicy compared with the blandness of traditional Dutch cooking. It is hard to beat a good "rijsttafel" (Dutch for "Rice Table"). This usually consists of a main rice dish with a combination of several smaller dishes of meat, chicken, fish, vegetable and egg, plus side dishes of hot sambals, satays, pickles and "soup" dishes traditionally all served at the same time. The other Dutch food I loved was raw herring, served as rich, creamy fillets that slid down your throat like raw oysters. My English colleagues disliked this delicacy so there was always plenty for me.

On one such visit I was in the company of Colin Richardson, possibly the only person I ever disliked but had to work with. Colin and I were walking together across a main square in The Hague looking for somewhere to eat, when our attention was caught by waiters dancing around outside one of the restaurants. We approached to be told they had just that moment received their first Michelin star. We were invited in to celebrate and sample the delights of the house at a knock-down price, drinks included. It was enjoyable but the meal could not have been that good because I can recall none of it.

I got on well with the Dutch scientists. There was Eric van Ballegooijen, an amiable extrovert who always got hopelessly drunk when we were in port on a sea trial, sometimes ending up in street brawls. On one occasion the police had to be called when he had an aggressive argument with guests in the foyer of our hotel in La Spezia, Italy, where we were attending a conference. Eric went on to become Director of Pensions at FELTNO in The Hague. Another person of particular note was Gerard van Mierlo, a gentle and clever man who apparently was a fine ballroom dancer, something I was unaware of at the time. He sadly died of a brain tumour a few years later.

In preparation for our joint UK/NL sea trials a large amount of engineering work had to be carried out on *Tydeman*'s afterdeck to enable her to take our twin arrays and winching equipment. Norman Field oversaw all this work involving several trips to *Tydeman*'s home port in Den Helder, Northern Holland.

The first UK/NL sea trial was in October 1984. I joined the *Tydeman* in Setubal, Portugal, the rest of the UK team having sailed with the ship from Portland where our equipment had been loaded. Life on board was very sociable. One day a week was Indonesian day when the cook prepared a "rijsttafel". Another day would be devoted to a barbecue on the helicopter deck to which scientists aboard the consort ships were invited. In the late evenings we retired to the wardroom where a favourite tipple was Genever, a Dutch sweet gin-flavoured spirit. In between these celebrations we did some useful work.

This sea trial was followed by a second in 1986/87 out of Madeira. At the end of the trial we berthed in Funchal and the Captain invited all the local dignitaries including the British consul and his wife to a grand party on board ship. There was much food, wine and merriment. We scientists took attractive young Portuguese ladies on hosted tours of the ship. The two trials were a great success generating huge amounts of useful data that kept us going for several years.

All our work with the Dutch on twin towed arrays was strictly bipartite. At the insistence of the Dutch we had agreed at the outset not to divulge anything about our work to third party nations. But after the success of our sea trials we decided the time was ripe to present our work to a NATO audience at a classified conference scheduled for October 1988, to be held at the NATO Undersea Research Centre in La Spezia, Italy. We jointly wrote three papers covering the work, two to be given by me and one by Gerard van Mierlo. Kees van Schooneveld, the Dutch leader was also to attend.

Kees suggested to me that it would be enjoyable to drive to La Spezia in his car. He invited me to stay at his house in Holland before setting off. Kees had a comfortable Volvo and he was good company. He was clever and meticulous in everything he said and did, and always willing to share his vast knowledge.

We set off on the long journey from The Hague over the German border, stopping for the night in Bavaria at a typical German hostelry. The next day we continued into Switzerland where we admired the Alps, and finally into Italy beside the northern lakes. Along the journey Kees told me at length about his poor and deteriorating eyesight. He explained that all the information was there, it was just scrambled. The solution was simply an exercise in signal processing. He was going to have an operation to improve things. All this time he was driving at breakneck speed in the outside lane. In an unrelated discussion he told me that his Volvo was the safest car in the event of an accident.

We had planned the journey to give us two days of sightseeing in Italy before turning up at the conference. Our first port of call was Florence. Kees had been there before and was an excellent tour guide. The city was dripping with magnificent churches and works of art. It seemed like people had Michelangelo sculptures in their back gardens. On our way to La Spezia we spent some hours in the ancient town of Lucca where, among other things, we saw the birthplace of Puccini.

The conference talks went well. We presented our twin array work as a world first. We were invited out one evening to a small restaurant in the nearby seaport of Lerici where we were staying. There I ate the most marvellous risotto I have ever tasted. It was a seafood risotto, smooth, creamy and rich. This was at variance with my many other experiences of Italian food, which generally has been disappointing.

After the conference we went home our separate ways. My own route took me through Pisa where I bought a large block of parmesan in a local market and climbed the Leaning Tower. Soon after my visit it was closed for safety reasons. It remained closed for several years while they surveyed it before injecting tons of supporting concrete into the foundations.

One outcome of our twin array work was an approach by the US to set up a tripartite agreement with ourselves and the Dutch to exploit twin arrays to the next stage of doing a full-scale demonstration. We had meetings in New London and The Hague to try and hammer out an agreement that satisfied all three parties, but the US were asking for a lot and eventually our side felt we would be giving too much away. The attempt to collaborate with the US was abandoned.

As it turned out, our Navy was never convinced of the feasibility of twin arrays, in spite of a successful submarine demonstration that Colin Richardson completed on HMS *Turbulent*. The MoD decided that the deployment and recovery issues were too expensive and risky. In the early 90s the programme was wound up through lack of funding. A lasting testimony to our collaborative work is a film, made largely by the Dutch with help from Colin Richardson, describing the experimental techniques and results of our work. It is now somewhat dated, but continues every year to entertain delegates when I show it on the Portland Sonar Course.

I was disappointed by the termination of the twin array programme, but in retrospect it was a sensible decision. There was a promising alternative approach offering all the advantages and none of the disadvantages of twin arrays. It involved the use of vector sensors that measured pressure gradient as well as pressure within a single array. This enabled the advantageous 3-dimensional beam patterns of a twin array to be achieved with a single array, without the handling difficulties of a twin array. This work continues to bear fruit.

The work with the Dutch lasted several years. During this time I became involved in another joint collaboration with the United States and the other Commonwealth countries of Canada, Australia and New Zealand. It was done under an umbrella agreement initiated in 1957 between the defence departments of the five governments. It is called TTCP - The Technical Cooperation Programme (or The Travelling Cocktail Party as some called it). The broad objective of this new collaboration was to improve the performance of our towed arrays. I made many trips to the US, to both East and West coasts, to visit labs and exchange information. I met one of my heroes, Dave Chase, at that time working as a private consultant out of his town house on Beacon Hill, Boston.

In about 1982 I was on a sea trial in the North Atlantic when I was called to the radio room to take a high priority call on the radio telephone. These phones were difficult to use. The reception was bad and you had to take turns to speak. It was Lucienne to tell me that Tonton Charly, her mother's brother, had died and she would shortly be on her way to the funeral in Roanne. There was no way I could attend.

A similar thing happened on another sea trial. I had been invited together with Peter Enoch, who worked for me, to be an observer on a US sea trial operating out of the Bahamas. We first visited the US Navy Research Labs in Washington, Orlando and Fort Lauderdale, taking a day out in Orlando to visit Disney World and the Epcot Centre (and O'Grady's Bar). We flew on to West Palm Beach where I found myself waiting on the tarmac to board the plane to Andros in the Bahamas. It was 24 June 1984. I heard an announcement over the tannoy calling for Dr Parsons to go immediately to the information desk to take a phone call. It was Lucienne who had been patched through via the Washington Embassy to tell me that her mother had just died. Funerals in France are never more than three days after the death, so Lucienne was on her way to France straight away. We discussed whether or not I should break off from my sea trial and travel to France. We decided against it and I continued with my schedule. I always seemed to be away when these things happened.

Madame Page's death cast a heavy cloud over the sea trial for me. The ship was the USS *Athena*, a converted gunboat formerly used in Vietnam. It had two screaming gas turbines stuck at $45°$ through the hull to propel the ship to a top speed of over 40 knots. It was difficult to sleep when the turbines were on, even when wearing the mandatory ear defenders. The purpose of the trial was for the US to test some experimental towed array modules. The trials site was Exuma Sound located some 100 miles East of Nassau. In view of the high security classification that the US applied to some of their work, Peter Enoch and I were confined below decks during some of the experiments.

The freezer on *Athena* was always stacked full of ice cream, and the fridge with cans of Pepsi, Dr Pepper, and the vegetable drink V8. This was my first experience of V8. Nobody else liked it so I had gallons of it to myself.

I recall a particular incident on this trial. I was in the miniscule shower room having finished my ablutions and trying to comb my hair in a pitching sea, when my comb caught in my glasses. They dropped to the floor breaking a lens. Fortunately I had brought my spare pair. I was successful in claiming £350 from MoD for the Zeiss replacement.

Although the *Athena* was a most uncomfortable ship with Spartan accommodation and poor food, I appreciated its value as an experimental platform because it was capable of high speed. Years later in 1995 and 1997 I was to hire the ship for our own high-speed trials.

On these trips to the States I always tried to meet up with my old friend Bill Moseley if I was passing through Washington. Bill was then working at the Navy Research Labs (NRL) in Washington where he was the US equivalent of a Group Leader. He often invited me to his house for an evening meal with his large family. On one occasion he and Sandra took me to a school soft-ball match in which two of their children were playing, and on another occasion to view a shopping mall. Bill seemed happy and doing well in his job.

In September 1985 on my way to a meeting in New London, Connecticut, I met Bill over the weekend in Washington. He picked me up early evening from my hotel, the Radisson in Crystal City, for the 20-minute drive to his house. Bill and his wife Sandra had always been heavy drinkers, but I was surprised to see in the drinks holder of Bill's car a large glass full of bourbon. He seemed to drive fairly well, narrowly missing a deer (or was it a bear) on the freeway. We arrived safely at his house to meet Sandra who was also heavily under the influence. The evening meal with his large family was strained and ended in a blazing row between Bill and Sandra with Bill walking out into the night. Sandra went outside to sit in her car where I joined her. She was angry and crying. She said that their marriage was over and Bill had a mistress, one of his secretaries. This was all very sad. Bill eventually returned home to drive me back to the Radisson. It was the last time I was to visit Bill and his family at his Washington house, and the last time I saw Sandra, although she sent us Christmas cards regularly from Tybee Island in Georgia where she was living with her new family.

We hear a lot today about the "special relationship" between the UK and the US. It has existed for a long time and to a certain extent is real, but in my experience it has always been skewed on account of the massive resources of the US compared with the UK. The US always "called the shots" if we were not careful. When it came to matters of research in support of defence, the UK would typically assign half a man and a boy to address a problem. The US strategy was to throw money at it. They would fund three or four contractors to work independently, then select the best solution for further development. In spite of the vast difference in resources we usually managed to hold our own against the US.

This difference in approach was brought home to me at a technology exchange meeting in New London, Connecticut, circa 1980. The meeting was headed on the UK side by Noel Reynolds, a procurement manager at AUWE who was much senior to me and regarded as a safe pair of hands. He was a talented pianist with a great skill for improvisation. At a pre-meeting briefing at our hotel in Mystic, Noel told me to tell the US everything about our recent research progress since this was the only way the US would open up on their own massive programme of work. I refused to comply because I believed I had made progress that put us ahead of the US. When it came to my talk I presented my results without explaining how they had been achieved. This turned out to be the right strategy.

The year 1984 saw yet another large reorganisation of the government research establishments. The name of the Portland site was changed from Admiralty Underwater Weapons Establishment to Admiralty Research Establishment (ARE). This was considered to be a less aggressive name in keeping with the national policy of defence rather than attack.

At AUWE we frequently had visits from high-ranking naval officers and I would be called upon to give talks on my work. We got quite used to doing this and it was not a big deal. One such occasion was on Friday, 14 November 1986, when we had a visit from the top man, Admiral of the Fleet Sir William Staveley, Chief of the Naval Staff and First Sea Lord. After I had given my spiel, a photographer walked in who I assumed was there to record the visit of our honoured guest. But Sir William arose from his chair, made a short speech and presented me with an award (Plate 42). It turned out to be an International Achievement Award for my work on towed sonars. It came as a total surprise and I was stunned.

A third collaboration in the 1980s was with the French, in some ways the most enjoyable of all my collaborations, though not the most fruitful. It came about through yet another international forum for technical exchanges called the Anglo-French-Netherlands Cooperation Programme (AFNCP). The first AFNCP meeting I attended was in 1986. It was held at the French naval research labs in Le Brusc on the Côte d'Azur. A French naval officer booked us into a hotel in the blue light district of Toulon. The hotel seemed to be doubling up as a brothel. The officer who had booked us in thought it was a great joke. I have always found that naval officers are blessed with a childish sense of humour.

The meeting itself was unexciting. It took the form that I was to get used to in the AFNCP forum, with the Dutch keen and always pressing the UK and French contingencies for more information, the UK sometimes complying and the French usually managing to give nothing away. But there were two significant outcomes from the sidelines of this meeting. Firstly, I discovered that the French had ambitions like us to build arrays of smaller diameter. The benefit would be smaller stowage volume, or longer arrays for a given volume. The topic was to form the basis of a bipartite collaboration between ourselves and the French. Secondly, I met Michel Boisrayon who was to become a good friend and collaborator.

Michel was an eccentric and fascinating character, atypical and unconventional in every sense. He was the manager of the French research effort on Sonar at Le Brusc, in which position he did not seem to fit comfortably. He was more interested in doing the basic science. He was articulate and spoke meticulous French and English. His writing was small and written in a beautiful calligraphic script. He was living in a rambling country house outside Toulon with his elderly parents whom he looked after. Michel was unmarried and at the time had no partner. He had a refined palate and was an expert cook, committing his exquisite recipes to memory in minute detail. Michel had ultra right-wing views. He once told me that he was not averse to slavery and believed that some races were sub-human. Michel should have been born into the Roman era. He kept a daily diary that ran to over two million words, probably much more by now.

Michel had a keen interest in fringe science and believed in extra-sensory perception, not only between humans but also between any living things, plant or animal. He diligently pursued these interests in his spare time performing practical experiments in his garden. This involved him in such extreme activities as the cultivation of bonsai trees and the entrapment of birds using glue sticks. There was much to disagree with in Michel's philosophy, but at the very least he was entertaining. Over the years Michel became a great support and inspiration to me in my work.

The collaboration between Michel's group and mine on small diameter towed arrays began in 1987. There followed frequent visits between our two laboratories with participation in each other's sea trials. During this time the French managed to tell us almost nothing of value about their work and we responded likewise. Yet on the surface a lot was going on.

On one occasion the French contingency joined my team for a sea trial on RMAS *Newton*. It was yet another enjoyable trial out of Madeira. Regrettably, I was unable to make the return invitation to a sea trial on the French research frigate, *Le Commandant Rivière*. Alan Morrell attended and reported on the fine food and wine that the French Navy enjoyed, like foie gras, caviar, champagne and lobster as standard fodder always available in the fridge for snacking.

When we visited Le Brusc, Michel's hospitality was generous. He liked to table a detailed agenda, with frequent breaks to drink his fine coffee and sample the local delicacies that he brought in specially for the occasion. He gave us lengthy lunches in a private room at the lab. I recall eating a huge plate of langoustines complete with shells and claws. It was a messy business because the langoustines were in a rich tomato sauce and we had to use our fingers.

In the evenings Michel usually entertained us at one of the local restaurants in Toulon or Sanary where everything would be provided. He had a curious habit of allowing the waiters to give us menus for us to study, then choosing on our behalf what all of us would eat. I never objected because Michel knew what he was doing. On one occasion he chose an enormous fish about three feet long that was presented by two waiters in the middle of a long table. I often wondered how he managed to be so lavish in his entertainment. Then I realised he had a tame contractor who always attended the meals to settle the bills.

Over the years I made about twenty visits to Le Brusc, usually flying to Nice or Marseille, and staying in various hotels in all the main seaside resorts between Marseille and Toulon, such as Cassis, La Ciotat, Bandol and Six Fours. It is certainly a beautiful part of France. My favourite venue was a mountain village called La Cadière d'Azur some 6km inland from Bandol. The hotel was called the Hostellerie Bérard set in idyllic surroundings with beautiful views across the valley. The chef was M. Bérard whom I never actually saw, only tasted his fine food. He was recently awarded a Michelin star. The restaurant was the domain of the proud Mme Bérard whom I got to know well. She limped badly from an injured leg, but I never saw her sit down.

With all these interconnected collaborations it might be thought that we would have difficulty preserving confidentiality and would get confused about what we could say to whom. In fact, we became used to erecting Chinese walls and it was not difficult to comply with the rules of the different agreements.

In November 1988 I went on another sea trial out of Madeira on RMAS *Newton*. I took this opportunity to invite the US hydrodynamicist Henry Bakewell as an observer. Henry was an old friend from the US Navy Underwater Systems Center in New London, Connecticut. The invitation was to reciprocate the US invitation that Peter Enoch and I had accepted four years earlier on the USS *Athena* out of the Bahamas. I told Henry to get his admin to book him into the Savoy Hotel in Funchal, at that time the best hotel excepting Reid's, and my favourite choice for the team. What with the sumptuousness of the hotel and the vastness of our trials ship compared with the USS *Athena*, Henry thought it was a holiday. As an observer he had little to do whereas the rest of us were working hard. In order to play the game I had Henry confined to his cabin during certain experiments, which I told him were sensitive and outside our collaborative agreement.

In 1988 my name was put forward by the then Head of Sonar Department, Dr Colin Pyckett (he who was to play the organ at Isabelle's wedding), to be considered for promotion to IM DCSO (Individual Merit, Deputy Chief Scientific Officer). I thought this was out of my league, there being less than a dozen IM DCSOs in the whole of the scientific civil service. There began the long process of refereeing, finding academic sponsors, submission of technical papers and preparation of supporting statements. I passed these hurdles and went for interview at the Treasury Office in Whitehall on Monday, 10 April 1989 at 14:30.

The Panel Chairman was Professor Sir Kenneth Blaxter FRS, Director of the Rowett Research Institute and President of the Institute of Biology. He was supported by: Sir Ronald Mason FRS, former Chief Scientific Adviser to the Ministry of Defence, Professor of Chemistry and former Vice-Chancellor at the University of Sussex and at that time Chairman of University College, London; Professor Cyril Hilsom CBE FRS, Director of Research at GEC and President of The Institute of Physics (also the inventor of liquid crystal displays, widely used in televisions and computer monitors); Professor R E D Bishop CBE FRS, Vice-Chancellor of Brunel University and former Professor of Engineering at University College, London; B Edmiston, Director of Civil Service Personnel. It was a most impressive panel. My interview lasted almost three hours and went quite well.

In May 1989 I ran a sea trial on RMAS *Newton* out of Santa Cruz in Teneriffe, finishing in Madeira. Gordon Murdoch, a reclusive character who had recently returned from a two-year exchange at NUSC, New London, also took part. I had passed on to him the responsibility for organising all our towed array sea trials. Gordon was good at his job but always gave the impression that he was carrying a heavy burden on his shoulders. He spent hours pacing up and down the corridor outside his office in deep thought, on matters probably unrelated to work. Gordon was married to Sandra who was also a bit strange. They had no children and lived self-indulgent lives of parties, expensive holidays and skiing. I never discovered Gordon's secret worries, but it may have been something to do with a boundary dispute at his home in Preston, Weymouth. Bob Weatherburn was also present on this sea trial, having taken over management of the Group on Peter Clynick's retirement in 1986.

At about this time I felt that Ameeco Hydrospace who built our towed arrays under contract were becoming too expensive and we had insufficient control over their work. At Gordon Murdoch's suggestion I made him responsible for setting up our own build facility on Portland (later transferred to Winfrith). We developed all the necessary in-house expertise to support the build process and never looked back.

In 1988 I became interested in the topic of the higher order spectra. The motivation for this was a seminar I had attended at Portland given by Professor Mel Hinich from the University of Austin, Texas. His talk was on the properties of the third order spectrum (the bispectrum) and on ways it might be used to aid signal detection. The higher order spectra are dependent on the third and higher order moments of a data sample, as opposed to the more conventional second order moments that define the power spectrum. For non-Gaussian signals the bispectrum and higher order statistics contain additional information not present in the power spectrum. Whereas the power spectrum describes the distribution of variance in a signal, the third and fourth order spectra describe the distribution of skew and kurtosis, respectively. A

Gaussian signal is completely described by its first and second order statistics. So the higher order spectra are necessarily concerned with non-Gaussian signals.

These ideas were all new to me and I resolved to investigate it further. In June 1989 I attended a conference on Higher Order Spectra in Vail, Colorado. This conference gave me some useful contacts. In particular, I was fortunate to meet Chris Dainty, Professor of Applied Optics at Imperial College, London. He was already using the special properties of the bispectrum to improve the optical imaging of stars. I resolved to team up with Chris if I ever managed to secure research funds. In the event, I obtained funding and we worked together for several years on applications of the bispectrum to underwater acoustics. I was also able to fund his PhD student, Mark Williams.

During this period I kept in close touch with Mel Hinich who was always encouraging and even arranged for his post-doc students to visit me if they were in UK. One that I recall was John Dalle Molle, an engaging character who stayed for a week on Portland visiting me each day. Mel Hinich was a modest man with time for everyone and a consuming interest in everything. His brain worked exceedingly fast making it difficult to keep up with his train of thought. On one of my US trips I met him in Austin, Texas, both at his place of work and at his house. Mel was addicted to red-hot chillies that he ate raw for breakfast. He was actually Professor of Local Government at the University of Texas as well as a research professor at the University's Applied Research Laboratories. He wrote seven books and published over 200 papers in such varied fields as statistics, signal processing, political science, economics, underwater acoustics, pharmacy, and biomedical engineering. He was also an accomplished musician. He was one of the few remaining polymaths in these days of specialisation. Mel died in 2010 aged 71 after falling down a staircase.

On 1 July 1989 I was promoted to IM DCSO. The Cabinet Office had tried unsuccessfully to contact me in Vail, so I got the news on my return. The position was akin to a personal chair at a university. It was prestigious with much influence but little authority. Within the hierarchy of the Scientific Civil Service my elevation attracted perks. I suddenly found I was entitled to a bigger office. This was duly provided by knocking two adjacent offices into one. I was entitled to wall-to-wall carpeting and a suite of office furniture made of wood rather than metal and plastic. My suite comprised a large oak desk, two conference tables with a set of six matching chairs, a bookcase, and most quirky of all, a hat-stand and coat rack. I was also provided with personalised headed letter paper and matching envelopes all embossed with the MoD crest. I was given my own parking space at work. When I visited an HM ship or travelled on an RAF flight I received VIP treatment. At Brize Norton I was provided with a staff car and driver and was always last on and first off the plane. This was all because my equivalent service rank was one star, or Rear Admiral.

I began to have requests from other institutions. I was asked to give key-note lectures at conferences. I was invited to become a Visiting Professor at Birmingham University in the Department of Engineering. I sat on several university Research Advisory Committees and Councils, for example, at the Royal Naval Engineering College (now closed) at Manadon in Plymouth, the Institute of Sound and Vibration Research at the University of Southampton, the Faculty of Science at the University of Bath, and the University Senate at Birmingham. I cannot say that I contributed much to any of these committees. I think my real function was to enable these institutions to tick a box to satisfy external scrutiny of their business strategies.

I was invited to become a Fellow of the Institute of Mathematics and Its Applications, and the Institute of Electrical Engineers (now the Institution *of* Engineering and Technology), even though I knew nothing about electronics. I became the UK editor of the US Journal of Underwater Acoustics. I committed to all these additional activities with enthusiasm, but their main effect was to interfere with my work.

The key thing about an Individual Merit promotion was that it gave the scientist a great deal of autonomy. Since the salary of an IM scientist came directly from the Treasury it was difficult for a manager to persuade an IM scientist to do something that he/she did not think was worthwhile. The system made explicit within the organisational structure the key role played by the scientists in the success of the whole enterprise. It was understood that the most important asset the organisation had was its scientists. This may seem strange to a present day culture in which the only opinions that matter are those of senior management. Regrettably, this recognition of scientific excellence was soon to end.

Meanwhile I was still making regular trips to the States visiting research laboratories, contractors, and attending the biennial classified US Navy Symposium on Underwater Acoustics (NSUA). In particular, in early November 1989 I presented two papers at the NSUA symposium in Biloxi, Mississippi, on the Gulf Coast, a 60-mile car drive from one of my favourite US cities, New Orleans, (along with San Francisco). One paper was on vibration isolation and the other was on the integral properties of low frequency pressure in a towed array. The organising host for the conference was my old friend Bill Moseley, now the Technical Director at the nearby Naval Ocean Research and Development Activity (NORDA) on the site of the Stennis Space Center. I was delighted to see Bill again and visited him at Stennis. He invited me round to his house in the evening. He was now divorced from Sandra and had a new partner, Margaret, although she was not present. Bill's house was open-plan. It had an enormous high-ceilinged living area that was open to the rooms on the top floor by way of a mezzanine landing, which strangely housed a large communal jacuzzi. We had drinks, and at the end of the evening said our goodbyes. I do not believe I have seen Bill since, but we have kept in close touch by email.

In 2005 Bill's house in Biloxi was totally destroyed by Hurricane Katrina, along with the rest of Biloxi and much of the Gulf Coast. Neither Bill nor his partner Margaret was hurt.

Now married, they share their time between houses in Lafayette in Indiana and New Port Richey, near Tampa, Florida. Although our meetings have been infrequent, I have always felt a strong friendship with Bill, cemented all those years ago by our time together at ARL Teddington.

# Change in the 1990s

In January 1991 my parents celebrated their 50$^{th}$ wedding anniversary at the Estuary Club in London Road, Leigh. My father was then aged 75 and my mother 79. Upwards of 100 people were present because both my parents had many friends. It was a poignant occasion particularly because, in order to attend, my father had to obtain special leave from Southend General Hospital where he returned at the end of the celebrations. This was a continuing problem with his bladder that he had over many years.

The backdrop to my work in the 1990s was the enormous change that took place in the organisation that employed me. In April 1991 we became the Defence Research Agency (DRA), a prelude to the privatisation which unknown to us workers was to follow. DRA was an amalgamation of all the MoD research centres: the Admiralty Research Establishment at Portland (ARE), the Royal Aerospace Establishment at Farnborough (RAE); the Royal Armament Research & Development Establishment (RARDE) at Fort Halstead in Kent; and the Royal Signals and Radar Establishment (RSRE) in Malvern. There was great sadness that these research centres, with their distinguished histories and many distinguished scientists, would lose their identity and eventually disappear along with their royal charters.

Very soon after DRA's formation John Chisholm was appointed CEO. He was a man who looked rather like a meerkat. Chisholm started with the controversial premise that as an organisation we were bankrupt. He made this declaration at Farnborough to a meeting of senior personel at which I was present. He vowed to sack all under-performing managers. It was Friday, 13 September 1991, and became known by us as Black Friday.

Chisholm closed several DRA sites and amalgamated the rest. His response to the argument that we were losing unique experimental facilities was that they were unique only because no other profit-making organisation considered them viable. It became clear that his main objective was to make money, not to serve the armed forces. He also had a hidden agenda. The managers became terrified of him.

Our site on Portland was closed at roughly the same time as the closure of the Navy Base. These events were a severe blow to the economy of Weymouth and Portland. Most of us scientists on Portland were moved to new premises at the former experimental atomic energy site at Winfrith, about 12 miles East of Weymouth. So there was no need for a house move.

Friday, 19 May 1995 was my last working day on Portland. On that day I committed a major security breach and a cardinal crime. I smuggled my video camera into the Navy Base and filmed my office and the empty building where I had worked for so many years. I then drove round the Navy Base filming everything for posterity out of my car window. I cannot say that the video is riveting but it does have some historic value, particularly now that many of the original buildings have been razed to the ground. The video still exists somewhere for those who are curious,

At about the same time that our site was closed we became The Defence Evaluation & Research Agency (DERA), the R&D branch of the Ministry of Defence. At that time we were the UK's largest science and technology organisation.

Chisholm began to make yet more sweeping changes. In a series of steps he reorganised the management structure from a multi-layered hierarchical one to a flat one, in itself a good thing. He achieved this by not replacing senior managers when they retired, retiring other senior managers early on special terms, organising the research into fewer but larger groups with fewer management layers, and placing relatively junior people at the head of these groups at lower salaries. The result was a leaner organisation, with managers who were paid much less than their predecessors and who were not sure what they were supposed to be doing. But the new managers were not displeased because for the most part they had been promoted from nowhere to positions of importance. We did not know that the real purpose of all these changes was to set us up for privatisation and eventual sell-off.

The main effect on us scientists of all the restructuring was the continual change in the way we had to run our projects. It also left me in a strange position. Because I was an Individual Merit scientist within a system that was still run by the Cabinet Office, my personal status was unaffected by all these reorganisations. So with the massive devaluation of all the management posts I found I was earning more than the managers above me, and even more than the top man at Winfrith.

Although the changes were painful I was able for several years to do whatever I liked, provided I could obtain supporting funds. Altogether I built up some 20 projects ranging across blue skies research, applied research, in-service naval equipment support, and commercial. I had an annual turnover of around £5M supporting a team of some 30 scientists and engineers. These included, in no particular order, Steve Pointer, Jon Davies, Nigel Apsey, Alan Morrell, Nick Goddard, Colin Mead, Andrew McLean, Linda Morrison, Peter Rapson, Steve Thompson, Bill Fagg, Steve Groves, Gary Davidson, Simon Gardner, Geoff Wadham, Steve Watts, Gary Dunham, Chris Chester, Richard Heal, Roy Baker, Norman Field, Shaun Dunn, Carol Tanner, Dawn Pridmore, Paul Masham, Chi-Kong Yeung, Jim Nicholson, Richard Brind, Hilary Green.

At this time I felt confident enough to ignore the ever-changing systems imposed by the new management structures. I set up my own project management system which I called ARA, standing for Accountability, Responsibility and Authority (rather "tongue-in-cheek"). The basic idea was that every node of the project plan had to address these three issues. One feature of the system was the monthly progress meeting at which I required everyone in the team to be present. The purpose of the meeting was to examine, record and update the progress on all the milestones for each project. I chose the chairman of each meeting by rotation, thus giving everyone the experience of running a meeting. It also meant that the most junior members of the team had the opportunity to hold to account the most senior members, a system that was actually Aristotelian, although I could not agree with much of his scientific philosophy. Because we made a profit and because my system was more effective than the official one, I got away with it.

In the mid-90s our organisation was subjected to an extensive external audit to assess quality control and position in the market. My team was ranked world class and my project management system was judged to be "beyond requirements". No other team at Winfrith achieved this. Our success enabled me to ignore the official management, at least for a time. The managers called me a prima donna. Lucienne told me this was not meant to be a compliment.

During this period the morale in my team was high and we were very productive. Everyone had interesting work from conception to implementation. We had stimulating contact with many universities and frequent sea trials in exotic places. Most of us had plenty of foreign travel both in Europe and as far afield as USA, Canada, Australia and New Zealand.

The other important change was the break-up in 1991 of the Soviet Union into fifteen independent sovereign states. This astonishing event followed rapidly after the fall of the Berlin Wall in 1989 and the reunification of East and West Germany in 1990. These events led to the end of the Cold War in 1992 and a complete rethink of UK defence strategy.

The catalyst for the fall of Soviet Russia was its bankruptcy brought about by its enormous defence budget. The West can take some credit for this in the pressure it put on the Soviets by its defence policy of technological advance. This was in spite of the many naval secrets given away to the Soviets in the 1980s by the American traitor John Walker, which went a long way to explaining why the Soviet Victor III and Typhoon (Akula) classes of submarine were so quiet and well-equipped.

During the Cold War things had been black and white. We knew who our opponents were and we had a clear-cut strategy of containment. The Soviets equally had their own strategy and, for the most part, there was a satisfactory balance of power.

After the Cold War the situation became blurred and complex with many potent forces emerging, including international terrorists. The "Peace Dividend" that had been predicted did not materialise. For our Navy the end of the Cold War meant a reduced requirement for deep water submarine operations because the perceived threat from Russian submarine activities in the North Atlantic was reduced (although in 2019 this is now changing again). The focus shifted to combined land-sea-air capabilities in littoral (territorial) waters with policies increasingly driven by the need to preserve commercial shipping routes and oil supplies. All this had a major impact on our work at DERA and our ability to obtain funds.

The 1990s were important for other reasons. Our daughters completed their degrees at Oxford, Isabelle some years earlier in 1988 at St Anne's College and Caroline in 1996 at New College. They each achieved 2:1 degrees just like Lucienne and me, so equality was maintained. Isabelle married Antony on 6 July 1991. This was followed in 1994 by the birth of our first grandchild, Emily.

Both Lucienne and I began to do things we had always dreamed of but had never had the time or money to achieve. We bought a flat in Bath in New King Street that we visited regularly as an escape. Bath is a beautiful spa city steeped in history. Lord Nelson's father had lived at the end of our road, and opposite us was the house where the astronomer William Herschel lived when he discovered the planet Uranus with the aid of a telescope in his back garden. We started ballroom dancing which was to become a lasting passion for both of us. Over the years we had many dance teachers: George Biles and Angela in Weymouth who started us off, Ron and Doreen Stone in Dorchester, Joyce Allen in Yeovil, Simon and Heidi Cruwys in Bournemouth, Ralph and Muriel Aird in Weymouth, Eileen Spracklen in Dorchester, Nigel Allen in Yeovil, Pauline Collison in Poole, Sue Cooper and Andrzej Mailkowski in Oxford, Jenny Marks in Swindon, David Pollard in Oxford, plus many master classes in Weymouth with visiting Strictly stars. We went on dancing holidays, usually over long weekends, in places like Torquay, Bournemouth, Brighton, Woolacombe, Paignton and Malvern with Sid and Roz Gateley, Ian and Ruth

Walker (former world Latin champions), Paul Eden, Philip Wylie, Sue Cooper, Gary Fleetwood, Jenny Marks, Pat Anslow, Alan and Ginny Newman (former world Latin champions), and several others. How much of all the steps and technique that we learned can we remember?

Lucienne learned to ride a motorbike and bought an 883cc Harley-Davidson Sportster, together with a complete set of black leathers. She later graduated to a beautiful pearlescent white Jaguar XJ12 which she occasionally used for weddings working for Gary Bartlett, then a Hyundai sports car. I bought a two-manual Wyvern electronic organ with full pedal board that we installed in our new "atrium", later to be replaced by a three-manual Johannus organ from Makin. I began occasional organ lessons with Michael Dover, formerly organist at All Saints church in Wyke Regis until he had an argument with the vicar, and then more regular lessons with John Wycliffe-Jones at St Mary's Church in Weymouth. I also attended art classes and watercolour painting at Weymouth College, and I began to take a keen interest in cooking. Lucienne had done most of the cooking up until the late 1980s after which she became too busy with translation work. It was before the internet, and each day she had to catch the last evening post in Weymouth town centre. So I began to take over the cooking as a relaxation after returning home from work.

Two fell-walking holidays in Cumbria (Lakeland) with daughters Isabelle and Caroline reawakened my school memories. The first was in Ambleside in 1988 when we stayed at the rather quaint Rysdale Guest House, later to be recommended by Hunter Davies, complete with its evening performances on the pianola. We were naive walkers in those days and ill-equipped wearing our ordinary shoes. After climbing Helvellyn via Dollywagon I was admonished as an irresponsible climber by the Ambleside cobbler when I asked him to stick my soles back on. The second holiday with Isabelle and Caroline was in Keswick in 1989 when we stayed at Greystones Guest House opposite the Parish Church, where one evening we attended an organ concert. One of several memorable climbs was beautiful Glaramara. We were still irresponsible walkers when it came to health and safety.

Soon after, I began making more regular visits to Lakeland that were to become an important feature of my life. I became intoxicated with the Lakeland Fells. The area measures no more than 30 miles from North to South and 20 miles from East to West. Yet in this small space lies the most beautiful, exciting and spectacular mountain scenery that one could wish for. I could not get enough of it, hot or cold, rain or shine. I loved the challenge of the climb, the sense of danger, the exhilaration of reaching the summit, the views in fine weather of lakes and valleys from the high ridges, and the wonderful feeling of complete exhaustion at the end of the day. This interest was not something I shared with Lucienne, who has no great love of walking or mountain scenery.

The first of my regular visits to Cumbria was in August 1995. Two weeks before that holiday I injured my right foot. I had been adjusting the heavy concrete counter-weight to the up-and-over door of our garage when I dropped it on my foot, removing the nail on my big toe. The A&E at Weymouth Hospital bound it up and I continued with my planned holiday. I stayed at the Wateredge Hotel by Derwent Water just outside Ambleside. On the first day in hot weather I repeated the school walk across Striding Edge that I had done almost 40 years previously. I then did Steel Fell from Wythburn via Greenup Edge, followed by the Langdale Pikes coming down North Rake, and the Wandsfell Pike Round via Troutbeck. Each evening I rebound the bandages on my foot and dined in the superb restaurant. The Wateredge Hotel was a high-class establishment with a colonial feel to it. Sadly, it has now been converted to a noisy theme pub with no charm, but probably attracts more punters.

In the early years I always visited Cumbria by myself, then later I teamed up with our daughters or with my brother Nick and cousin John. I found myself driving up there four times a year. The wonderful thing about the Fells is that they are always there and welcoming whatever the weather. After almost 70 Lakeland holidays I know the area intimately having completed over 220 different walks. I have a photographic record of every walk, which helps me recall each one in some detail. All this must qualify as an obsession. I have surely made a significant contribution to the Cumbrian economy.

The photos that record my Lakeland walks give the mistaken impression that the weather is always fine. The fact is that there is little incentive to extract the camera from zipped-up waterproofs to take a photo in thick mist, howling gale or driving rain. The camera is poor at capturing these conditions when there are no views anyway.

We made regular visits to my parents' bungalow in Lympstone Close in Southend. They had moved there from Henley Crescent in 1972. They were always welcoming and my mother was never happier than when she was surrounded by family. I knew my parents were growing older but I somehow clung to the illusion that they would live for ever.

In April 1992, as a professor at Birmingham University, I began giving my series of lectures on the intensive Portland Sonar Course organised by the University. The course was then run by Professor David Creasey, Head of the Department of Electrical & Electronic Engineering at Birmingham. David was a generous man and a great encouragement to me. In July 1995 he died tragically from a blood clot on the brain following a fall from a ladder at his house while repairing the roof of his porch. I went to his funeral. Several people I know have died or injured themselves falling off ladders. After David's passing the Portland Sonar Course was organised by the estimable Phil Atkins, an excellent lecturer and a talented scientist. My contribution was to give additional context to the theoretical aspects of the course by way of practical examples taken from my own experience. I have now contributed to this annual event for over 25 years and have cherished every minute of my involvement.

In 2004 I was invited to lecture on the biannual Bournemouth University Sonar Course, which I continued to do with enthusiasm until 2019. One reason these courses remain viable is that a high proportion of the delegates are now non-EEC, many from Asia and the Far East where there is money and a growing eagerness to develop sonar technologies.

In December 1993 I was invited to FELTNO in The Hague to give a guest lecture in honour of my friend and colleague Kees van Schooneveld on the occasion of his retirement. I was free to choose the subject. I chose to talk about the higher order spectra. The material was based on my previous work and some new results that hopefully Kees and the large audience enjoyed. This was the last time I was to see Kees.

Unfortunately, my funding for this fascinating topic of the higher order spectra was not to last much longer. With the help of some Scottish universities (Aberdeen and Strathclyde) I was able to give it an extended lease of life for a few years with a novel application to machine condition monitoring, but as far as sonar is concerned the fact remains that signals of interest tend to be dominated by second order (Gaussian) statistics, not higher order. In other words, most of the information is in the variance. With all the current interest in non-Gaussian statistics I feel that there must somewhere be an important place for the "higher order". Nevertheless, it does not appear to have been fully appreciated that there is no such thing as a signal without second order statistics, unless it is identically zero, whereas the higher order statistics are just useful additions.

At about this time the Dutch asked me if I would like a job in The Hague working at their research labs. When I declined they suggested that maybe, with the agreement of MoD, I could work for them part-time commuting to The Hague for a few days each month. For a short while I was tempted but it all seemed too complicated and they were not prepared to pay me enough. I turned it down. It occurred to me that perhaps I could have been more adventurous in my career. I had declined interesting jobs in the US, Holland, France and Italy, always arguing to myself that it would interrupt our daughters' education.

In March 1994 Lucienne's father died. He had been in a care home for some time but died in hospital following a stomach problem. We had been planning to visit him but were just a bit too late. Lucienne, Caroline and I travelled to Roanne for the funeral, one of the few that took place when I was not away at sea. He had been a man of few words and in my experience had never showed much emotion, but he was always most kind and generous towards me. I respected him greatly for what he had achieved in life.

In November 1994 I made the first of several trips to New Zealand and Australia. This particular trip was to a conference and exchange meeting organised by the afore-mentioned TTCP forum (The Technical Cooperation Programme). I gave an obscure paper entitled "The Quotient and Quimp methods of signal detection and bearing estimation", which was poorly understood, not surprisingly.

My immediate and lasting impression of New Zealand is that it would be a fine place to live as long as you are not too interested in the latest fashions. I liked New Zealand for the generosity of its people, its climate, scenery and simplicity. I particularly enjoyed the sights of Rotorua, with its bubbling mud pools and spouting geysers. By contrast, my immediate and lasting impression of Australia is that the people are common and brash, many of them in the Les Patterson mould. I found little to commend Australia and its people except for those who like swearing, steak and beer. The climate is hot and sticky, and far less comfortable than New Zealand.

In mid-1993 I became interested in a new sonar concept. The catalyst was an old colleague, Shane Heaney, whom I knew from my days at ARL Teddington where he had worked in the operational assessment group. Shane had transferred from Teddington to an MoD headquarters post in Whitehall where he had customer oversight of our work in Dorset. I liked Shane and always found his comments constructive and incisive. One day during a routine visit Shane suggested to me that in addition to hull-mounted and towed sonars there was a third possibility - sonars left behind in the water. It was an idea that could be traced back to Homer Bucker in San Diego in the early 1980s. Traditional air-dropped sonobuoys had been around for a long time, but this idea was on a more ambitious scale and opened up new possibilities.

The basis of the idea was for a submarine or ship, or even an aircraft, to leave behind in the water a large acoustic array, or a network of interconnected arrays. They would free-float in the water column or sit on the sea bed transmitting data back to the mother vessel on demand. They would be distributed systems with a limited lifetime and could be used in passive or active receive modes for surveillance, sanitising an area, port exit monitoring, or intelligence gathering.

With the help of Smith Systems Ltd in Guildford I began a short assessment study to determine the best options. This led to some MoD funding to kick-start the work. I put a bright electronics engineer, Steve Pointer, in charge of the project after which it went from strength to strength.

Part of the project plan was to try for international collaboration. Our MoD customer encouraged this because it increased the overall return (gearing) and demonstrated that other nations thought the work important. Personally, I liked collaboration because it greatly increased the work interest and incentive and, most important, improved our chances of funding. It turned out that other nations were indeed interested in the idea. The concept became known as RDS (Rapidly Deployable Systems).

I became the Technical Leader for the collaborative RDS effort. I negotiated a joint programme of work between ourselves, the US, Canada, Australia, and New Zealand under the afore-mentioned umbrella of TTCP. It took several months to put together a programme that satisfied the defence requirements of each nation. In producing the collaborative work plan I employed my trusted ARA management system, making sure that responsibilities were spread between the participating nations in proportion to their contributions. My system was later adopted as the template for all future collaborations under TTCP.

One of the RDS planning meetings was in San Diego. It was June 1995. I arranged it so that Steve Pointer and I had a free weekend in California after the meeting. We drove in our hire car through Los Angeles, along Route One to Santa Barbara, stayed the night in Santa Maria, then continued along the coast road through Big Sur stopping off in Monterey, Steinbeck country, and then on to a meeting in Silicon Valley before visiting San Francisco, my favourite US city alongside New Orleans. I have always loved San Francisco. The hilly streets, redwood parks, seafood and waterside, and the excitement of visiting Alcatraz are all intoxicating.

Key to the RDS programme was three full-scale sea trials in which all the TTCP nations participated: June 1997 on the Scotian Shelf off the East coast of Canada; November 1998 in the Timor Sea out of Darwin, NW Australia; and September 2000 in the Adriatic out of Bari in southern Italy. There was later added a fourth sea trial in the NW Approaches in July 2002.

The sea trial on the Scotian Shelf was out of Halifax, Novia Scotia. It involved several ships. In spite of bad weather, some difficult boat transfers and several equipment losses we did some useful work and collected a lot of data to take the programme forward. When the weather prevented us working some of us did fishing. I was successful in helping to catch several fine haddock that the cook served up at dinner. Over the years my trips to Halifax always included a memorable team meal of fresh lobster at Salty's waterfront restaurant.

The sea trial in the Timor Sea, out of Darwin, was particularly interesting. I saw my first flying fish and a water spout, the aquatic equivalent of a tornado. It is here I first experienced a total flat-calm. It was hot with the sun beating down like an anvil. The sea was like a sheet of glass extending to infinity in all directions and perfectly reflecting the blue sky, making the sea and sky indistinguishable. It felt like our ship was floating in a hot void. On the technical front this sea trial was also very successful.

Unfortunately, in September 2000 I had to cancel at the last minute my participation in the third RDS sea trial in the Adriatic, owing to a bout of flu.

The RDS collaboration successfully achieved all its objectives. I enjoyed it immensely. But such is the slowness of the procurement process in MoD that, at the time of writing, RDS systems are still not in service except as demonstration equipments.  Possibly our Navy is still not quite sure how best to use them.

The RDS systems we developed had no physical connection with the mother vessel.  They had to communicate acoustically.  There was therefore a requirement for large amounts of data to be transmitted reliably over acoustic underwater channels.  The existing technology was not up to it, the only means of underwater communication available being the underwater telephone which was pretty hopeless and most of the time unintelligible.  What we needed was something that would work in a multipath environment and in situations with high Doppler.  The long-term vision was for an underwater acoustic internet.  So a key component to the success of RDS was a parallel project on underwater acoustic communications, also led by Steve Pointer, using spread spectrum techniques.  This developed into an important project in its own right with many commercial spin-offs: diver detection systems, communication with autonomous underwater vehicles, video image transmission, remote monitoring, control of subsea well-heads for oil and gas, and monitoring competitive swimmers.

Another important project was the Thin Towed Array, a follow-on from the French collaboration.  The objective was to develop a high performance array of smaller diameter than its predecessors, with increased length, strength and channel capacity.  It had to be robust enough to be winched on to a submarine.  I christened the design "Crustacean" because the strength members were round the outside.  By this time we had an impressive towed array assembly line at Winfrith that made us self-sufficient.  We also had full-scale facilities for winch-testing our arrays, under the direction of Norman Morris and Richard Williams.  The overall aim was to persuade MoD to choose our array as the preferred design for the new nuclear submarine class Astute.

Towards the end of February 1997 I rang my parents prior to another foreign trip.  For some reason my mother recalled holidays when she was young, staying with relatives on a farm in Hemel Hempstead and riding from the station in a pony and trap. At the end of this recollection she said, "Those were happy days" [20]. These were to be her last words to me.

On 26 February 1997 I flew with several of my team to the US where we had arranged an important sea trial to test our Crustacean array modules at high speed on a rented US vessel, the Athena II, out of Andros in the Bahamas. This was the second of our two trials on Athena, the first in 1995 being out of Panama City. It was a key stage in the demonstration of our technology.

On our way to the Bahamas we stayed at the sumptuous Holiday Inn in West Palm Beach where we met the Spice Girls checking in. The next day we flew from West Palm to Andros in blistering heat to prepare the ship. Equipment malfunctions delayed our departure on Athena until Tuesday, 4 March. We then did four days of test runs.

On Saturday, 8 March at 04:00 a signal arrived saying that Andros needed to speak to me urgently on the ship's radio. They said they had to speak to me personally so I guessed it had to be some sort of bad news. The radio reception was poor and I was not able to speak to Andros for another five hours during which time I was going through mental torture. When I finally spoke with them I learned that my mother had died at Southend General Hospital in the morning of Friday, 7 March 1997.

The ship returned to Andros making 40 knots to drop me off. The UK Navy looked after me with their usual efficiency in emergencies. I phoned Lucienne who told me my mother had

---

[20] The relatives my mother referred to were her cousins Joan and Molly and her Uncle Eddie Besent and Aunt Milly. Eddie Besent was my paternal grandfather's younger brother. In 1911 he succeeded my great-grandfather Thomas Besent as farm bailiff on the Halsey Estate at Great Gaddesden, near Hemel Hempstead, where the Besents lived in the spacious Home Farm House. After Eddie's death in 1949 the family moved into an estate cottage at nearby Water End. I recall visiting Molly and Milly at Water End Cottage with my parents. It was one of the first trips that we made in my father's first car, his blue Ford Anglia, and so must have been circa 1958/9.

fallen in the garden and broken her hip.  She had been operated on and appeared to be recovering but had died the next day.  I later learned she had been left for twelve hours on a trolley in a hospital corridor before her operation.  The death certificate said she had died of pneumonia, the usual get-out clause.  She was 85 years old.  I feel that she received poor treatment and should have survived her injury.  She was basically a fit person, much fitter than my father.  The sea trials continued in my absence and were a success.

In order to complete the Thin Towed Array work we did two more sea trials.  The first was in March 1998 on RMAS *Newton* out of Funchal, Madeira.  This was to be my last visit to this much-loved island.  The second trial required a high-speed platform.  With little money left in the budget I could not afford to rent the US vessel Athena again.  Instead I rented a much cheaper US facility on Lake Pend Oreille, Idaho.  The lake is in wild country, surrounded by mountains and forests.  The trials were in September 1999.  It was a good experience that worked out well - quite different from working at sea and more civilised.

I assembled a final report on all the work we had done on the new generation of thin towed arrays.  It was well-received by MoD.  Unfortunately we were hit by the familiar problem of risk transfer and resistance to "pull-through".  The Astute contractors were not persuaded by our achievements and had no real incentive to take up the technology.  The eventual contract for the Astute sonar system went to BaE Systems who used their own designs, which happened to be very similar to ours.  They were assisted by our results to which they had access.  As a consolation I was asked by the contractors to head one of their "Red Team Review" panels.  Interestingly, the Crustacean design of towed array achieved a new lease of life in 2010 after the acquisition of the sonar work at Winfrith by the German company Atlas Elektronik.

I should say that it was becoming increasingly difficult to improve the performance of our sonar systems against quieter targets.  All the easy things had been done.  In broad terms there are only three ways to improve performance:  bigger arrays with more sensors to provide better discrimination against noise; data processing that is better matched to local environmental conditions; and networking between cooperating platforms. The work to date had concentrated mainly on the first of these but, cost issues aside, there is a limit to useful array size set by the length scale over which propagating signals remain coherent. In many cases this limit was already being approached. So future improvements will increasingly rely on the other two approaches. Conventional array processing makes rather unrealistic and over-simplified assumptions about the nature of the signal. With the increasing improvements in real-time knowledge of environmental conditions, processing methods can be employed that make assumptions better matched to the environment and the effect it has on the received signal, with a corresponding improvement in performance. These methods require a lot of computer power. The networking approach between platforms is all about making better use of coordinated information, based on the maxim that "the power of the sum is greater than the sum of the powers". This requires good data communication both above and below the air-water interface.

I have never been superstitious, so I surprised myself at being drawn into an area of speculative, fringe science. The seed was sown by an unlikely encounter in February 1993. I had a phone call from Professor Tony Christer, an expert on operational research at Salford University, asking if I would be the external examiner at the *viva* of one of his PhD students. There was a second external examiner, Paul Sutcliffe, who had worked at Portland before transferring to a senior management position at MoD in London, where he had some responsibility for allocating research funds. The oral exam went well and the student received his PhD. But the interesting thing was a side-meeting that Paul had set up with a Dr Cyril Smith at Salford who had recently written a book called "Electromagnetic Man". It was all about the electromagnetic aura surrounding a person, and how to tap into it in order to influence the behaviour of the brain and other vital bodily processes, including healing. What impressed me was that Paul, a level-headed and rational man, was considering funding some work in this area. I stored this experience away.

In July 1997 I examined another of Tony Christer's PhD students. I failed this one because I thought his thesis lacked originality. Instead of a reaction of shock followed by a desire to know what needed to be done to salvage his thesis, the student confronted me with uncontrolled anger. When I left the building he was waiting at the bottom of the stairs. I thought he was going to attack me so I made a rapid exit to the safety of my car. The student eventually rewrote his thesis and I approved his PhD 18 months later.

Our parallel collaboration with the French on thin towed arrays had come to a conclusion some years earlier with the production of a final report that I compiled in 1995. Meanwhile I had become increasingly interested in Michel Boisrayon's biological experiments at Le Brusc. I began independent discussions with people at the Chemical & Biological Defence Establishment, Porton Down (CBD). My main contacts were Paula Gibson and Ric Titball. Paula had a general interest in bacterial infections. She worked for Ric who was a world authority on "The Plague". Ric was responsible for the secure isolation of one of the last known examples of this bacterium, sometimes known as the Black Death.

Over the years I made many visits to Porton Down. Ric would serve us coffee with cookies prepared on site. On two occasions I felt ill on the drive back to Weymouth. Once I had to stop while Steve Pointer was sick in a lay-by. We put it down to the cookies and supposedly clandestine experiments that Ric and Paula were carrying out on us.

From my talks with Ric I drew up an innovative research proposal on biological acoustic sensors based on a certain deep-sea bacterium (*Photo-bacterium* SS9) that was weakly sensitive to pressure. The idea was to generate mutations and select those with increased pressure sensitivity. In parallel I made contact with Prof George Lunt and Dr Adrian Rogers at Bath University who suggested examining certain mecano-sensory neuronal cells such as Merkel cells (rat whiskers) and dorsal root ganglia cells that might be employed as vibration sensors of pressure. My proposal was bolstered by a collaboration I was planning with Michel Boisrayon in France. Michel had an ambitious programme to develop bacterial acoustic sensors by a process of conditioning and training similar to the methods used much earlier by Pavlov.

I sold these ideas to our MoD on the basis of the novel concept of growing living acoustic sensors that could regenerate themselves in the event of failure. I argued that this was the first step towards intelligent sensors and biological computing. I obtained funding for a 3-year programme from 1996 to 1999 thus enabling me to continue my regular trips to the Côte d'Azur.

The programme included funding for a biologist to carry out experiments at Winfrith. This required a biological laboratory on site. I obtained in-house funding of £100K from the DERA capital budget to build a facility. But people on the Winfrith site were wary of having a bio-lab in the building in which weird experiments would be conducted. Would we use live animals? How would we stop the bugs escaping? Were we using dangerous pathogens? So I had to do a PR job to get the locals on side. The self-contained bio-lab was completed in September 1997. I recruited a neuroscientist, Richard Heal, who oversaw the specification and installation. Richard was to become a tower of strength in our biological research. We were now fully equipped for bacterial and neuronal cell culturing.

I have mentioned earlier Michel Boisrayon's interest and fervent belief in extra-sensory perception. One of his premises was that if such a mode of communication existed between living creatures, then it should also exist between the cells that constitute those creatures. He made the point that it was much easier to perform controlled experiments on the constituent cells than on the creatures themselves. Now that we had our own bio-lab I thought we should explore this area.

Richard began a series of experiments and soon came up with an astonishing result. He found that when a healthy bacterial culture was placed near another culture of the same bacteria that had been doped with a level of antibiotic normally enough to kill it, the doped culture survived. The healthy culture was evidently providing some sort of "assistance" to the stressed culture. This had nothing to do with extra-sensory perception. It appeared to be a new type of signalling, or communication system that transmitted through air. If we could understand how it worked there were many potential applications, a topical one at the time being the control of MRSA (Methicillin-Resistant Staphylococcus Aureus) in hospitals.

On the basis of these results I applied for MoD funding for a second bio-project that I called Cellular Communications. There was great competition for MoD funding on innovative topics such as this, and my initial attempts failed. The panel that judged the presentations from proposers comprised stakeholders and academics. The latter were typically elderly university professors who are notoriously conservative, and my proposal was just too way out. So I resorted to a different strategy. I rewrote my proposal and classified it at a level that was above the security clearances of the academics on the panel. I justified the classification level on the grounds that the new communication channel we had discovered had potentially important applications for the intelligence services. This persuaded MOD to allow me to give them a private presentation. In June 1999 I travelled to London for an appointment with Mike Markin who was Director General, Research & Technology. Mike Markin had worked at Portland in his early career. He had actually rented a flat for a short period right next door to us in Carlton Road North. After my short presentation he authorised funding on the spot.

This enabled me to recruit a second biologist Chi-Kong Yeung (known as Kong). The later publication of some of this work prompted media attention with live radio interviews and a 10-minute slot on Geoff Watts' "Leading Edge" science programme on BBC's Radio 4. The publicity upset my MoD sponsors because I had not properly consulted them, and I probably damaged my relationship with them.

Not surprisingly, I enjoyed every minute of my several collaborations with the French. In September 1998 I had a biological sensors meeting with Michel at Le Brusc, I and my biologist Richard Heal stayed at La Cadière d'Azur together with Ric Titball and Paula Gibson from CBD, Porton Down. We were eating in the hotel restaurant when we noticed a noisy group of English-speaking people at a nearby table. At their head was the TV personality Monty Don, presiding over several toadies who seemed programmed to laugh at every stupid joke he made. Monty was making a BBC film about French food with the assistance of our chef. It soon became apparent that my colleague Paula, a vivacious but normally well-balanced young lady, was besotted with Monty Don. She strained to overhear every word he said at the neglect of her meal. Monty and his sycophants departed the restaurant before we had finished. Paula promptly jumped up and went to sit in Monty's chair where she had an orgasmic experience drinking the dregs from his coffee cup.

The next morning after breakfast in an unrelated incident we walked to our rental cars for the drive to Le Brusc. I was astonished to see my Mercedes supported by rocks under the axles with all four wheels missing. I waited for the police and a replacement rental car while the other three drove off to Le Brusc in Ric's car. I was late for the meeting.

The work on biological acoustic sensors did not produce any convincing devices, but it did spawn several interesting areas of work. The key was a new technology of micro-electrode arrays (MEAs) that Adrian Rogers at Bath had brought to our attention. The device comprised an 8 x 8 array of miniature electrodes mounted in an area less than 1mm square. Networks of various types of cell could be kept alive in a closed and controlled environment on the plate surface allowing recordings of their electrical activity to be made non-invasively over long periods of time.

In 1999 we began using MEAs to examine the effects of different drugs on heart cells from rat that Richard obtained from Porton Down. He dissected the rats himself, and developed the protocol for keeping the cells alive on the MEAs for long periods of several days or even weeks. I developed a range of processing techniques, drawing on ideas from sonar classification, to extract the essential temporal and spatial features from the electrical action potentials. I designed some 40 features which, taken together, provided a "fingerprint" for each drug. The aim was to demonstrate a system capable of characterising in real-time the effects of drugs on cardiac physiology in order to reduce the time it took for new drugs to come to market. One of the selling points was the reduced dependency on live animal testing. The success of the work attracted funds from a private venture company who eventually bought the technology.

In parallel we obtained funding for a three-year research programme to investigate the use of our MEA plates for growing neural networks (brain cells). My objective was to develop neuronal sensors that took advantage of the living nature and the computing powers of neurons. This led to interesting collaborations with the Medical Schools at Nottingham and Newcastle, where I worked with the inspirational Gary Green, an unassuming polymath who excelled as a neuroscientist, mathematician and practising clinician.

By this time I had a strong vision for the direction in which all this biological work was going. It involved the fusion of physics, biology and signal processing leading to the laboratory growth of intelligent biosensors and bio-computers using live biological material such as mecano-sensory cells and neurons. The timescale I envisaged for the research and development was 10 to 15 years. My ideas turned out to be somewhat naive and optimistic for the low levels of funding I was receiving, but only the timescale has turned out to be wrong.

I continued to have other interesting university contacts through a number of smaller projects. I worked for several years with Asoke Nandi at the University of Strathclyde in Glasgow and with Jim Penman and his PhD students at Aberdeen on the use of higher order spectra for machine condition monitoring. Glasgow is a sad city of drunks.

Aberdeen is an expensive, dismal city of wet granite where it is impossible to find Aberdeen Angus beef on any restaurant menu. From my experience of many visits to Scotland I find there is little to commend it, except whisky and possibly the mountains of which I have only dim schoolboy memories.

In parallel I was doing work with Chris Dainty at Imperial College on the use of higher order spectra for signal detection. I had a series of contracts with Roger Pinnington and Joe Hammond in the Department of Sound and Vibration (ISVR) at Southampton University, and with Phil Atkins at Birmingham on novel methods of signal detection using manipulations of the data covariance matrix for large arrays.

I had several projects with the "special-fit" submarine service to improve the performance of their in-service sonars. Some of this work involved novel methods of data analysis that was ably led by Andrew McClean.

In addition to my neuronal work at Bath University with George Lunt and Adrian Rogers. I worked with Keith Walton at Bath on the theory of damping of vibration in fluid-saturated open-cell foam. My visits to Bath led me to meet Alan Rayner, the expert on fungi and their growth. It was he who made the sharp observation that the most interesting activities in any system occur at its boundaries. This assertion was made in the context of fungal colonies, but I believe it applies more widely to both organisms and organisations. It even applies to land masses. For example, seaside resorts tend to have bigger populations and much more going for them than inland towns of similar size.

Within my project on vibration damping using fluid-saturated foam I had obtained funding to support one of my staff, a bright young engineer called Colin Mead, to do a PhD on the topic. I had planned to link up with Cambridge University where I introduced Colin to Professor David Crighton, Head of Department at DAMPT and Master of Jesus College, but Colin seemed overawed. (Sadly, it was not long after our visit that David died of a recurrent cancer). So we finally went with Southampton University. Colin completed the work but never wrote up his PhD thesis, a great disappointment to me. In contrast, John Davies, a clever electronics engineer, successfully completed his PhD as an external student at Birmingham.

I did further work at Southampton with Geraint Price and Owen Tutty on the turbulent boundary layer on ultra-thin cylinders. Unfortunately, I never did any work with my *almer mater*, Bristol University.

With so many projects, my team and I were producing a growing number of technical papers and reports, most of which I reviewed personally. I was always struck by the poor standard of report writing regardless of the scientific abilities of the author. Many found it difficult to express themselves concisely in writing and to assemble a report within a logical framework, unlike my experience of US scientists who were more organised. I became quite obsessed with achieving a uniformly high standard of report writing. Some people joked about my ability to scan a page and identify in a few seconds any double spacings between words, a strange capability I have for this particular type of pattern recognition.

My last work trip in 1999, the end of the millennium, was my annual December "Christmas Lecture" to the $2^{nd}$ year engineering students at Birmingham. I stayed at Lucas House the night before after the obligatory curry in Selly Oak.

# The Noughties and Retirement

We were warned that on 1 January 2000 the millennium bug would crash our computers and the world as we knew it would come to an end. The day turned out to be uneventful.

In August 2000 Caroline married Simon, a union that was short-lived. The ceremony took place in Oxford's New College Chapel, and afterwards at Milton Hill House near Abingdon, all at incredible organisation and expense. We loved Simon, but for reasons we have never understood Caroline and Simon were soon divorced. We could not have guessed that a few years later we would be moving from Weymouth to Oxfordshire to live close by to Caroline and her new husband Pete.

My father was too weak to attend Caroline and Simon's wedding. In fact, he seemed disinterested. He died on 29 June 2001. With failing health he had been in a retirement home in Westcliff for the previous two years. His last months were made bearable by the regular daily visits of his close and faithful friend Rene. My father left a letter to Nick and me, asking that we give a sum of money from his estate to Rene. It was not legally binding but I complied with his request. Nick preferred not to comply, a decision that I have never entirely understood. Rene died some years later taking many secrets to the grave. I attended her funeral in Southend in September 2011. Up to the last my father remained a philanthropist through his connections with the Masons and his gifts to the nurses on his many visits to Southend General Hospital. It was a mixture of generosity and an inability to keep money in his pocket.

My uncle Frank died a few months later on 27 November 2001, leaving my Auntie Margaret as the last surviving member of my father's generation of Parsons. Sadly, Margaret's health was to deteriorate rapidly. She suffered bravely from Parkinson's for several years. She died on 7 July 2013. So I became the senior member of the Parsons clan.

Shortly after the death of my father we had 9/11. Most of us remember clearly what we were doing when we heard that

President Kennedy had been shot dead. The same goes for 9/11. I was hosting a visiting Admiral for part of a tour round our labs at Winfrith. I recall I was giving him examples of "blow-back" in our research as opposed to the more conventional "pull-through". While engaged with the Admiral some of my team in a neighbouring lab saw the first reports of the planes crashing into the Twin Towers on a small TV screen. I heard some commotion from this lab, and while somebody else was entertaining our visitor I went to see what was going on. The news and film reports were so horrific we could not believe it was for real. By the time the Admiral's tour had finished everyone except him knew about it.

DERA remained within the UK MoD until 2 July 2001 when it was divided into two organisations: a new commercial company named QinetiQ, and a government arm called the Defence Science & Technology Laboratory (Dstl).

John Chisholm, the chief executive of DERA, had convinced Blair's government of the benefits of floating DERA on the stock market. At a stroke it would dramatically reduce the number of civil servants and generate cash for the Treasury. When QinetiQ was created in 2001 Chisholm bought £129,000 worth of preferential shares at a gearing of 1 to 180. Upon flotation these shares became worth £23m. The QinetiQ directors each became multi-millionaires and even the lowly project managers were richer by up to £1.8m. Scientists were not able to profit in this way, being limited to buying share options up to £2500. The Carlyle Group, a US private equity company fronted in Europe by former Prime Minister John Major, invested £42m. This translated to over £370m when they sold their stake four years later following full flotation. Meanwhile, Blair gave Chisholm a knighthood.

The National Audit Office investigated the privatisation of DERA. They were critical of the directors' share deals and the Carlyle Group investment, saying that the UK taxpayer had lost tens of millions of pounds. QinetiQ inherited expertise, real estate and hundreds of valuable patents previously belonging to MoD and hence to the British taxpayer. After privatisation the US Department of Defense, which had been happy to share classified information with DERA, shut QinetiQ out deciding that working with a plc was an unacceptable risk to security. The share issue price on the 2003 vestment day was £2. Ten years later after many ups and downs the shares were trading at around £1.50, rising to £2.10 in 2018. QinetiQ is kept buoyant through its foreign acquisitions, the UK division failing despite having undergone repeated restructuring. The former government sites, once centres of scientific excellence, have all been decimated or sold off. For example, what was left of the sonar and underwater systems component of QinetiQ was sold to the German company Atlas Elektronik in 2009. The Blair legacy is that over half of British companies are now owned by foreigners who do not have UK's best interests at heart.

At the split, QinetiQ was formed from the majority of DERA, about 75% of its staff and most of its facilities. Dstl assumed responsibility for those aspects that were judged best done in government, basically contract management and the sensitive work on nuclear, chemical, and biological research. I was one of those who moved into the new company, QinetiQ, together with all my team. At this point the morale of the workers began to deteriorate rapidly.

Nevertheless, on the basis of their unique contributions I successfully supported both Steve Pointer and Nick Goddard in their achievement of Individual Merit promotions. It was remarkable to have three IMs in such a small team.

In June 2002 I made my last serious foreign trip. It was to an underwater acoustics conference in La Spezia, Italy. I was determined to make it a memorable "swan song". Steve Pointer accompanied me and suggested we travel by car. This turned out to be an excellent idea and something of a nostalgia trip.

We crossed the Channel on the overnight ferry from Portsmouth to Le Havre. The next day we drove through France to our favourite haunt on the Côte d'Azur. Madame Bérard in La Cadière welcomed us with open arms when we checked in. In the balmy heat of the evening we strolled down the narrow main street of this mountain village to a nearby bar beneath the platanes where we each drank an apéritif of pastis. The owner must have sensed it was a special occasion and served us another free of charge. We returned to the Hostellerie Bérard where we ate a wonderful meal followed by digestifs that Madame Bérard offered on the house.

The next day we drove the short distance to Ollioules to meet up with my old friend Michel Boisrayon. He had not yet retired but was not working that day. He took us to his art studio in a converted wine cellar that he shared with Michèle, his former secretary with whom he was now partnered. They both had a passionate interest in Chinese art and calligraphy producing several fine works, some of them commissioned. They also ran art courses for school children. Michel and Michèle later produced a personalised Chinese painting for me on my retirement. It forms the cover of this book. We had lunch with Michel in a nearby restaurant before continuing our journey along the coast road to Italy.

Steve presented a paper at the conference in La Spezia but otherwise our contributions to the proceedings were minimal. We spent most of the time talking to old friends and eating nice meals. I met up with Chris Harrison who had shared my office all those years ago in Teddington. We talked about old times.

We left La Spezia before the conference finished to take our time on the long drive back. We drove over the Alps and back into France where we visited some vineyards in Burgundy, staying overnight at an excellent hostelry, Le Château de Sainte Sabine, that had been converted into a hotel. We returned home stocked up with French food and wine. This was the last and possibly the most extravagant of my many foreign trips paid for by Her Majesty.

Our esteemed CEO, John Chisholm, pretended that the scientists were still important to the business. To demonstrate this he instituted an annual event for all the QinetiQ Fellows and Senior Fellows (formerly Individual Merit scientists like me) at prestigious venues like Blenheim Palace, at which newly appointed Fellows were invested like film stars winning an Oscar. It was a tremendous publicity stunt complete with lavish food and entertainment, citations and videos describing each new Fellow's technical achievements, followed by presentations and inauguration by a distinguished personage. I went to several of these events, including Blenheim Palace and one at the Abbey Hotel in Malvern. I enjoyed the hospitality but found the formalities quite hollow and excruciating.

My last funded project before my formal retirement in 2003 was my most controversial and right on the edge of respectable science. It was a theoretical investigation into a novel method of electromagnetic communication that exploited the vector potential field of Maxwell's equations. Conventional wisdom has always been that the vector potential is just a useful mathematical device with no physical reality. This was challenged in 1959 by an intriguing "thought experiment" that became known as the Aharanov-Bohm effect. I examined the equations more closely and identified the existence of an apparently new type of wave that might be called a scalar EM wave. The important property of this wave was that it could exist even in the absence of any associated electric or magnetic fields. I came up with a conceptual design for a device capable of generating and receiving this new type of wave. Phil Atkins at Birmingham tried building the device but obtained no useful results. There were several philosophical problems with my formulation and I never received further funding to proceed to the next stage. I am sure there was something fundamentally wrong with my approach, otherwise I should be receiving a Nobel Prize.

Lucienne and I were now doing a lot of ballroom dancing. When not having lessons we went frequently to evening dances at the Weymouth's Royal Hotel, Weymouth Pavilion, the Riviera at Bowleaze, Lulworth, Dorchester Ceroc, Dorchester Trinity Club, Charlton Down, and others. Sometimes we went further afield to venues in Yeovil or Bournemouth, the latter being a principal centre for professional ballroom dancing. There was somewhere to go every night. Many of these venues were for holidaymakers, but the organisers got to know us and let us in because we gave encouragement to shy holidaymakers to get up and dance. Sometimes there was nobody else on the floor so we did a demonstration of our latest moves. We occasionally received applause and were asked to our embarrassment if we were professional dancers. Dorset was a great centre for dancing. We immensely enjoy dancing together.

My retirement was now rapidly approaching. One weekend in July 2002 during a visit to London to see Caroline and Simon, we visited the former house of the Victorian historian and philosopher Thomas Carlyle in Cheyne Row, Chelsea. We were struck by a book on display that contained hundreds of short entries comprising compliments and anecdotes about Carlyle, all written by his friends and acquaintances. The particular point of interest was that the book was a celebration of Carlyle's life that was compiled and completed before he died. Lucienne and I both remarked that these things were usually said at a person's funeral, which was of course too late for the dead person to appreciate. After that I forgot all about it.

But Lucienne did not forget and she set about doing a similar thing for me. She somehow managed to find the addresses of a good proportion of all the people I had ever worked with around the world. She sent each of them a small card asking them to write down a few words about me and to return the card to her. This was a mammoth task that she managed to complete without me knowing anything about it. It must have helped that she was able to use her business address in Caroline Place for all the correspondence.

As my retirement day of 5 June 2003 approached I agreed reluctantly to host a lunch time retirement party for everyone at Winfrith. As it turned out it was a memorable occasion. Several people from outside took the trouble to attend, including our daughter Caroline and Phil Atkins from Birmingham. The QinetiQ CEO failed to attend. Instead the Winfrith site manager, John Wade, made a generous speech to which I replied. I was given various presents and plaques, including one from my old US friend and colleague Bill

Moseley. Lucienne presented me with the book she had prepared containing the kind remarks she had collected from all those people I had worked with. I could not believe that she had gone to all this trouble. She made me very proud and happy. Everyone admired the book and thought what a brilliant idea it was.

We got home late that afternoon and I thought it was the end of the day. But it turned out that Lucienne had prepared another surprise for me. At about 6 p.m. she told me to look smart and walk obediently with her down to the seafront. When we reached the Prince Regent hotel she steered me inside and there confronting me in the bar was a crowd of people numbering at least 100. I was taken aback. Not only were all my current work friends there but also former work colleagues, together with a contingency of dance friends. Even my cousin John Besent and his wife Sheila were present.

After pre-dinner drinks we sat down to a semi-formal meal that I think everyone enjoyed (Plate 45). After the meal I stood up and made an impromptu speech. I recall that when I said I had been totally ignorant of all Lucienne's preparations, somebody quipped about what else Lucienne might be up to that I did not know about. Then John Besent, who can never resist an opportunity to make a speech, stood up and spoke off the cuff. I cannot remember what he said but it raised lots of laughs. The floor was then cleared for dancing to the live piano music and singing of Ed Hinze.

So ended my career in full-time employment that was both exhilarating and fulfilling. I had worked on many exciting projects and with many interesting and talented people enabling me to see much of the world. There were some bad moments but I would not have changed any of it. My work is covered in 235 technical papers, conference papers and publications that I have written.

My plan had always been to retire *in situ* (an expression that I can ascribe to Michel Boisrayon), that is, to carry on doing part-time work and phasing it out until I was no longer useful. The plan worked well. I was fortunate to have a series of contracts with QinetiQ and bio-companies for another fifteen years, initially working two or three days a week, then on a reducing scale. For the first five years I kept my old office at Winfrith before I was eventually thrown out (literally, for health and safety reasons) when the acoustic interests of QinetiQ were sold off to the German company Atlas Elektronik with whom I had occasional contracts.

I remain an Honorary Professor at Birmingham University and continue with my duties, notably, lecturing on the annual Portland Sonar Course and every year processing at the degree congregation at Birmingham, always a happy event. I also continued to lecture on the biannual Bournemouth University Sonar Course until 2019. I have enjoyed all this enormously. When I have been contacted of late on a professional basis it has often been to give personal references, a task in which I have probably been over-generous in my responses.

So what happened to all my projects? Some have disappeared while others have thrived, or at least stayed alive. Many have resulted in some form of in-service equipment.

My biggest disappointment is that my biological work fell on stony ground. I won only limited funding for new bio-projects after I retired, for example, Live Neuronal Learning & Computing for which I obtained funding in November 2003. This was about growing live neuronal networks on micro-electrode arrays (MEAs) and training them to perform simple tasks. I saw this as a step towards bio-computing. I think this project was too ambitious and way ahead of its time.

Generally I failed to obtain critical mass with the biological work. In particular, we were never able to isolate the mechanism for the biocommunication phenomenon that Richard Heal had discovered. I made several other proposals that were unsuccessful. For example, the use of the MEA platform to test shellfish meat for the presence of toxins to replace the current method using live mice, but unfortunately we missed the boat even though our method was superior. Another unsuccessful proposal was about the control of the underwater bio-chorus, an idea that I developed together with

Phil Bardswell.  We had both experienced the chorus of noise at dawn and dusk from biological activity in shallow waters, analogous to the dawn chorus of birds.  A sonar equipment in the vicinity could easily be swamped by this biological noise, rendering it incapable of detecting targets of interest.  The idea was to understand what triggers the dawn and dusk choruses so that we might be able to turn it on and off at will.  This would enable a submarine to "hide" in the high level of background noise so providing it with useful cover.  The proposal was never funded.  Yet another proposal was to extend the study of bio-communication to cultures of the "super-bug" MRSA (*meticillin-resistant Staphylococcus aureus*).  The idea was to develop a sensor for detecting MRSA in hospital environments.  The proposal went through several iterations with the Hospital Infection Society and Dorset County Hospital before eventually failing as a result of red tape.  For example, you had to obtain a patient's permission before experimenting on samples of his MRSA.  The fundamental problem was our lack of track record and credibility.  Nevertheless, for a time I had sufficient work to support three biologists at Winfrith.

Eventually, my biological work no longer fitted with QinetiQ's overall business plan.  The management felt that anything to do with biology was the business of Porton Down.  So in 2007, with the pending sell-off of QinetiQ's acoustic interests, the management decided to wind up my bio-work.  My bio-lab closed in September 2007, exactly ten years to the day after it was opened.  One biologist moved out to a post at Hong Kong University, another stayed with QinetiQ but retrained to become an accountant, and the lead biologist Richard Heal to whom I was so close left to start a teachers training course.  After a successful period of several years as a biology and chemistry teacher at Bryanston Public School in Blandford, Richard became a research scientist at the Centre for Environmental Fisheries and Aquaculture Science (CEFAS) in Weymouth.

At the time of writing, the bio-work is not entirely dead.  The drug characterisation work is being pursued by the company that bought the technology. After 2008 I carried on working from home on bio-projects until 2011 when I sold my software.

In February 2014 I was awarded a three-year part-time contract with Atlas Elektronik to help develop and predict the performance of a new generation of passive sonar systems for military applications, an exciting project. Curiously, it has its roots in work on vector sensors that began all those year ago in 1979 with the Canadian exchange scientist, Barrie Franklin. One thing I have learned is that there are very few original ideas. They go round in circles invigorated each time by advances in the enabling technologies. I continue with occasional work as a part-time consultant on sonar projects.

My partial retirement allowed me to indulge in my other interests: fell walking, ballroom dancing, watercolour painting, organ and guitar playing.

On the organ front I went on several dedicated holidays. The first was in July 2002 when I travelled to Paris with my brother Nick on an organ extravaganza to celebrate his 50th birthday. The trip was organised by the IAO (Incorporated Association of Organists), and the detailed schedule was arranged by its then President, Ian Tracey, the virtuoso resident organist at Liverpool Cathedral. Ian used his many contacts developed while he was studying organ in Paris to come up with a stunning programme of recitals at most of the major churches, all housing magnificent organs mostly by Cavaillé-Coll. The great French organ composers of the late 19$^{th}$ and early 20$^{th}$ century had titulaire positions at these churches where they could be heard playing and improvising every Sunday, a tradition that continues. This was a holiday never to be repeated. The Paris venues I have visited for organ recitals include St Sulpice (Lefébure-Wély, Widor, Dupré, Roth), La Trinite (Guilmant, Messiaen), Notre Dame (Vierne, Cochereau), Saint Augustin (Gigout), Sainte-Clotilde (Franck, Tournemire, Langlais), La Madeleine (Saint-Saëns, Dubois, Fauré), Sacré-Cœur (Marghieri, Hakim), Saint-Étienne-du-Mont (Duruflé), Saint Eustache (Bonnet, Marchal, Guillou), and the Basilica of Saint Denis (Heurtel).

The Paris trip in 2002 encouraged Nick and me to go on other organ holidays together: Harrogate (2003), Cologne (2006), Wirral (2009), Paris (2011), Toulouse (2012), Nuremburg (2013), Bath (2016). This, together with fell-walking, has brought me closer to my brother in recent years.

I have only two regrets in my life. The first is that I did not discover earlier the things that I am good at, so that I could have applied myself to them and become proficient when I was young. Life is just too short. The second is that, while my ancestors were alive, I did not ask them about their experiences when they were young. It is now too late. Hence, this account of my life to save my daughters the trouble.

Printed in Great Britain
by Amazon